Cellular and Molecular Effects of Mineral and Synthetic Dusts and Fibres

NATO ASI Series

Advanced Science Institutes Series

A series presenting the results of activities sponsored by the NATO Science Committee, which aims at the dissemination of advanced scientific and technological knowledge, with a view to strengthening links between scientific communities.

The Series is published by an international board of publishers in conjunction with the NATO Scientific Affairs Division

A Life Sciences B Physics	Plenum Publishing Corporation London and New York
C Mathematical and Physical Sciences D Behavioural and Social Sciences E Applied Sciences	Kluwer Academic Publishers Dordrecht, Boston and London
F Computer and Systems Sciences G Ecological Sciences H Cell Biology I Global Environmental Change	Springer-Verlag Berlin Heidelberg New York London Paris Tokyo Hong Kong Barcelona Budapest

NATO-PCO DATABASE

The electronic index to the NATO ASI Series provides full bibliographical references (with keywords and/or abstracts) to more than 30000 contributions from international scientists published in all sections of the NATO ASI Series. Access to the NATO-PCO DATABASE compiled by the NATO Publication Coordination Office is possible in two ways:

- via online FILE 128 (NATO-PCO DATABASE) hosted by ESRIN, Via Galileo Galilei, I-00044 Frascati, Italy.
- via CD-ROM "NATO Science & Technology Disk" with user-friendly retrieval software in English, French and German (© WTV GmbH and DATAWARE Technologies Inc. 1992).

The CD-ROM can be ordered through any member of the Board of Publishers or through NATO-PCO, Overijse, Belgium.

Series H: Cell Biology, Vol. 85

Cellular and Molecular Effects of Mineral and Synthetic Dusts and Fibres

Edited by

John M. G. Davis

Institute of Occupational Medicine
8 Roxburgh Place
Edinburgh, Scotland, EH8 9SU

Marie-Claude Jaurand

INSERM
Unité 139
CHU Henri Mondor
Creteil, Cedex, France

Springer-Verlag
Berlin Heidelberg New York London Paris Tokyo
Hong Kong Barcelona Budapest
Published in cooperation with NATO Scientific Affairs Division

Proceedings of the NATO Advanced Study Institute on Cellular and Molecular Effects of Mineral and Synthetic Dusts and Fibres, held in Paris, France, October 11–13, 1993

ISBN-13:978-3-642-79043-0 e-ISBN-13:978-3-642-79041-6
DOI: 10.1007/978-3-642-79041-6

CIP data applied for

Softcover reprint of the hardcover 1st edition 1994

Typesetting: Camera ready by authors
SPIN 10122375 31/3130 - 5 4 3 2 1 0 - Printed on acid-free paper

PREFACE

The Fifth International Workshop on "Cellular and Molecular Effects of Mineral and Synthetic Dusts and Fibres" was held on October 11-13, 1993 in Paris, France. The emphasis of the NATO Advanced Research Workshop was the use of cell and organ culture and lavage cell populations obtained from man and laboratory animals to elucidate cellular and molecular events occurring after their interaction with fibrous and non-fibrous particulates including metal compounds. In four sessions, an international representation of scientists from 18 countries (Australia, Belgium, Canada, Finland, France, Germany, Hungary, India, Israel, Italy, Japan, Netherlands, Norway, South Africa, Sweden, Switzerland, UK, USA) presented research findings in the following areas:

- Cellular and metabolic changes caused by mineral dusts.
- Molecular changes and DNA alterations produced by mineral dusts.
- In vivo dust-related pathological processes. Correlations between in vitro and in vivo data.
- Physico-chemical properties of minerals in relation to their biologic effects.

The organising committee for this workshop was:
Co-Chairs: M.C. Jaurand (France) and J.M.G. Davis (United Kingdom)

G. Barlovatz-Meimon (France)	R. Bégin (Canada)
J. Bignon (France)	R.C. Brown (UK)
B. Fubini (Italy)	B.T. Mossmann (USA)
Q. Rahman (India)	

The workshop was sponsored by the Advanced Study ARW Program of NATO. Additional financial sponsorship was also obtained from the following:

- Agence de l'Environnement et de la Maîtrise de l'Energie
- Communauté Economique Européenne
- Ministère du Travail, de l'Emploi et de la Formation Professionnelle
- Université Paris Val de Marne
- Association Française de l'Amiante
- Council for Tobacco Research
- Du Pont de Nemours
- Joint European Medical Research Board
- European Ceramic Fibres Industries Association
- North American Insulation Manufacturers Association
- Procter & Gamble Company

The editors wish to thank Springer-Verlag for their interest in publishing this state of the art volume on cellular and molecular interaction between mineral and organic dusts and cells.

The Sixth International Workshop on "Cellular and Molecular Effects of Mineral and Synthetic Dusts and Fibres" will be held in the USA in 1997 and co-chaired by K.E. Driscoll and G. Oberdorster.

John MG Davis
Marie-Claude Jaurand

TABLE OF CONTENTS

The Role of Active Oxygen Species in Uptake of Mineral Particles by Tracheobronchial Epithelial Cells*

Andrew Churg, MD
Department of Pathology and University Hospital
University of British Columbia
Vancouver, BC V6T 2B5

Introduction

Inhaled mineral particles that are not removed by macrophages or the mucociliary escalator tend to be taken up by pulmonary epithelial cells (Adamson and Bowden 1981; Bowden 1987). This process of is considerable pathologic importance. Depending on the exact mineral species, particle uptake may be associated with a variety of mutagenic events (Mossman et al 1990; Hei 1989; Hesterberg and Barrett 1985; Hei et al 1992). As well, some mineral particles are directly cytotoxic and their uptake leads to cell death and epithelial ulceration. Particles may also be transported through the cell cytoplasm to the underlying interstitial tissue (airway wall, alveolar interstitium), and in this location they can interact with interstitial macrophages to cause release of fibrogenic mediators, eventually leading to airway wall or interstitial fibrosis (Adamson et al 1989; Ferin et al 1992; Brody and Hill 1982).

The factors that determine when and to what extent a particle will be taken up by an underlying epithelial cell are poorly defined, although there is evidence that particle type, shape, size, and possibly charge all play a role (Adamson et al 1989; Ferin et al 1992; Brody and Hill 1982; Adamson and Bowden 1981; Bowden 1987). We have been investigating an additional mediator of particle uptake, namely active oxygen species. This paper briefly reviews some of our current observations in this area.

Materials and Methods

The experiments reported here were all carried out using rat tracheal explant cultures as originally described by Mossman et al (1979). The general approach has been to prepare explants, expose them to suspensions of mineral dust for 1 hour, and then maintain them in organ culture in air/CO_2 with basal agarose/culture medium feeding for periods up to one week to allow particle uptake to occur. Dust uptake has been evaluated by morphometric methods on 5 micron paraffin sections, and lipid per-

---

* Supported by grants MA8051 and MA7820 from the Medical Research Council of Canada

NATO ASI Series, Vol. H 85
Cellular and Molecular Effects
of Mineral and Synthetic Dusts and Fibres
Edited by J. M. G. Davis and M.-C. Jaurand

oxidation measured by the method of Wong et al (1987). The effects of scavengers were evaluated by adding the scavenger to the dust suspension, or, for the iron chelator, deferoxamine, by preincubating the dust with the chelator. To evaluate the effects of added iron, dust suspensions were incubated overnight with mixtures of ferrous and ferric chloride before use. To test the effects of cigarette smoke, explants were first exposed to whole smoke or air (control) and then to dusts. Details of the preparation methods can be found in Hobson et al (1990) and Churg et al (1990).

Results and Discussion

Different types of mineral particles show rather markedly different relative uptake by tracheal epithelium in organ culture. Table 1 illustrates this effect for a variety of dusts. Even among nonfibrous dusts of almost exactly the same particle diameter applied at the same concentration (for example, titanium dioxide, nonfibrous

Table 1: Basal Particle Uptake and Effects of Cigarette Smoke on Particle Uptake by Tracheal Epithelial Cells after 7 Days in Organ Culture (Values as Mean (SD), from Keeling et al 1993)

	Basal Uptake	**After Cigarette Smoke Exposure**
Nonfibrous dusts (Values as Volume Fraction of Epithelium)		
Iron oxide	0.31 (0.29)	0.26 (0.25)
Silicon carbide	0.26 (0.17)	0.36 (0.15)
Titanium dioxide	0.87 (0.13)	7.7 (4.4) *
Fibrous and platy dusts (Values as Particles/MM Epithelial Length)		
Amosite asbestos	13.5 (4.8)	28.9 (6.0) *
Silicon carbide	14.4 (1.2)	22.9 (20.8) *
Talc	6.5 (2.7)	22.9 (20.8) *
Wollastonite	3.8 (3.3)	2.2 (1.9)

* Indicates significant effect of smoke on dust uptake

iron oxide, and nonfibrous silicon carbide in Table 1), relative uptake varies by a factor of almost 4 fold. Similarly, for fibrous or platy dusts in which there is more variation in size, there is again an almost a 4 fold difference in particles/mm of epithelial length (Table 1). This is clearly not a size effect and probably not a charge effect, since all of the minerals in question have a negative surface charge.

Basal uptake can be markedly reduced by including catalase, which scavenges hydrogen peroxide, or superoxide dismutase, which scavenges superoxide anion, in the medium with the dust suspension, or by preincubating the dust with deferoxamine. Table 2 illustrates this effect for catalase or deferoxamine and amosite asbestos. Diminution of particle uptake is catalase or deferoxamine dose related and, with sufficient concentrations of catalase or deferoxamine, reaches about 1/2 to 1/3 the basal level. Inactivating the catalase by boiling destroys the protective effect. Essentially identical decreases in uptake are observed with superoxide dismutase. These findings suggest that, at least for amosite asbestos, particle uptake is related to the generation of active oxygen species by the dust, an effect which is now well documented for asbestos and many other types of mineral particles (Pezerat et al 1989; Mossman and Marsh 1989). However, the fact that uptake cannot be completely eliminated, no matter how much scavenger is used, suggests that basal particle uptake is also mediated by other, non-oxidant related, mechanisms.

Table 2: Effects of Deferoxamine (DFX) and Catalase (CAT) on Basal Asbestos Fiber Uptake After 3 Days in Organ Culture (Values are Mean (SE) Fibers/MM Epithelial Length, from Hobson et al 1990)

Effects of Deferoxamine		Effects of Catalase	
Asbestos only	37 (8)	Asbestos only	60 (7.0)
Plus 0.01mM DFX	26 (2.8)	Plus 325U/ml CAT	46 (13)
Plus 0.1mM DFX	13 (2.8)*	Plus 650 U/ml CAT	34 (6.1)*
Plus 1.0mM DFX	9.7 (3.0)*	Plus 900 U/ml CAT	41 (3.7)
Plus 10mM DFX	11 (4.5)*	Plus 1300 U/ml CAT	14 (3.6)*
		Plus 1300 U/ml Boiled CAT	80 (18)**

* Indicates significantly less than asbestos only
** Indicates significantly greater than 1300U/ml active CAT

The role of surface iron in the formation of active oxygen species by mineral particles is a topic of current interest. For example, Fontcave et al (1990) demonstrated that, for many different minerals, measured surface iron correlated well with lipid peroxidation of artifical membranes. Ghio et al (1992) noted that artificially increasing the level of surface iron leads to increases in the generation of active oxygen species, leukotriene B_4 release from macrophages, and lavageable inflammatory cells. Goodglick and Kane (1986) pointed out that titanium dioxide, which is essentially nontoxic to macrophages, becomes toxic after preincubation with ferric chloride. Lund and Aust (1992) observed that the number of crocidolite-induced DNA single strand breaks could be increased if iron was mobilized from the surface, in which case uptake was not augmented.

These reports and the observations made above on the role of deferoxamine in reducing basal particle uptake suggested to us that surface iron might also be important in the determining the extent to which epithelia take up mineral particles. To test this hypothesis, we preincubated titanium dioxide or amosite asbestos with increasing concentrations of a 1:1 mixture of ferrous and ferric chloride overnight, washed the dust to remove excess iron, and then exposed tracheal explants to the iron loaded dust. After 7 days in culture, some explants were used for histologic evaluation of particle uptake, and duplicate samples for malonaldehyde (MDA) analysis. We found (Table 3 and Keeling et al, unpublished data) that increasing levels of iron loading were associated with increasing amounts of particle uptake for both amosite and titanium dioxide. Increases in MDA content of the tissue paralleled particle uptake. Varying the ratio of ferric to ferrous iron had little effect on amosite fiber uptake unless only pure ferrous or pure ferric chloride was used to load the fibers (Table 3).

These observations suggest to us that cationic iron may be an important mediator of particle uptake by pulmonary epithelia, and, further, that this process is related to lipid peroxidation of the apical cell membrane. Although lipid peroxidation in this system might be expected to be caused by formation of hydroxyl radical from hydrogen peroxide using surface iron as a Fenton catalyst, the requirement for both ferrous and ferric iron noted above may provide some evidence for the suggestion made by Braughler et al (1986) and Minotti et al (1987) that initiation of lipid peroxidation actually requires a ferrous-ferric iron complex.

Table 3: Effects of Iron Loading with Various Ratios of Fe(III):Fe(II) on Uptake of Amosite Asbestos Values as Mean (SD) Asbestos Fibers/MM Epithelial Length

Fe(III):Fe(II) Ratio*	Fibers/MM Epithelium
Amosite asbestos alone	21.3 (7.5)
Plus 1:1 Fe(III):Fe(II)	49.9 (14.4)**
Plus 2:1 Fe(III):Fe(II)	64.6 (18.7)**
Plus 1:2 Fe(III):Fe(II)	45.4 (16.1)**
Plus Fe(II) only	21.3 (10.3)
Plus Fe(III) only	16.6 (11.9)

* Total iron concentration is held constant at 10mM
** Significantly greater than asbestos alone

Although it is difficult to exactly translate results from in vitro explant culture models to whole animals, iron mediated increases in particle uptake and lipid peroxidation are phenomena of potential pathologic importance. Intracellular iron is normally kept tightly bound to protein, but can be released by a variety of insults, including oxidant exposure. Mere exposure of the lung to dusts results in adsorption of iron onto the dust particle surface (Ghio et al 1992), and increased iron loads can be found in the lungs of workers exposed to some types of mineral dusts, particularly asbestos (Guest 1978). Moreno et al (1992) have shown that cigarette smoke causes release of iron from ferritin by an oxidant mechanism, and the lungs of cigarette smokers contain significantly greater amounts of iron than the lungs of nonsmokers (Thompson et al 1991). Thus mineral dust exposure may establish a positive feedback loop in which the dust progressively adsorbs iron and progressively increases cell damage by increased particle uptake and increased oxidant release.

Cigarette smoke is also a rich source of oxidants including a variety of active oxygen species (39). This fact lead us to evaluate the role of cigarette smoke in particle uptake, a question of considerable interest since, in vivo, smoke enhances whole lung retention of many mineral dusts in man and animals (Keeling et al 1993). As shown in Table 1, a 10 minute pre-exposure of the tracheal explants to cigarette smoke increases the uptake of asbestos, titanium dioxide, talc, and fibrous silicon carbide by several fold over a period of 7 days. For asbestos and titanium dioxide this effect can by be blocked by catalase, superoxide dismutase, and deferoxamine, indicating that enhanced uptake is mediated by active oxygen species (Keeling et al 1993). Interestingly enough, these scavengers do not block smoke-enhanced talc uptake. Also of interest is the fact that iron oxide, wollastonite, and nonfibrous silicon carbide uptake are not increased by cigarette smoke (Table 1). The latter findings again suggest that the uptake of mineral particle by tracheobronchial epithelium occurs via a variety of different mechanisms.

Summary

We have used a tracheal organ culture system to investigate the role of active oxygen species as mediators of particle uptake. Using this system, uptake of asbestos fibers can be decreased in a dose response fashion by treatment of the explants with catalase or superoxide dismutase, or treatment of the dust with deferoxamine. Conversely, uptake can be increased by loading asbestos fibers or titanium dioxide particles with iron chloride, and iron-enhanced particle uptake parallels tissue malonaldehyde levels. Brief exposure to cigarette smoke before dust exposure enhances the uptake of some, but not all, types of mineral particles, and this effect can generally be prevented by scavengers of active oxygen species. These findings suggest that active oxygen species, and in particular hydroxyl radical formed by Fenton catalysis using iron on the particle surface, play an important role in particle uptake. Lipid peroxidation of the apical membrane may be one way in which the ability of the cell to take in particles after oxidant damage is increased.

References

Adamson IYR, Bowden, DH (1981). Dose response of the pulmonary macrophagic system to various particulates and its relationship to transepithelial passage of free particles. Expt Lung Res 2, 165-171.

Adamson IYR, Letourneau HL, Bowden DH (1989). Enhanced macrophage-fibroblast interactions in the pulmonary interstitium increases fibrosis after silica injection to monocyte-depleted mice. Amer J Pathol 134, 411-418.

Bowden DH (1987). Macrophages, dusts, and pulmonary diseases. Exptl Lung Res 12, 89-107.

Braughler JM, Duncan CA, Chase RL (1986). The involvement of iron in lipid peroxidation. J Biol Chem 261, 10282-10289.

Brody AR, Hill LH (1982). Interstitial accumulation of inhaled chrysotile asbestos fibers and consequent formation of microcalcifications. Am J Pathol 109, 107-14.

Church DF, Pryor WA (1991). The oxidative stress placed on the lung by cigarette smoke. In, Crystal RG, West JB, editors: The Lung: Scientific Foundations. New York, Raven Press. Pages 1975-1980.

Churg A, Hobson J, Wright JL (1990). Effects of cigarette smoke dose and time after smoke exposure on uptake of asbestos fibers by rat tracheal epithelial cells. Amer J Resp Cell Molec Biol 3: 265-269.

Ferin J. Oberdorster G, Penny DP (1992). Pulmonary retention of ultrafine and fine particles in rats. Amer J Resp Cell Molec Biol 6, 535-542.

Fontcave M, Jaouen M, Mansuy D, Cost D, Salma R, Pezerat H (1990). Microsomal lipid peroxidation and oxy-radicals formation are induced by insoluble iron-containing minerals. Biochem Biophys Res Comm 173, 912-918.

Guest L (1978). The endogenous iron content, by Mossbauer spectroscopy, of human lungs. Lungs from various occupational groups. Ann Occup Hyg 21, 151-157.

Ghio AJ, Kennedy TP, Whorton AR, Crumbless AL, Hatch GE, Hoidal JR (1992). Role of surface complexed iron in oxidant generaiton and lung inflammation induced by silicates. Amer J Physiol 263, L511-L518.

Goodglick LA,Kane AB (1986). Role of reactive oxygen metabolites in crocidolite asbestos toxicity to mouse macrophages. Cancer Res 46, 5558-5566.

Hei TK (1989). Oncogenic transformation by asbestos fibers and radon-stimulated alpha particles. In, Effects of Mineral Dusts on Cells, edited by Mossman BT, Begin RO. Berlin, Springer-Verlag, pages 389-398.

Hei TK, Piao CQ, He ZY, Vannais,D, Waldren CA (1992). Chrysotile fiber is a strong mutagen in mammalian cells. Cancer Res 52, 6305-6309.

Hesterberg TW, Barrett JC (1985). Induction by asbestos fibers of anaphase abnormalities: mechanism for aneuploidy induction and possibly carcinogenesis. Carcinogenesis 6, 473-475.

Hobson J, Wright JL, Churg A (1990). Active oxygen species mediate asbestos fiber uptake by tracheal epithelial cells. FASEB J 4, 3135-3139.

Keeling B, Hobson J, Churg A (1993): Effects of cigarette smoke on tracheal epithelial uptake of non-asbestos mineral particles in organ culture. Amer J Respir Cell Molec Biol 1993; 9: 335-340.

Lund LG, Aust AE (1992). Iron mobilization from crocidolite asbetos greatly enhances crocidolite-dependent formation of DNA single-strand breaks in PX174 RFI DNA. Carcinogenesis 13, 637-642.

Minotti G, Aust SD (1987). The requirement for iron(III) in the initiation of lipid peroxidation by iron(II) and hydrogen peroxide. J Biol Chem 262, 1098-1104.

Moreno JJ, Foroozesh M, Church DF, Pryor WA (1992). Release of iron from ferritin by aqueous extracts of cigarette smoke. Chem Res Toxic 5, 116-123.

Mossman BT, Marsh JP (1989). Evidence supporting a role for active oxygen species in asbestos-induced toxicity and lung disease. Environ Hlth Perspect 81, 91-94.

Mossman BT, Bignon J, Corn M, Seaton A, Gee JBL (1990). Asbestos: Scientific developments and implications for public policy. Science 247, 294-300. .

Mossman BT, Craighead JE. Use of hamster tracheal organ cultures for assessing the cocarcinogenic effects of inorganic particulates on the respiratory epithelium. Prog Exptl Tumor Res 24, 37-47.

Pezerat H, Zalma R, Guignard J, Jaurand MC (1989). Production of oxygen radicals by the reduction of oxygen arising from the surface activity of mineral fibers. In, Biological Effects of Mineral Fibers in the Nonoccupational Environment, edited by J. Bignon et al. Lyon, IARC. Pages 100-111.

Thompson AB, Bohling T, Heires A, Linder J, Rennard SI (1991). Lower respiratory tract iron burden is increased in association with cigarette smoking. J Lab Clin Med 117, 493-499.

Wong SHY, Knight AJ, Hopfer SM, Zaharia O, Leach CN, Sunderman FW (1987). Lipoperoxides in plasma as measured by liquid-chromatographic separation of malondialdehyde-thiobarbituric acid adduct. Clin Chem 33, 214-220.

OXYGEN RADICAL GENERATION BY ASBESTOS AND ITS CORRELATION TO CYTOTOXICITY

Val Vallyathan, Pathology Section, National Institute for Occupational Safety and Health, Morgantown, WV 26505, USA

INTRODUCTION

Etiologic identification of occupational minerals as pathogenic in humans usually requires epidemiologic, pathologic, toxicologic and animal experimental studies. Epidemiologic studies are often initiated by potential health concerns prompted by cytotoxicity, animal experimentation and/or case reports of overt human health effects. Because we need to know in advance the biosafety of the materials we work with, bioassays and animal experiments to predict the health effects in humans would be of great value. Animal experimental studies are time consuming, relatively expensive and the results are often controversial and difficult to extrapolate to humans due to differences in the responsiveness between species. Disease prediction by cytotoxicity tests would be beneficial because in vitro bioassays are relatively inexpensive, rapidly concluded for the valuable depictions of disease mechanisms and biologic interactions.

In this context during the last few years, molecular toxicology has gained substantial support as a valuable tool in risk evaluation and prediction of adverse health effects of toxic substances in humans. The recognition of mechanisms involved in the initiation of disease through molecular reactions has become relevant to the evaluation of pathogenic risk . Therefore in risk evaluations of chemicals, minerals and possibly in future studies of man-made mineral fibers and untested materials, it is important to pursue some advances in molecular toxicology to enhance our understanding of disease initiation. To test the validity of a possible correlation of cytotoxicity we have attempted, in

NATO ASI Series, Vol. H 85
Cellular and Molecular Effects
of Mineral and Synthetic Dusts and Fibres
Edited by J. M. G. Davis and M.-C. Jaurand

this study, to evaluate asbestos minerals with known fibrogenic potential and fiberglass, using a battery of in vitro bioassays.

There is considerable debate concerning the mechanism of iron-stimulated biologic generation of oxygen radicals and the associated lipid peroxidation in the pathogenicity and carcinogenicity of asbestos (Craighead et al.1982; Mossman et al.1989; 1990). All types of asbestos have been shown to cause fibrotic and malignant pulmonary diseases in humans and animals (Craighead et al.1982). Several investigators have proposed that amphiboles are more fibrogenic and carcinogenic than serpentines because of their higher iron content and potential to produce more oxygen radicals (Craighead et al.1982; Mossman et al.1988; 1989; Kamp et al.1992). On the other hand several studies have suggested that dissimilar physical and chemical characteristics of amphiboles and serpentines are important determinants of cytotoxicity, pathogenicity and carcinogenicity (Mossman et al.1990). In conventional cytotoxicity assays these dissimilar physical and chemical characteristics of minerals play an important role giving false positive and false negative results. We have shown in a recent study that in addition to the physical characteristics, the ability of asbestos to generate oxygen radicals through surface based iron and by frustrated phagocytosis may play an important role in cell injury (Vallyathan et al.1992).

The objective of the present study was to assess the potential role of several asbestos fiber types in the generation of hydroxyl radicals ($\cdot OH$) from hydrogen peroxide (H_2O_2) and to correlate, if possible, in vitro cytotoxicity, lipid peroxidation potential and in vitro generation of oxygen radicals during phagocytosis. Known mass concentrations of asbestos were mixed with H_2O_2 in the presence of a spin trap to capture the $\cdot OH$ generated. The stable radical adduct produced was measured in an electron spin resonance (ESR) spectrometer. The signal peak heights of the radical adduct provided a quantitative measure of the $\cdot OH$ generated. For correlation, cytotoxicity was measured as red blood cell hemolysis and release of alveolar macrophage (AM) enzymes after exposure to minerals. Lipid peroxidation, as a measure of radical-induced initial cellular membrane injury, was determined using a non-cellular in vitro assay.

MATERIALS AND METHODS

Asbestos used in this study originated from different geographic areas. They were obtained from: chrysotile (Quebec, Canada); amosite (Transvaal, South Africa); crocidolite (Kalahari Desert, South Africa). In addition, Code 100 fiberglass was obtained from Johns Manville, Denver, CO. All types of asbestos were originally obtained by the National Institute of Environmental Health Sciences (Research Triangle Park, NC) and was chemically and physically characterized in detail (Campbell et al.1980). Scanning electron microscopic analyses with the aid of x-ray spectrometric analyses on Nucleopore filters were made to confirm the size characteristics of fibers. Surface area measurements were made by the nitrogen adsorption technique.

Electron Spin Resonance (ESR) Measurements:

Spin trapping of radicals generated from the minerals was made in a total volume of 1 ml containing 1 mg dust, 100 μl 1 M 5,5,-dimethyl-1-pyrroline-1-oxide (DMPO) and 100 μl 0.1M H_2O_2. The reaction was initiated in a plastic syringe by the addition of H_2O_2, mixed well for 10 sec and filtered through a 0.45 μm nylon Acrodisc filter attached to the syringe. ESR measurements were made on 250 μl samples using a flat quartz cell. The measurements were completed within 7 min after mixing H_2O_2, using a Varian E 109 ESR operating at x-band ($\sim$9.4 GHz) frequency. A microwave power of 50 mW and modulation amplitude of 2 G were found to be adequate for the optimal development of peak heights without compromising on spectral resolution.

Cytotoxicity Measurements:

Hemolysis of red blood cells was measured as an index of cellular membrane damage. The test system consisted of 2 ml PBS, pH 7.4, containing 1 mg mineral and 2 ml of a 4% suspension of sheep erythrocytes. The percent hemolysis was measured according to the method described earlier (Vallyathan et al. 1988).

Alveolar Macrophage enzyme Studies:

For enzyme studies, AM were harvested from pathogen free male Fischer 344 rats by repetitive pulmonary lavages using a calcium and magnesium free Hank's balanced salt solution. AM were sedimented by centrifugation and suspended in HEPES buffered solution containing 10 mM glucose. Cell viability counts were made by the trypan blue dye exclusion method. Approximately 95% of the cells obtained by this technique were AM.

For enzyme studies, AM (2×10^6) were incubated with 1 mg dust at 37° C for 2 hr in a water bath with gentle mixing. Percent of LDH, β-NAG and β-GLUC released from the AM into the medium were determined according to the method described earlier (Vallyathan et al. 1988).

Lipid Peroxidation:

Mineral-induced lipid peroxidation, of the polyunsaturated lipid linoleic acid, was measured by monitoring the production of malondialdehyde generated in one hour according to the method of Hunter et al (1963). The reaction mixture, in a total volume of 1 ml, contained 1 mg mineral and 10 μl emulsified linoleic acid in phosphate buffer at pH 7.4. The reaction was terminated at the end of one hour by the addition of 0.3 ml 5 N HCl and 0.625 ml 40% trichloroacetic acid. The tubes were mixed and 0.625 ml of 2% thiobarbuturic acid was added and heated in a water bath at 95° C for 20 min. The thiobarbuturic acid reactive substances developed a color which was measured at 535 nm after cooling and centrifugation for 10 min at 600 x g. Malondialdehyde produced was calculated from a standard curve. Control experiments were carried out without minerals and in the presence of an antioxidant, butyl hydroxy toluene, to inhibit lipid peroxidation.

ESR Measurements of Oxygen Radicals During Phagocytosis:

Asbestos stimulated generation of oxygen radicals by rat AM was studied in the presence of a spin trap PBN. AM (1 X 10^6) were incubated with 250 ug dust and 0.1M PBN for 30 min. Reaction was terminated by the addition of 5 ml chloroform:methanol (3:1) and the total lipids extracted. The chloroform was evaporated and the lipid reconstituted in 0.5 ml chloroform and read in a ESR spectrometer as described earlier (Vallyathan et al.1992).

RESULTS

Physical and Chemical Characteristics of Dusts:

The physical and chemical characteristics of the minerals used in this study are presented in Table 1.

Mineral	Length (um)	Diameter (um)	Surface Area (m^2/G)	Iron %	MnO %
Fiberglass	>10	>0.25	4.3	>0.1	ND
Amosite	>41	0.53	4.0	34.61	2.66
Crocidolite	>10	0.27	9.8	18.32	0.08
Chrysotile	>21	0.17	24.9	2.93	0.06

ND-Not detected

The particle size and the size distributions differed considerably among the various types of minerals. Surface area measurements, by nitrogen adsorption, also showed wide differences in surface area reflecting the variability in size (Table 1).

·OH Radical Generation:

Minerals reacted with H_2O_2 in the presence of an ·OH radical trap DMPO to produce DMPO-

OH radical adducts showing a typical 1:2:2:1 quartet signal. All the ESR signals were similar except for differences in peak intensities (Fig.1a-1d). The ESR spectra obtained for amosite was 375% bigger than that of chrysotile. Crocidolite asbestos produced a signal 166% bigger than that of chrysotile. Fiberglass generated signals barely detectable above the background level produced by DMPO and H_2O_2. The signals produced by minerals in the presence of DMPO was centered around g= 2.0059 + 0.0003 with a splitting constant of 14.9 G, very characteristic of the DMPO-OH adduct (Buettner,1987). To further confirm the generation of ·OH radicals, a competitive ·OH radical scavenger, ethanol, was added in increasing concentrations to the reaction mixture. Addition of ethanol to the dust-DMPO-H_2O_2 mixtures resulted in the generation of a six-line spectrum characteristic of DMPO ethanolyl radical adducts. Metal chelator deferoxamine and ·OH radical scavengers such as catalase and sodium benzoate inhibited ·OH radical generation significantly (80-90%).

The relative potential of all minerals in equal mass concentration (1 mg/ml) to generate ·OH radicals from H_2O_2 is illustrated in Figure 2. It is clearly evident that amosite and crocidolite asbestos, which contained higher concentrations of iron, generated greater amounts of ·OH radical. However, iron concentration of minerals showed no direct correlation to the ·OH radical generation. This non-correlation was more evident when the ·OH radical generation by the minerals was expressed per unit surface area. On a unit surface area basis ($0.1 m^2$), amosite generated the strongest signal, whereas crocidolite and chrysotile produced much weaker signals. The ESR signal generation thereby showed no direct correlation with total iron content.

Cytotoxicity Studies:

The hemolysis potential of all the minerals in equal mass concentration (1 mg/ml) is presented in Figure 3. Chrysotile was the most hemolytic followed by amosite and crocidolite. Hemolytic potential on an equal surface area basis ($0.1 m^2$) showed that amosite was the most hemolytic followed by crocidolite, chrysotile and fiberglass. Release of the cytosolic enzyme LDH and lysosomal enzymes β-GLUC and β-NAG in equal mass concentrations (1 mg/ml) are shown in Figures 4-6. Amosite, crocidolite and chrysotile induced the release of the cytosolic enzyme LDH in

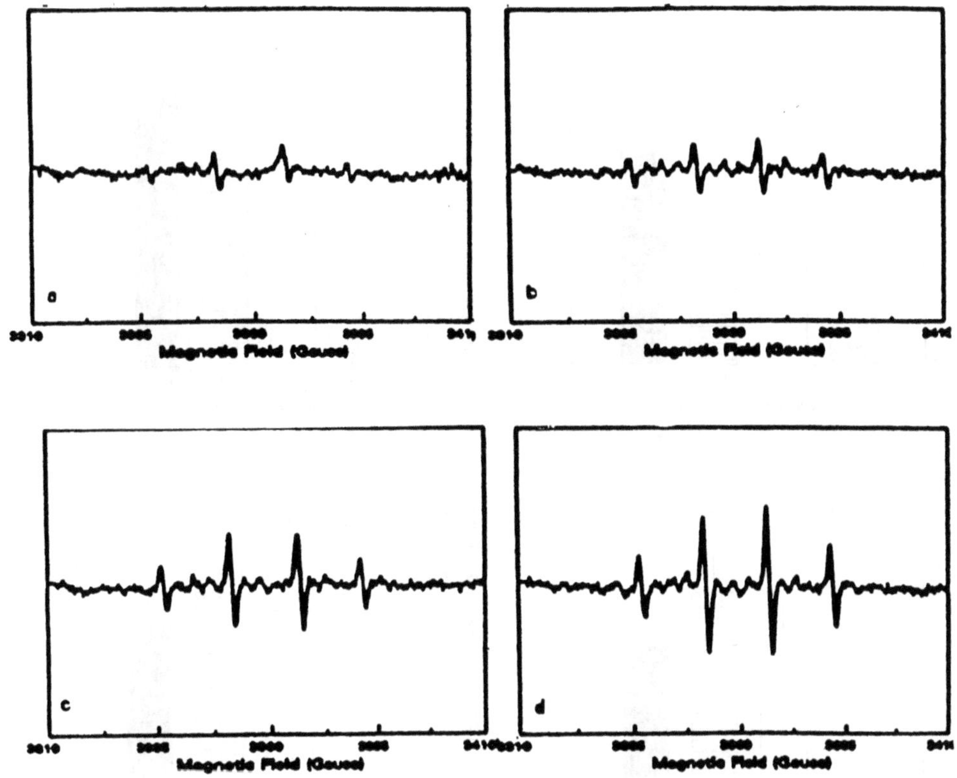

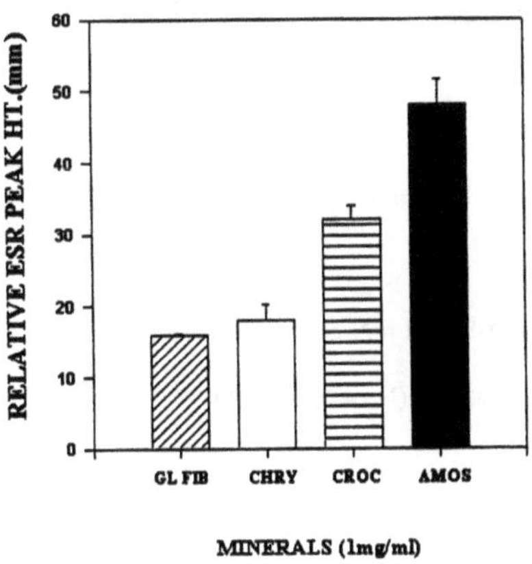

__Fig.1__ Electron spin resonance spectra obtained from mixtures of minerals, H_2O_2, and spin trap DMPO. Fig.1a. Spectrum obtained from 1mg fiberglass. Fig 1b. Spectrum obtained from 1 mg chrysotile. Fig.1c. Spectrum obtained from 1mg crocidolite. Fig.1d. Spectrum obtained from 1mg amosite.

__Fig.2__-Graph showing the relative intensities of DMPO ·OH adducts formed from 10 mM H_2O_2 and 1mg dust samples in the presence of 100mM DMPO.

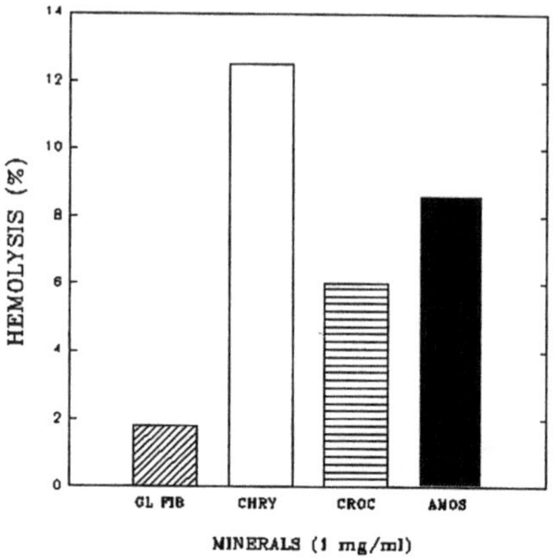

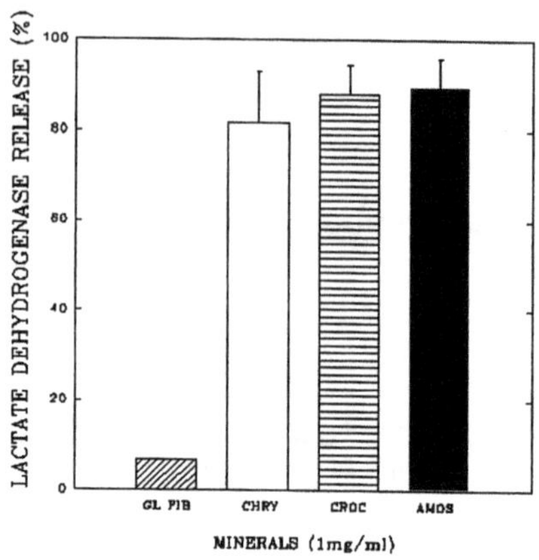

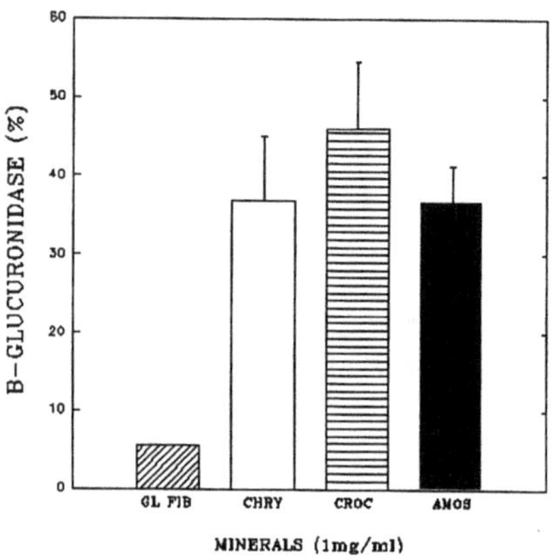

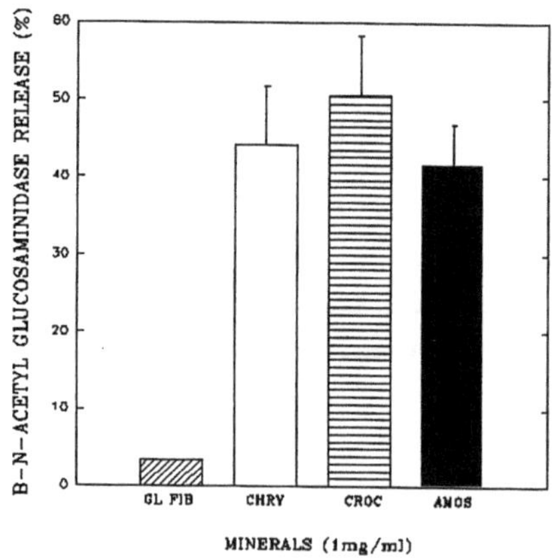

Fig.3 Hemolytic potential of minerals in equal mass concentration. Results presented are the means of (N=5).

Figs.4-6 Release of lactate dehydrogenase (Fig.4) β-glucuronidase (Fig.5) and β-N-acetyl glucosaminidase (Fig.6) from alveolar macrophages ($2x10^6$)/2hr. Results presented are means (N = 5) and SD.

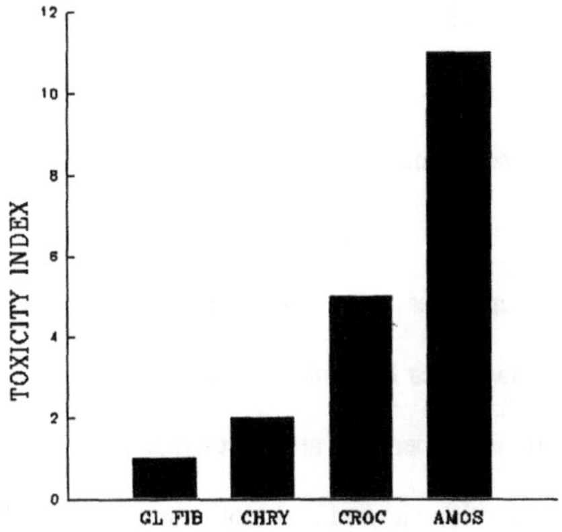

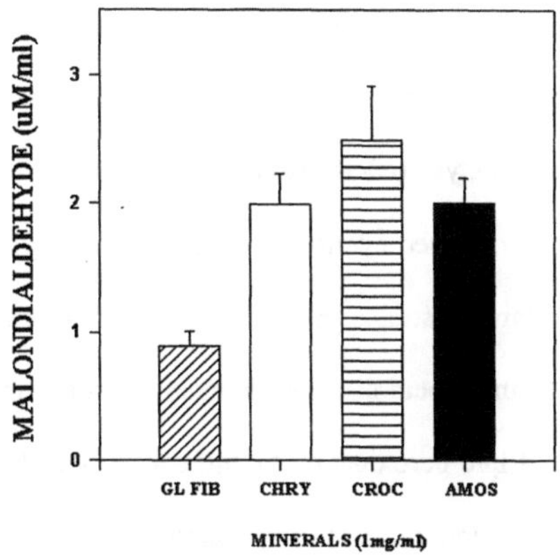

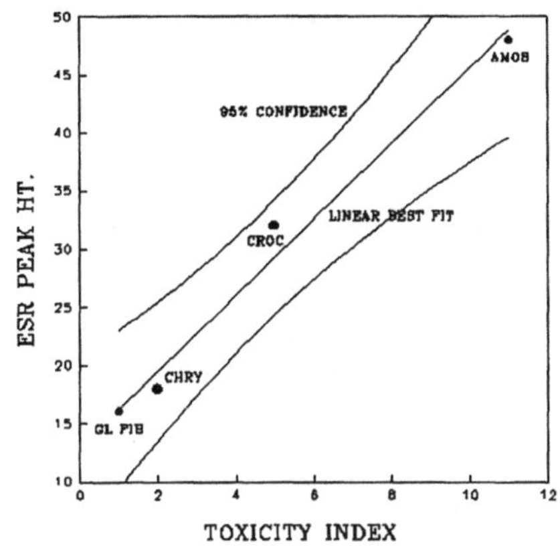

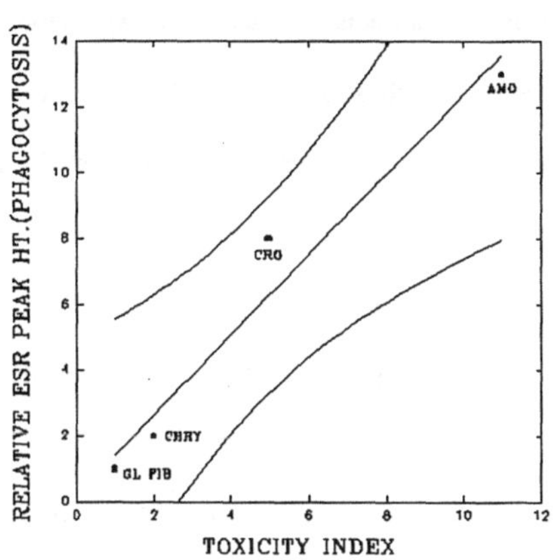

Fig.7 Combined cytotoxicity index of minerals on an equal surface area basis ($0.1m^2$). Cytotoxicity index was derived by adding the data on hemolysis and enzymes and normalizing to 1 for inert dust fiberglass.

Fig.8 Lipid peroxidation potential of minerals in equal mass (1mg) concentrations.

Fig.9 Cytotoxicity index of minerals based on equal surface areas ($0.1m^2$) vs ESR peak heights based on 1 mg mass. Toxicity index showed a good correlation with •OH radical generation.

Fig.10 Relative ESR peak heights vs toxicity index. ESR peak and toxicity index normalized to 1 for fiberglass on an equal surface area basis. The data indicates good correlation.

comparable levels. However, the selective releases of lysosomal enzymes β-GLUC and β-NAG were more pronounced by crocidolite compared to amosite and chrysotile. On an equal surface area basis amosite was the most cytotoxic followed by crocidolite, chrysotile and fiberglass.

Combined Cytotoxicity Index:

Combined cytotoxicity index of hemolysis and three enzymes on an equal surface area basis ($0.1m^2$) are illustrated in Figure 7. The data on an equal surface area basis clearly indicates that amosite and crocidolite are the most cytotoxic and chrysotile and fiberglass are least cytotoxic.

Lipid peroxidation by minerals in equal mass concentration showed crocidolite as a strong inducer of lipid peroxidation followed by amosite and chrysotile (Figure 8). However, when compared on an equal surface area basis, amosite was found to be a stronger inducer of lipid peroxidation followed by crocidolite and chrysotile.

Cytotoxicity VS ·OH Radical Generation by Minerals:

Figure 9 illustrates a direct relationship between the combined cytotoxicity index calculated on an equal surface area basis ($0.1\ m^2$) and ESR peak heights generated from minerals using equal masses ($r^2 = 0.99$, $p<0.01$). There was also no correlation between cytotoxicity index and ESR data calculated on an equal mass basis.

Cytotoxicity VS Oxygen Radical Generation during Phagocytosis:

The oxygen radicals generated during the phagocytosis of minerals is presented in Figure 10. The cytotoxicity index and the ESR peak heights are based on equal surface area ($0.1\ m^2$) basis. The data was normalized to 1 for fiberglass. The ESR peak generated as a result of the combined contributions of mineral and cells correlated well with cytotoxicity ($r^2=0.98$, $p<0.02$).

DISCUSSION

The goal of this study was to investigate the potential use of existing in vitro bioassays to evaluate their applicability to predict fibrogenicity of mineral dusts. This is particularly important because of the proposed ban on the use of asbestos by the United States Environmental Protection

Agency and the manufacturing and introduction of several asbestos substitutes for commercial use. Therefore an attempt was made to identify bioassays that would provide an alarm and reasonable justification for initiating further investigations. A group of well characterized asbestos particles was tested using a battery of cytotoxicity assays, including an assay for the ability of minerals to generate oxygen radicals by themselves and during phagocytosis.

The results suggest that several mechanisms are likely to be involved in the generation of oxygen radicals during asbestos cell interactions. Results with the iron chelator deferoxamine indicate that bioavailability of iron on the surface of asbestos is a major factor in the production of oxygen radicals. These data are also supported by our previous studies showing the augmentation of oxygen radical production when asbestos was treated with chelators such as DETAPAC thereby making iron available for the Fenton reaction (Vallyathan et al.1992).

Asbestos was shown to generate ·OH radicals from H_2O_2 by Weitzman and Gracefa (1985). Potential of asbestos to induce cytotoxicity and lipid peroxidation was shown by several investigators (Weitzman and Weitberg,1985; Rahman and Casciano,1985; Mossman et al.1988). From these studies, it is evident that asbestos particles cause the generation of oxygen free radicals and that these radicals are known to cause cell injury and disease. It is not clear what mechanisms are involved in the generation of oxygen free radicals during the interactions of asbestos and phagocytic cells. First it is evident that iron (Fe^{3+}) present in the asbestos is a major contributing factor in the generation of oxygen radicals (Mossman et al.1988). Another major source of oxygen radical generation is the "respiratory burst" resulting from the phagocytosis of asbestos fibers. The ability of asbestos to perpetuate the generation of oxygen free radicals as a result of cell death and recycling of fibers from repetitive phagocytosis may be a major factor. Finally, frustrated phagocytosis by several phagocyte commonly observed on long asbestos fibers may be a major source of oxygen free radicals.

From the results of these studies made on reference asbestos, we consider it is more valuable to study lipid peroxidation potential and generation of oxygen radicals during phagocytosis as predictive bioassays for fibrogenicity screening. The value of these predictive bioassays depends on their ability to estimate the potential fibrogenicity of untested materials. In order to establish that goal these bioassays must be tested and validated with several known fibrogenic reference materials.

REFERENCES

Buettner GR. Spin trapping: ESR parameters of spin adducts. Free Rad Biol Med 3: 259-303, 1987.

Cambell WJ, Huggins CW and Wylie AG.Chemical and physical characterization of amosite, chrysotile, crocidolite and non-fiborous tremolite for oral ingestion studies by the National Institute of Environmental Health Sciences. U.S.Department of Interior,R18452,pp1-63,-1980.

Craighead JE, Abraham JL, Churg A, Green FHY, Kleinerman J, Pratt PC, Seemayer TA, Vallyathan V and Weill H. Asbestos-Associated Diseases. The pathology of asbestos asoociated diseases of the lungs and pleural cavities: Diagnostic criteria and proposed grading schema. Arch Path Lab Med 106: 544-596,1982.

Hunter FE Jr, Gebicki JM, Hoffsten PE, Weinstein J and Scott A. Swelling and lysis of rat liver mitochondria induced by ferrous ions. J Biol Chem 238: 828-832, 1963.

Kamp DW,Graceffa P, Pryor W and Weitzman S. The role of free radicals in asbestos induced diseases. Free Rad Biol Med 12:293-315,1992.

Mossman BT, Marsh JP, Shatos MA, Doherty J, Gilbert R, and Hill S. Implication of active oxygen species as second messengers of asbestos toxicity. In: Asbestoc Toxicity, Editors Fisher GL and Gallo MA, Dekker, New York, NY, 1988, pp 157-180.

Mossman BT, Gee J. Asbestos related diseases. NEJM 320:1721-1730, 1989.

Mossman BT, Bignon J, Corn M, Eaton A and Gee J. Asbestos. Scientific developments and implications for public policy. Science 247:294-301, 1990.

Rahman Q and Casciano DA. Involvement of superoxide radical in the toxicity of mineral fibers. In: In Vitro Effects of Mineral Dusts. Springer-Verlag, Berlin, Germany, pp483-488, 1985.

Vallyathan V, Mega JF, Shi X and Dalal NS. Enhanced generation of free radicals from phagocytes induced by mineral dusts. Am J Respir Cell Mol Biol 6:404-413, 1992.

Vallyathan V, Schwegler D, Reasor M, Stettler L, Clere J and Green FHY. Comparative in vitro cytotoxicity and relative pathogenicity of mineral dusts. Ann occup Hyg 32:279-289, 1988.

Weitzman SA and Graceffa P. Asbestos catalyzes hydroxyl and superoxide radical generation from hydrogen peroxide. Arch Biochem Biophys 228:373-376, 1984.

Weitzman SA and Weitberg AB. Asbestos-catalyzed lipid peroxidation and its inhibition by deferoxamine.Biochem J 225:259-262, 1985.

Fiber-Induced Hydroxyl Radical Formation and DNA Damage

Neil F. Johnson and Kirk R. Maples
Inhalation Toxicology Research Institute
P. O. Box 5890, Albuquerque
New Mexico 87185
USA

Summary

We have measured the phagosomal pH (ppH) in tracheal epithelial cells, alveolar type II cells, and mesothelial cells. The ppH was measured using a flow cytometric procedure employing fluorescein-isothiocyanate-labeled, amorphous silica particles. The ppH for these cells ranged from 7.0 to 7.6, with no statistically significant difference in ppH between the cell types. We also compared the ability of a series of fibers to initiate hydroxyl radical formation in an acellular system at pH 4.0, an acidic pH similar to the reported ppH for rat alveolar macrophages, and at pH 7.0, the ppH determined in this study for epithelial and mesothelial cells. In general, for those fibers capable of initiating hydroxyl radical formation, the capacity of the fiber to generate hydroxyl radicals was dependent on the pH. Most of the fibers tested were more effective at initiating hydroxyl radicals at pH 7.0 than at pH 4.0. These results suggest that the ability of fibers to induce varying amounts of injury to different cell types may be due to differing pH effects within cellular regions. The capacity of crocidolite to induce DNA damage was determined by measuring p53 expression. Crocidolite induced expression of p53 in cultured lung epithelial cells. The induced p53 expression was extinguished by salicylic acid suggesting that hydroxyl radicals play a role in crocidolite-induced DNA damage.

NATO ASI Series, Vol. H 85
Cellular and Molecular Effects
of Mineral and Synthetic Dusts and Fibres
Edited by J. M. G. Davis and M.-C. Jaurand

Introduction

The differences in toxicity of various fibers to the same cell population have been ascribed to the differences in the physical and structural properties of the fibers. These properties include fiber dimensions (Stanton and Wrench, 1972; Lippman, 1988), surface charge (Light and Wei, 1977a;b), mineral content (Timbrell, 1970; Lund and Aust, 1990; 1991), and solubility (Johnson, 1993). However, these properties alone cannot account for the differences in toxic response found among different cell types that receive equivalent doses of the same fiber. Cells must have certain distinguishing characteristics that make them more or less susceptible to fiber-induced toxicity.

Recently, we investigated whether differences existed among the phagosomes from different cell types. We found that the phagosomal pH (ppH) in cultured rat alveolar macrophages and the ppH in cultured nasal epithelial cells are quite different (Johnson, 1993). The ppH in alveolar macrophages had a mean value of 5.4, while the ppH in nasal epithelial cells had a mean value of 7.2. This variation in pH could make a major difference in the surface chemistry associated with the fibers and could help explain the differences in susceptibility of cells to fiber-induced injury.

This study represents an extension of the previous work and examines other target cells at risk from fibers, since fibers and particles have been detected not only in alveolar macrophages (Suzuki *et al.*, 1972; Brody *et al.*, 1981; 1982; Jaurand *et al.*, 1984; Nyberg *et al.*, 1989; 1991), but also in alveolar type I epithelial cells (Suzuki *et al.*, 1972; Brody *et al.*, 1981; 1982; Johnson *et al.*, 1984), alveolar type II epithelial cells (Corrin, 1969; Suzuki *et al.*, 1972; Johnson *et al.*, 1984), tracheobronchial epithelial cells (Mossman *et al.*, 1977; Haugen *et al.*, 1982), and pleural mesothelial cells (Jaurand *et al.*, 1979; 1981; 1984). We have used flow cytometric, fluorescence-emission-ratioing techniques with fluorescent-labeled, amorphous silica particles to measure the ppH in rat tracheal epithelial, alveolar type II, and mesothelial cells. We have also examined the effects of differing pH conditions on the ability of a series of fibers to initiate hydroxyl radical (•OH) formation in an acellular system. In the previous work, we showed that at pH 7.4 significant •OH formation was detected for erionite, glass fibers, crocidolite, amosite, and chrysotile, but not for anthophyllite, in an acellular system (Maples and Johnson, 1992). In this study, we examined the ability of these same fibers to initiate •OH formation at pH 4 and studied one additional glass fiber and three silicon carbide whiskers at both pH 4 and 7. In addition, we have investigated the involvement of •OH in the induction of DNA damage by using the expression of p53 as an indicator of DNA damage. Transient expression of p53 after exposure to various different DNA damaging agents has been observed in many cell types (Kastan *et al.*, 1991; Fritsche *et al.*, 1993).

Material and Methods

Fibers and Chemicals: We examined the following fibers: anthophyllite (U.I.C.C.), amosite (U.I.C.C.), chrysotile (U.I.C.C.), crocidolite (U.I.C.C.), erionite (provided by Dr. R. Brown, MRC Toxicology Unit, Leicester, UK), JM Code 100 (fine glass fibers, Johns Manville Corporation, Denver, CO), glass wool (GFB; Owens-Corning Fiberglas, Toledo, OH), specialty glass fibers (GFA; Owens-Corning Fiberglas, Toledo, OH), silicon carbide whisker sample #1 (SiC1; Alcan Aluminum Corporation, Clarks Summit, PA), silicon carbide whisker sample #2 (SiC2; Advanced Composite Materials Corp., Greer, SC), and silicon carbide whisker sample #3 (SiC3; American Matrix Inc., Knoxville, TN). Salicylic acid, 2,3-dihydroxybenzoic acid (2,3-DHBA), 2,5-dihydroxybenzoic acid (2,5-DHBA), and 2,6-dihydroxybenzoic acid (2,6-DHBA) were purchased from Aldrich (Milwaukee, WI). We purchased hydrogen peroxide (30%), sodium acetate, sodium citrate, propidium iodide, phosphate buffered saline (PBS), RPMI medium 1640, alpha MEM medium, newborn calf serum, and fetal calf serum from Sigma (St. Louis, MO).

Methanol (HPLC Grade) was purchased from Burdick and Jackson (Baxter, McGraw Park, IL). Chelex 100 resin (sodium form, 100-200 mesh) was purchased from Bio-Rad (Richmond, CA). The p53 monoclonal antibody was purchased from Oncogene Science (Uniondale, NY). Horse serum, biotinylated antimouse IgG, and streptavidin-fluorescein was purchased from Vector (Burlingame, CA). The isotype monoclonal antibody was purchased from Dako (Carpinteria, CA). A cell line of normal lung epithelial cells (LEC) established by Li *et al.* (1983) was cultured between passages 17 and 25 in RPMI 1640 media containing 10% heat-inactivated newborn calf serum.

Phagosome pH Determination Procedure: The pH of the phagosome was determined using a flow cytometric procedure employing 3-μm diameter fluorescein-isothiocyanate (FITC)-labeled, amorphous silica particles (Nyberg *et al.*, 1989; 1991); the fluorescence of fluorescein is pH sensitive. The cells were cultured in the presence of the FITC-coated particles for 20 hours after a 20-hour plating period and 20 days of culture. A target concentration of five particles per cell was used. The pH was analyzed using a dual-laser flow cytometer (Becton Dickinson, San Jose, CA); the FITC was excited at 457- and 488-nm wavelengths. The emissions from the particles were measured using a 515-nm, narrow-band pass filter. The stored fluorescence data from the cell studies were edited using forward and side-scatter signals to eliminate non-cell-associated events, i.e., those events relating to free particles and debris. The pH was estimated by taking the difference between the mean fluorescence channel number of the 457/515- and 488/515-edited histograms (Nyberg *et al.*, 1989; 1991). This value was then used to obtain the pH of the phagolysosome from its intersection with the regression lines obtained from the pH calibration curve. This curve was derived from determining the fluorescence ratios of coated particles measured in buffers of premeasured pH.

Hydroxyl Radical Assay: The ability of the fibers to initiate •OH formation was determined using the salicylate assay (Floyd *et al.*, 1984; Grootveld and Halliwell, 1986). In this assay, an HPLC (Spectra-Physics, model SP8700XR, San Jose, CA) equipped with an electrochemical detector and a glassy carbon electrode (BAS model LC-4B, W. Lafayette, IN) was used to separate and quantitate the •OH products derived from •OH attack on salicylic acid. The primary products are 2,3-DHBA and 2,5-DHBA. In the experiment, either the fiber (1 mg/mL, final concentration) or ferrous iron (0-10 μM) was added to a solution of salicylic acid (1 mM) in chelexed Milli-Q water. All regents were prepared using chelexed Milli-Q water, and the salicylate solution was stored over chelex 100 resin. The salicylate solution was carefully adjusted with small amounts of HCl (6M) or NaOH (6M) to either pH 4.0 or 7.0 before the addition of any other reagents. Once either the fiber or iron was added, the mixture was incubated at 37°C for 5 minutes. At the end of this time, hydrogen peroxide (20 mM) was added, and the solution was re-incubated for 5 minutes. 2,6-DHBA (1 μM) was added as an internal standard to the solutions immediately after the second incubation. The pH of the mixtures did not change during the incubations. The samples were then filtered using 0.2-μm LV PVDF acrodisc filters (Fisher Scientific, Pittsburgh, PA) and immediately run on the HPLC. The concentrations of 2,3-DHBA and 2,5-DHBA were determined by comparison of their peak heights to that for the internal standard. Chromatography was performed on an Ultrasphere ODS 5-μm, 4.6 mm x 25 cm column (Beckman, San Ramon, CA). The mobile phase was run isocratic and consisted of 15% methanol/85% 0.03 M sodium acetate, 0.03 M sodium citrate, and pH 3.6 buffer. The HPLC flow rate was 0.9 mL/minute, and the electrochemical detector potential was +0.9V. Under our conditions, 2,5-DHBA eluted at 7 minutes, 2,3-DHBA eluted at 9 minutes, 2,6-DHBA eluted at 21 minutes, and untreated salicylate eluted at 28 minutes.

We express our units for •OH formation capacity in terms of μM Fe(II) equivalents per mg fiber. An Fe(II) equivalent is equal to the extent of the hydroxylation product formed as compared to what was produced by 1 μM Fe(II). The unit conversion used to convert from peak-height ratios for 2,3-DHBA:2,6-DHBA and 2,5-DHBA:2,6-DHBA to μM Fe(II) equivalents was based on the results obtained from at least four replicate trials using ferrous iron in place of the fibers in the various experiments. We chose to normalize our data to the weight of the fiber to allow comparison to our published data (Maples and Johnson, 1992).

Cell Isolation and Culture Techniques: Rat tracheal epithelial cells (Johnson *et al.*, 1987), alveolar type II cells (Johnson and Jaramillo, 1989; Johnson *et al.*, 1991), and mesothelial cells (Jaurand *et al.*, 1981) were isolated and cultured as previously described. In addition, nasal epithelial cells were obtained as follows. The head of the rat was removed; the lower jaw, skin, and musculature were removed; the head was sectioned longitudinally in half from the nares to the pharynx. The nasal septum was removed, and the epithelium from the lateral wall was stripped and placed in RPMI 1640 media. The epithelium was rinsed, placed in alpha MEM media containing protease type XIV (3.3 mg/mL), and incubated for 30 minutes at 37°C. The resultant cell suspension was mixed with media containing fetal calf serum and filtered. The cells were enumerated using a hemacytometer, and their viability was assessed by a trypan blue exclusion method. The cell suspensions were adjusted to a concentration of 5.0×10^5 cells/mL by dilution with RPMI 1640 media.

p53 Expression: Lung epithelial cells which were originally derived from normal F344 rats as described by Li *et al.* (1983) were cultured for 48 hours so that the LEC were 70-80% confluent prior to treatment. Cells were cultured in 225 cm^2 tissue culture flasks in RPMI 1640 medium containing 10% newborn calf serum. The culture flasks were divided into three groups. One group was incubated with 20 μg/mL UICC crocidolite (3.1 $\mu g/cm^2$) for 4 hours after which the cells were washed in PBS and harvested by trypsinization, pelleted by centrifugation, and resuspended in PBS. Another group of cultures was treated with 10 mM salicylic acid for 30 minutes prior to the addition of crocidolite; thereafter they were treated as above. The third group of cultures received no treatment and were harvested as above. The cell suspensions were fixed in 70% methanol at -20°C for at least 20 minutes after which they were washed in PBS and permeabilized with 0.05% Triton 100 in PBS. Nonspecific immunoreactivity was blocked with 1.0% horse serum and stained overnight at +4°C with a p53-specific monoclonal antibody at a dilution of 1:100. The cells were washed and incubated with biotinylated IgG (preabsorbed with 1.0% rat serum) for 60 minutes at room temperature. The cells were washed again and labeled with a streptavidin/fluorescein conjugate for 30 minutes in the dark at room temperature. Stained cells were analyzed by flow cytometry; forward light scatter, side angle light scatter, and fluorescent signals were collected and stored for analysis.

Statistics: Statistical significance was based on a Student's *t* test number using a Bonferroni correction for multiple comparisons. The criterion for significance was set at $p < 0.05$.

Results

ppH measurements: The measurement of the ppH showed that the values for the tracheal epithelial, alveolar type II, and mesothelial cells were equivalent to that obtained for nasal epithelial cells (Table 1). As with other cell types, nasal epithelial cells in culture took up particles into membrane-bound cytoplasmic phagosomes. Electron microscopic techniques have been used to establish that the organelle within which particles or fibers are found are phagolysosomes in the case of macrophages (Nyberg *et al.*, 1989; 1991), tracheal epithelial cells (Mossman *et al.*, 1977), and mesothelial cells (Jaurand *et al.*, 1979). Thus, we believe the organelle in which the pH was measured in these cells was also a phagolysosome. However, to our knowledge there is no evidence to suggest in the case of the nasal epithelial cells and the alveolar type II cells that the organelle measured was a phagolysosome rather than a phagosome. Therefore, in the case of these two latter cell types, we refer to the organelle measured as a phagosome.

Cell Type	pH	pH Range	Reference
Nasal Epithelium	7.2 ± 0.3[a]	7.0 - 7.5	Johnson, 1993
Tracheal Epithelium	7.3 ± 0.5[b]	7.2 - 7.5	This work
Alveolar Type II Cells	7.4 ± 0.2[a]	7.2 - 7.6	This work
Alveolar Macrophages	5.4 ± 0.3[c]	5.2 - 5.8	Johnson, 1993
		4.5 - 5.1[d]	Nyberg *et al.*, 1989, 1991
			Kreyling *et al.*, 1991
Mesothelial Cells	7.1 ± 0.5[a]	7.0 - 7.4	This work

[a]Mean ± SD, n = 3
[b]Mean ± SD, n = 4
[c]Mean ± SD, n = 5
[d]Range of pH values obtained from *in vitro* and *in vivo* studies for macrophages from various species of animals.

Table 1. Measured values of ppH for pulmonary epithelial cells, alveolar macrophages, and mesothelial cells.

●OH formation: The capacities of the various fibers to initiate ●OH formation at pH 4.0 and 7.0 are shown in Figure 1. As we have shown (Maples and Johnson. 1992), erionite is the fiber most capable of initiating ●OH formation at pH 7.0. Erionite and crocidolite, amosite, chrysotile, JM Code 100, GFA, and GFB all yielded significantly elevated capacities to initiate ●OH at pH 7.0, as compared to the no-fiber control. In contrast to the results at pH 7, erionite was unable to stimulate ●OH formation at pH 4 over control levels. However, amosite, JM Code 100, GFA, GFB, and crocidolite could initiate all capable of initiating ●OH formation at pH 4 (Fig. 1).

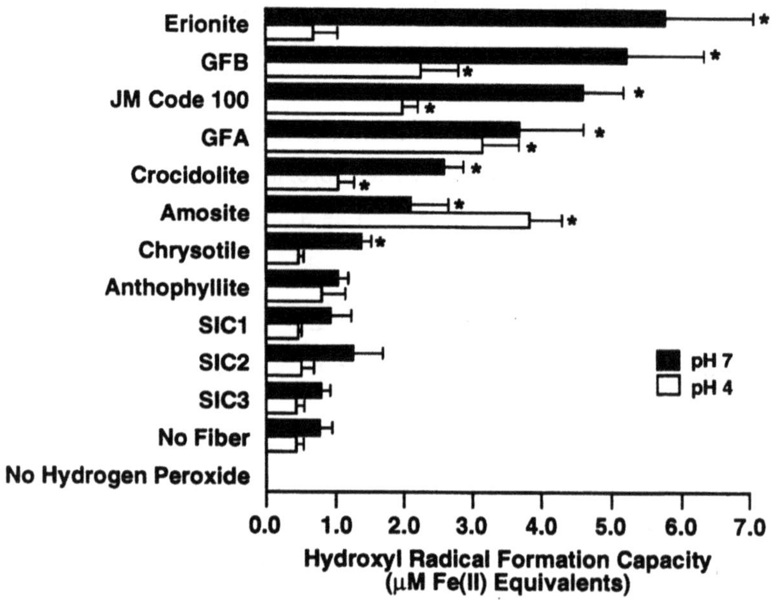

Figure 1. Hydroxyl radical formation capacity of fibers (1 mg/mL) incubated at 37°C with salicylate (1 mM) and hydrogen peroxide (20 mM) at pH 4.0 or 7.0. The data are expressed as mean ± SE for three to six determinations. The asterisks represent a value significantly different from the no particle control.

The three silicon carbide whiskers did not evince any significant ●OH-forming capacity at either pH. With the exception of amosite, the fibers' capacities to generate •OH were favored at the higher pH. The background capacity shown for the no-fiber control was likely due to catalysis by trace-metal contamination on the surface of the reaction flasks during the 37°C incubation. If hydrogen peroxide was not included in the incubations, 2,3-DHBA and 2,5-DHBA could not be detected (our limit of detection was 0.01 μM Fe(II) equivalents). Likewise, we found the signals for these hydroxylation products to be dependent on the inclusion of salicylate.

We examined the capacity of erionite and crocidolite to initiate •OH formation in solutions of varying pH. As shown in Figure 2, the capacity of erionite to initiate •OH formation was more greatly influenced by the pH of the solution than was the case for crocidolite.

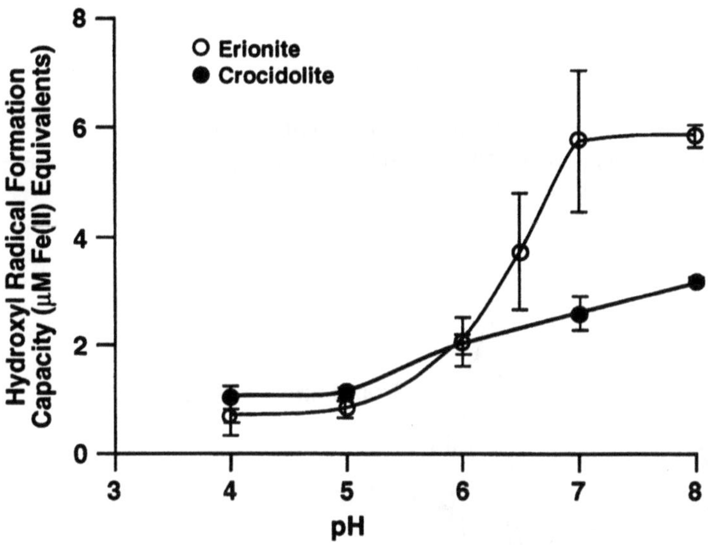

Figure 2. The effect of pH on the capacity of 1 mg/mL of either crocidolite or erionite to initiate •OH formation in mixtures containing hydrogen peroxide (20 mM) and salicylate (1 mM). The data are expressed as the mean ± SE for three to six determinations.

We examined the effect of crocidolite to induce the expression of p53 protein and determined whether this expression was associated with the formation of •OH. As shown in Figure 3, crocidolite induced a transient expression of p53

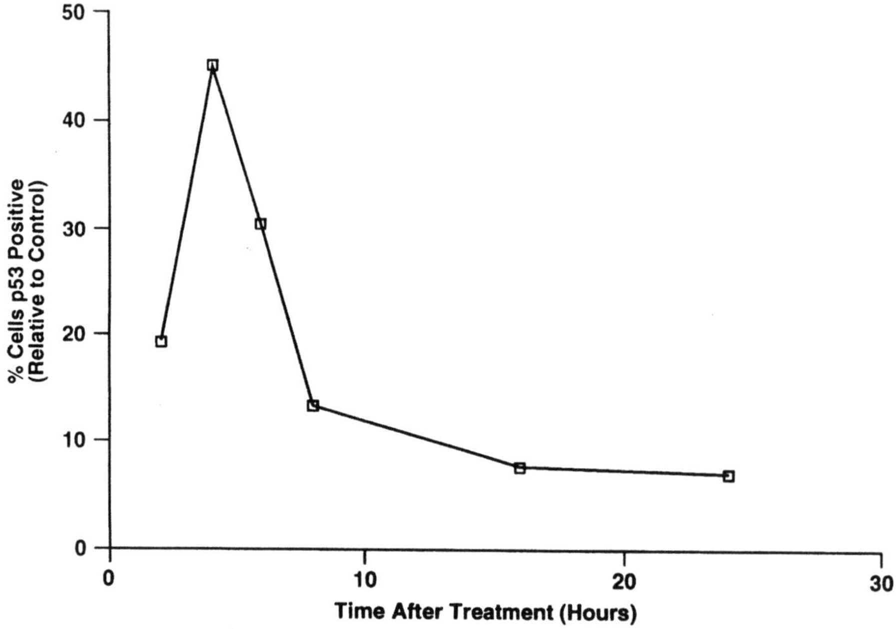

Figure 3. The expression of p53 protein in LEC following treatment with crocidolite. The data are from a single experiment.

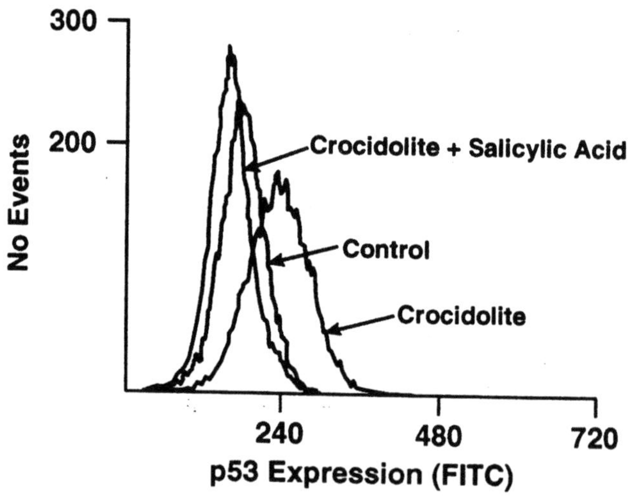

Figure 4. Expression of p53 protein in LECs treated with crocidolite, salicylic acid and crocidolite, or salicylic acid (control). The data are from a single experiment.

protein in LECs which was maximal 4 hours after treatment. This response was extinguished by treating the cells with salicylic acid prior to the addition of the crocidolite.

Discussion

Fibers can be phagocytized by many cell types within the lung. In rats, the majority of inhaled fibers are found within alveolar macrophages; however, a small fraction (< 0.2%) of the inhaled fibers can be found within the epithelial cells lining the respiratory tract (bronchial and type II cells), in fibroblasts within the interstitium, and in cells in the subpleural connective tissue layer (Johnson *et al.*, 1984; Johnson and Wagner, 1989). The interaction of fibers within epithelial cells and mesothelial cells can result in cancer, while fiber/alveolar macrophage interactions may be important in the pathogenesis of pulmonary fibrosis (Dunnill, 1982).

As we and others have shown, the ppH in alveolar macrophages is acidic with a pH of 4-6 (Nyberg *et al.*, 1989; 1991; Kreyling *et al.*, 1991; Johnson, 1993). Our results suggest that the threat of •OH formation from an erionite fiber engulfed in an alveolar macrophage phagolysosome under these ppH conditions would be minimized, whereas the threat of •OH formation in an epithelial or mesothelial phagolysosome wherein the pH is neutral would be much greater. Some evidence supports this concept. It has been shown that when erionite, crocidolite, or JM Code 100 are incubated in culture with various cell types, the ED_{50}, the effective dose to reduce cell survival by 50%, is much less in rat tracheal cells than in alveolar macrophages (Johnson *et al.*, 1992). This finding indicates that these three fibers were less toxic in the macrophages than in the tracheal epithelial cells. Thus, for some fibers, the differences in toxicity to different cell types may be partially attributed to the ppH influencing the capacity of the engulfed fiber to participate in redox reactions such as •OH generation.

For the majority of the fibers tested, if the fiber was capable of initiating •OH formation, the pH of the solution affected this capacity. Erionite, GFB, JM Code 100, crocidolite, and chrysotile were more effective initiators at pH 7.0, whereas amosite was more effective at pH 4.0. It is interesting to note that GFA yielded statistically equivalent results at both pH values. GFA is a specialty glass fiber with a rapid dissolution rate (Potter and Mattson, 1991; Johnson, 1993). The lack of a pH effect for GFA may possibly be attributed to its solubility, if solution chemistry is not pH dependent and plays a larger role in GFA •OH formation capacity than does surface chemistry. The striking differences in •OH formation capacity for both erionite (most effective at pH 7.0) and amosite (most effective at pH 4.0) under the different pH conditions suggest different surface chemistries for these fibers.

In addition to affecting the capacity of a cell to initiate •OH formation, the differences in ppH could influence fiber dissolution rates (Potter and Mattson, 1991; Johnson, 1993). Although pH has been shown to be important for glass fiber dissolution in an acellular system (Potter and Mattson, 1991), this point could not be confirmed in studies using cultured cells (Johnson, 1993). This discrepancy in results was explained on the basis that in the cells, protein binding to the fibers could alter the dissolution rates sufficiently to prevent detection of differential dissolution due to ppH effects (Johnson, 1993).

Our preliminary results demonstrate that crocidolite like other agents can induce the expression of p53 protein. The induction of p53 has been associated with the ability of various agents to induce DNA damage (Kastan *et al.*, 1991). This suggests that crocidolite may also cause DNA damage in cultured cells. The time course of induction is similar to other agents such as UV and sparsely ionizing radiation (Maltzman and Czyzyk, 1984; Kuerbitz *et al.*, 1992). Crocidolite-induced p53 expression appears to be mediated by •OH since the expression is extinguished in cells treated with salicylic acid, which has been shown to be a free-radical scavenger in biological systems (Sagone and Husney, 1987). The increased levels of p53 following exposure to DNA-damaging agents have been shown to result from post-translational stabilization of the protein (Fritsche *et al.*, 1993). This post-translational modification of the p53 protein may result from changes in the redox status of the cell (Hainaut and Milner, 1993). Our preliminary investigations of fiber-induced p53 expression are being continued to determine the dose-response relationship for crocidolite treatment and to determine whether p53 expression is common to any fiber treatment of LECs or whether it is restricted to fibers known to induce pulmonary disease.

Acknowledgements: This research was sponsored by the U.S. DOE/OHER under

Contract No. DE-AC04-76EV01013 in facilities fully accredited by the American Association for the Accreditation of Laboratory Animal Care.

References

Brody, AR, Hill, LH, Adkins, B, and O'Connor, RW. (1981) Chrysotile asbestos inhalation in rats: Deposition pattern and reaction of alveolar epithelium and pulmonary macrophages. *Am. Rev. Respir. Dis.* 123: 670-679.

Brody, AR, Roe, MW, Evans, JN, and Davis, GS. (1982) Deposition and translocation of inhaled silica in rats. Quantification of particles distribution, macrophage participation, and function. *Lab. Invest.* 47: 533-542.

Corrin, B. (1969) Phagocytic potential of pulmonary alveolar epithelium with particular reference to surfactant metabolism. *Thorax* 24: 110-115.

Dunnill, MS. (1982) Industrial lung disease. In: *Pulmonary Pathology.* Churchill Livingstone, New York, NY pp. 399-438.

Floyd, RA, Watson, JJ, and Wong, PK. (1984) Sensitive assay of hydroxyl free radical formation utilizing high pressure liquid chromatography with electrochemical detection of phenol and salicylate hydroxylation products. *J. Biochem. Biophys. Methods* 10: 221-235.

Fritsche, M, Haessler, C, and Brandner, G. (1993) Induction of nuclear accumulation of the tumor-suppressor protein p53 by DNA-damaging agents. *Oncogene* 8: 307-318.

Grootveld, M and Halliwell, B. (1986) Aromatic hydroxylation as a potential measure of hydroxyl-radical formation *in vivo.* *Biochem. J.* 237: 499-503.

Hainaut, P, and Milner, J. (1993) Redox modulation of p53 conformation and sequence-specific DNA binding. *Cancer Res.* 53: 4469-4473.

Haugen, A, Schafer, PW, Lechner, JF, Stoner, GD, Trump, BF, and Harris, CC. (1982) Cellular ingestion, toxic effects, and lesions observed in human bronchial epithelial tissue and cells cultured with asbestos and glass fibers. *Int. J. Cancer* 30: 265-272.

Jaurand, MC, Kaplan, H, Thiollet, J, Pinchon, MC, Bernaudin, JF, and Bignon, J. (1979) Phagocytosis of chrysotile fibers by pleural mesothelial cells in culture. *Am. J. Pathol.* 94: 529-538.

Jaurand, MC, Bernaudin, JF, Renier, A, Kaplan, H, and Bignon, J. (1981) Rat pleural mesothelial cells in culture. *In Vitro* 17: 98-106.

Jaurand, MC, Gaudichet, A, Halpern, S, and Bignon, J. (1984) *In vitro* biodegradation of chrysotile fibers by alveolar macrophages and mesothelial cells in culture: comparison with a pH effect. *Brit. J. Indust. Med.* 41: 389-395.

Johnson, NF. (1993) Intraphagosomal pH glass fiber dissolution in cultured rat nasal epithelial cells and alveolar macrophages: A preliminary study. *Environ. Health Perspect.* (in press)

Johnson, NF, Griffiths, DM, and Hill, RJ. (1984) Size distribution following long-term inhalation of MMMF. In: *Biological Effects of Man-Made Mineral Fibers.* Vol. 2; WHO, Copenhagen, pp. 102-125.

Johnson, NF, Margiotta, EA, Wilson, JS, Sebring, RJ, and Smith, DM. (1987) Preparation of viable single cell suspensions of tracheal epithelial cells. *Br. J. Exp. Path.* 68: 157-165.

Johnson, NF, and Jaramillo, RJ. (1989) Flow cytometric isolation of rat alveolar Type II cells. In: *Inhalation Toxicology Research Institute Annual Report,* LMF-126, National Technical Information Service, U. S. Department of Commerce, Springfield, VA, pp. 123-124.

Johnson, NF, and Wagner, JC. (1989) Effects of erionite inhalation on the lungs 6 rats. In: *Biological Interaction of Inhaled Mineral Fibers and Cigarette Smoke.* Battelle Press, New York, NY, pp. 325-345.

Johnson, NF, Jaramillo, RJ, McLeod, MK, and Lechner, JF. (1991) Matrix driven differentiation and proliferation of rat alveolar type II cells. In: *Inhalation Toxicology Research Institute Annual Report,* LMF-134, National Technical Information Service, U. S. Department of Commerce, Springfield, VA, pp. 145-148.

Johnson, NF, Hoover, MD, Thomassen, DG, Cheng, YS, Dalley, A, and Brooks, A. L. (1992) *In vitro* activity of silicon carbide whiskers in comparison to other industrial fibers using four cell culture systems. *Am. J. Ind. Med.,* 21: 807-823.

Kastan, MB, Onyekwere, O, Sidransky, D, Vogelstein, B, and Craig, RW. (1991) Participation of p53 protein in the cellular response to DNA damage. *Cancer Res.* 51: 6304-6311.

Kreyling, WG, Nyberg, K, Nolibe, D, Collier, CG, Camner, P, Heilmann, P, Lirsac, PN, Lundborg, M, and Matejkova, E. (1991) Interspecies comparison of phagolysosomal pH in alveolar macrophages. *Inhal. Toxicol.* 3: 91-100.

Kuerbitz, SJ, Plunkett, BS, Walsh, WV, and Kastan, MB. (1992) Wild-type p53 is a cell cycle checkpoint determinant following irradiation. *Proc. Natl. Acad. Sci.* USA 89: 7491-7495.

Li, AP, Hahn, FF, Zamora, PO, Shimizu, RW, Henderson, RF, Brooks, AL, and Richards, R. (1983) Characterization of a lung epithelial cell strain with potential applications in toxicological studies. *Toxicology* 27: 257-272.

Light, WG, and Wei, ET. (1977a) Surface charge and asbestos toxicity. *Nature* 265: 537-539.

Light, WG, and Wei, ET. (1977b) Surface charge and hemolytic activity of asbestos. *Environ. Res.* 13: 135-145.

Lippman, M. (1988) Asbestos exposure indices. *Environ. Res.* 46: 86-106.

Lund, LG, and Aust, AE. (1990) Iron mobilization from asbestos by chelators and ascorbic acid. *Arch. Biochem. Biophys.* 278: 60-64.

Lund, LG, and Aust, AE. (1991) Mobilization of iron from crocidolite asbestos by certain chelators results in enhanced crocidolite-dependent oxygen consumption. *Arch. Biochem. Biophys.* 287: 91-96.

Maltzman, W, and Czyzyk, L. (1984) UV irradiation stimulates levels of p53 cellular tumor antigen in nontransformed mouse cells. *Mol. Cell Biol.* 4: 1689-1694.

Maples, KR, and Johnson, NF. (1992) Fiber-induced hydroxyl radical formation: Correlation with mesothelioma induction in rats and humans. *Carcinogenesis* 13: 2035-2039.

Mossman, BT, Kessler, JB, Ley, BW, and Craighead, JE. (1977) Interaction of crocidolite asbestos with hamster respiratory mucosa in organ culture. *Labor. Invest.* 36: 131-139.

Nyberg, K, Johansson, A, and Camner, P. (1989) Intraphagolysosomal pH in alveolar macrophages studied with fluorescein-labeled amorphous silica particles. *Exp. Lung Res.* 15: 49-62.

Nyberg, K, Johansson, U, Johansson, A, and Camner, P. (1991) Phagolysosomal pH and location of particles in alveolar macrophages. *Fund. Appl. Toxicol.* 16: 393-400.

Potter, RM, and Mattson, SM. (1991) Glass fiber dissolution in a physiological saline solution. *Glastech. Ber.* 64: 16-28.

Sagone, AL, and Husney, RM. (1987) Oxidation of salicylates by stimulated granulocytes: Evidence that these drugs act as free radical scavengers in biological systems. *J. Immunol.* 138: 2177-2183.

Stanton, MF, and Wrench, C. (1972) Mechanisms of mesothelioma induction with asbestos and fibrous glass. *J. Natl. Cancer Inst.* 48: 797-821.

Suzuki, Y, Churg, J, and Ono, T. (1972) Phagocytic activity of the alveolar epithelial cells in pulmonary asbestosis. *Am. J. Pathol.* 69: 373-388.

Timbrell, V. (1970) Characteristics of the International Union Against Cancer standard reference samples of asbestos. In: *Pneumoconiosis: Proceedings of the International Conference, Johannesburg.* (Shapiro, H. A., ed). Oxford University Press, Cape Town, pp. 28-36.

Involvement of the Oxidative Stress in the Toxicity of Iron-Containing Particles on Tracheal Epithelium in Primary Culture.

Catherine Guilianelli, Armelle Baeza-Squiban, Odile Houcine, Henri Pezerat, Francelyne Marano*

Laboratoire de Cytophysiologie et Toxicologie Cellulaire
Université Paris 7 Denis Diderot, case 7073
2 Place Jussieu
75 251 Paris Cédex 05, France

and
*Laboratoire de Réactivité de Surface et Structure,
Université P. et M. Curie
4 Place Jussieu
75 252 Paris Cédex 05, France

Mineral particles are known to be responsible for many respiratory diseases and especially cause lung cancer among exposed populations. However the physico-chemical interactions of particles with biological tissues are unclear. In addition to particle morphology and size, the nature of the material and particularly its surface reactivity are involved in particles toxicity. Indeed it has been shown that some mineral particles, including asbestos, can induce the production of activated oxygen species (AOS) in relation with the presence of divalent iron at the interface solid-liquid (Zalma et al, 1987a, Mossman and Marsh,1989).
Moreover the cellular and molecular mechanisms leading to particles induced-respiratory pathologies are not yet understood. In order to determine how iron-containing particles are involved in these respiratory diseases, three naturally-occurring mineral particles: hematite, nemalite and chrysotile were tested on the tracheal epithelium in primary culture. These particles have a different iron II content which allows to test the hypothesis of a toxicity by an iron-induced oxidative stress. Nemalite is the richest Fe^{2+}-containing mineral, Canadian chrysotile is an asbestos form naturally contaminated by nemalite and magnetite, and hematite only contains Fe^{3+}. Measurements of oxidizing power for formate anion revealed that nemalite was the most AOS-productive mineral (Guilianelli et al.,1993).

NATO ASI Series, Vol. H 85
Cellular and Molecular Effects
of Mineral and Synthetic Dusts and Fibres
Edited by J. M. G. Davis and M.-C. Jaurand

The mucociliary epithelium of the respiratory system is the first target of inhaled particles. The development of *in vitro* models allows to investigate in controled conditions the reaction of target cells to the particle-induced stress. The rabbit tracheal epithelium obtained *in vitro* by the explant technique proved to be representative with respect to morphology (Baeza-Squiban et al.,1991), and physiology (Romet et al.,1991) of its *in vivo* conditions and has been described as a suitable model to study the toxicity of various inhaled pollutants (Romet-Haddad et al.,1992, Blanquart et al., 1991). In a previous study the effects of iron-containing particles on rabbit tracheal epithelial (RTE) cells were assessed by morphological studies, by cytotoxicity tests and by the induction of squamous metaplasia (Guilianelli et al.,1993). According to these three parameters, nemalite, the richest iron II-containing material appeared as the most active suggesting an involvement of iron II.
In the present paper a thorough morphologic study was performed at the ultrastructural level and the implication in particles toxicity of the oxidative stress linked to iron II content, was investigated by the ability of the 3 different particles to induce lipid peroxidation. Moreover an eventual protection by an iron chelator, desferrioxamine and also by antioxidant enzymes such as superoxide dismutase and catalase was determined. It was performed on two assays already shown to be sensitive to particles activity: the measurement of culture growth and the counting of cells forming cross-linked envelopes. The formation of cross-linked envelopes is the ultimate stage of squamous metaplasia which is a transformation of the mucociliary epithelium in a squamous, stratified one, and which is considered as the first step towards neoplasia .

Materials and methods

Culture conditions

Primary cultures of rabbit tracheal epithelial (RTE) cells were obtained by the explant technique as already described by Baeza-Squiban et al. (1991). Briefly, tracheas were removed from 1-month old fauve de Bourgogne rabbits. 2-mm^2 explants were deposited on a thick collagen gel supplemented with Minimum Essential Medium (MEM) and serum, and covered with MEM containing serum, epidermal growth factor, glutamin, insulin, hydrocortisone and fungizone.

Mineral particles

Chrysotile (UICC) B, $Mg_3Si_2O_5(OH)_4$ is contaminated by nemalite (about 5%) and magnetite (Fe_3O_4). The proportion of Fe^{2+}, expressed in FeO is 1.44%. Nemalite (Asbestos, Québec) is a fibrous brucite $Mg(OH)_2$ in which about 8% of the Mg^{2+} are substituted by Fe^{2+}. Hematite (αFe_2O_3) is a commercial compound (Merck) which contains Fe^{3+}.

Minerals at different concentrations were suspended in serum-free culture medium by vortexing and immediately added to cultures. In some experiments, particles were preincubated with 10 mM desferrioxamine (Desferal, Ciba-Geigi) in MEM for 4 hr at 37°C. The particles were then spun down and excess desferrioxamine removed by washing with MEM. Before the addition of the particles to the cultures, they are resuspended in MEM supplemented with growth factor. In the case of experiments with antioxidant enzymes, they were added simultaneously with mineral particles. The superoxide dismutase (SOD: from bovine erythrocytes, Sigma Chemicals Co, St Louis, Mo; Activity: 3900 U/mg prot), a scavenger of $O_2{\cdot}^-$ was used at 400 U/ml medium whereas catalase (from bovine liver; Activity: 3400 U/mg prot, Sigma), a scavenger of H_2O_2 was tested at 1400U/ml medium.

Ultrastructural studies.

Cultures were fixed in 2% glutaraldehyde and post-fixed in 1% OsO_4 according to the method previously described (Guilianelli et al., 1993). Mineral particles were applied at 10 µg/cm^2 on 3-days old cultures and the fixation was performed after 7 days of treatment.

Growth assay

Culture growth was evaluated by an image analysis system (Guilianelli et al.,1993). The culture growth was determined by a measurement of the ratio of outgrowth surface to explant surface. Surface ratios of treated cultures were expressed as a percentage of control culture values. Particles were applied on day 3 of culture and inhibition of growth was determined 24 hrs later.

Cross-linked envelopes (CLE) determination

Cells were removed from the culture dish by trypsination and counted to determine total cell number. Cells were then centrifuged, the pellet resuspended in 20mM dithiothreitol (DTT) and 2% sodium dodecyl sulfate (SDS), heated to 100°C for 10 min and the remaining cornified envelopes counted in hemocytometer. The data is expressed as number of CLE divided by the total cell number times 100. The determination of CLE formation was realized on 10 days-old cultures treated for 7 days with particles at 10 µg/cm^2 pretreated or not with desferrioxamine.

Measurement of lipid peroxidation

Lipid peroxidation was measured as malonaldehyde (MDA) production. MDA formation was determined by thiobarbituric acid (TBA) coloured reaction (Ohkawa et al., 1979). Cultures were treated for one hour with 100 µg/cm^2. At the end of incubation, butylated-hydroxytoluene (BHT) at 10% (v/v) was added to the supernatant and the cellular pellet of each culture. Samples were frozen until use at -20°C. Then they were supplemented with 1ml TBA reactant. The mixture was then heated at 80°C for 15 min. After cooling, MDA was extracted with n-butanol, and the absorbance was measured at λ: 550 nm. Results are expressed as nmol of MDA/ outgrowth surface unit.

Results and discussion

In order to determine the behaviour of mineral particles on the ciliated tracheal epithelium *in vitro,* morphologic investigations were first performed. Previous scanning electron microscopy studies had revealed dramatic lesions of RTE cells. These lesions were mainly located in peripheric cells of the outgrowth where particles are concentrated because of the ciliary beat. Moreover semi-thin sections had demonstrated the phagocytosis of the three types of particles (Guilianelli et al., 1993). Such results are now confirmed at the electron microscopic level which in addition reveals cellular modifications linked to particle treatment. Endocyted particles are seen in apical cells (Fig. 1d and Fig. 2a) as well as in basal cells (Fig. 1b). By contrast, intracellular particles are never observed in ciliated cells (Fig. 2c). In phagocytic cells particles are only found in the cytoplasm and they are surrounded by a membrane (Fig. 1c,d and Fig. 2b,f). The presence of an associated membrane is the most common case resulting from the phagocytosis process of particle. However it has already been observed asbestos fibers lacking this membrane (Rieder et al., 1991). Some particles are also present in intercellular spaces (Fig. 2e). Thus mineral particles are able to penetrate the tracheal epithelium reconstituted *in vitro* either by the mechanism of phagocytosis or by transport between epithelial cells as in tracheal organ culture (Woodworth et al., 1983a). In comparison to control cultures (Fig. 1a), treated cultures exhibit ultrastructural lesions such as the development of numerous lamellar bodies

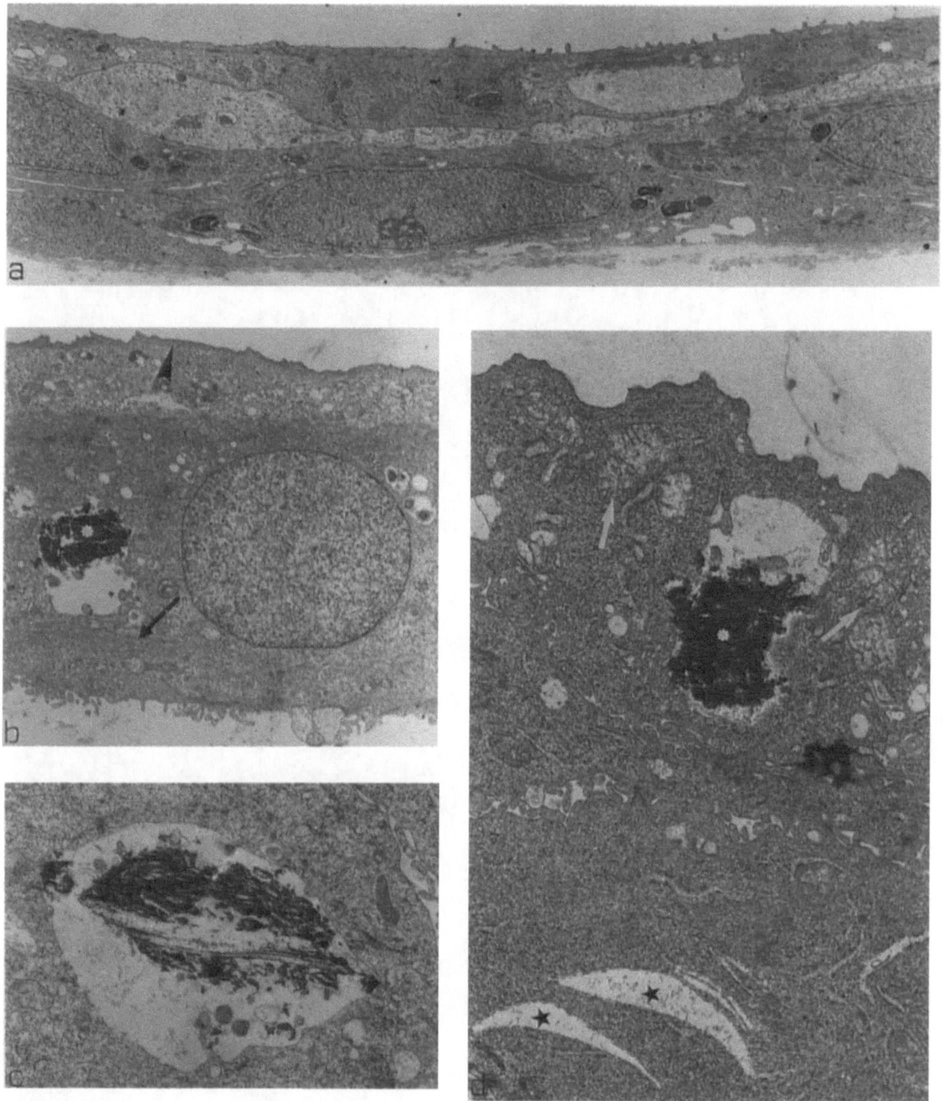

Fig. 1
a: control culture of RTE cells after 10 days of culture (x 6930)
b, c, d: Nemalite-treated culture, (b): the basal cell exhibits a nemalite-filled vacuole (asterisc) and numerous keratin filaments (arrow), whereas the apical cell shows a dark-staining marginal band (arrow head) beneath its plasma membrane (x 5900). (c): a detail of a nemalite-filled vacuole surrounded by a membrane (x 11150). (d):high magnification revealing ultrastructural lesions such as swollen mitochondria (arrow) and swollen cisternae of endoplasmic reticulum (star)(x 18800).

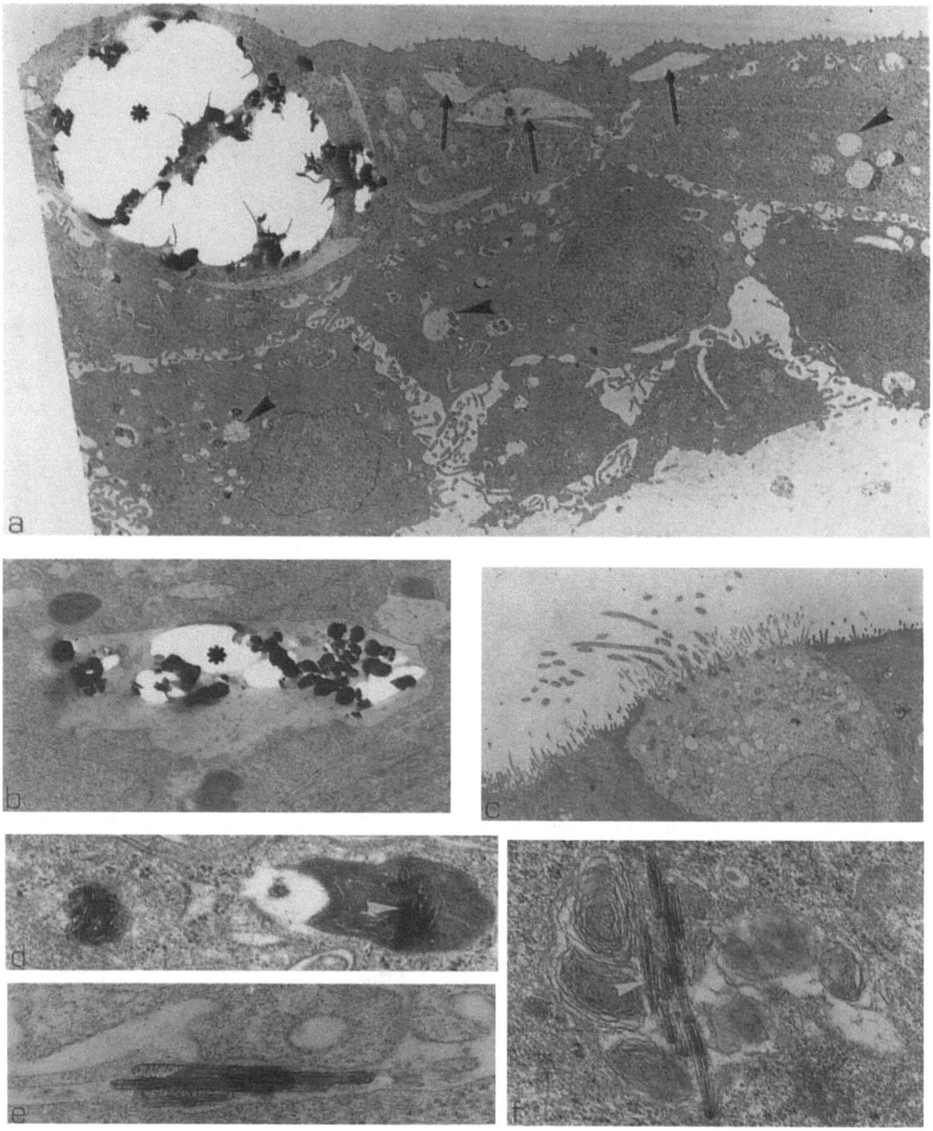

Fig. 2:
a, b, c: Hematite-treated culture, (a) general view of the culture showing the stratification of the epithelium, a large hematite- filled vacuole (asterisc), swollen cisternae of endoplasmic reticulum (arrow) and numerous vacuoles (arow head) (x 4940). (b) a detail of a hematite-filled vacuole (x 11750). (c): a detail of a ciliated cell without any lesion under hematite treatment (x 2520).
d, e, f: Chrysotile-treated culture, (d, f): a detail of lamellar bodies containing chrysotile fibers (arrow head)(d,x 21900; f,x 27200), (e): chrysotile fibers in intercellular spaces (x 34600).

(Fig. 2d,f) in which particles are sometimes present (Fig. 2d,f). Moreover treated cells show some swollen mitochondria (Fig. 1c) and swollen cisternae of endoplasmic reticulum (Fig. 1c, Fig. 2a). Some cells have numerous keratin filaments bundles in their cytoplasm (Fig. 1b). Beneath the plasma membrane of apical cells, a dark-staining marginal band lies (Fig. 1b) which could correspond to the appearance of the cross-linked envelope. These two last observations assume an evolution of treated cultures towards squamous metaplasia as demonstrated, especially with nemalite, in a previous article (Guilianelli et al., 1993).

These three mineral particles have already been shown to have an effect on the culture growth of RTE cells and on the induction of squamous metaplasia, but the implication of oxidative stress in these effects was not clearly established. Nevertheless it is now strongly suspected that besides particle size, reducing surface sites of mineral particles can act via the generation of AOS (Zalma et al., 1987b, Mossman et al., 1990). Furthermore a previous determination of the oxidizing power for formate anion of these 3 particles in serum-free culture medium has shown that hematite was totally inactive whereas a significative signal of a radical adduct of the carboxylate radical could be detected by electron-spin resonance for chrysotile and nemalite, this last being the most productive mineral (Guilianelli et al.,1993).

In addition to this direct method, now indirect methods were used to determine if the 3 tested particles produce AOS. First the ability of particles to induce lipid peroxidation was investigated since it is a well known AOS-dependent process. The TBA assay, which detects the appearance of MDA, a product resulting from lipid peroxidation, was performed. Nemalite and chrysotile

	nmol MDA/ outgrowth surface unit
Control	1.78 ± 0.62
Hematite	1.79 ± 0.85
Nemalite	3.06 ± 0.59 *
Chrysotile	3.44 ± 1.03 *

Table 1: Effect of iron-containing mineral particles at 100 $\mu g/cm^2$ on lipid peroxidation in RTE cells after 1 hour of treatment. Each data represents the mean of 6 values ± SE. (*) Significative difference with control culture at $P<0.05$ (Student's *t*-test).

have a significant effect on MDA production whereas hematite is inefficient (Table 1). These results are in agreement with those obtained by Fontecave et al. (1990) on liver microsomes.

Second, to assess the implication of AOS in particle effect on RTE cells, the culture growth assay representative of an effect on proliferation and the cross-linked enveloppe assay representative of the induction of squamous metaplasia (Banks-Schlegel et al., 1985), were carried out with particles in addition to antioxidant enzymes such as catalase and superoxide dismutase or after a pretreatment with an iron chelator, desferrioxamine.

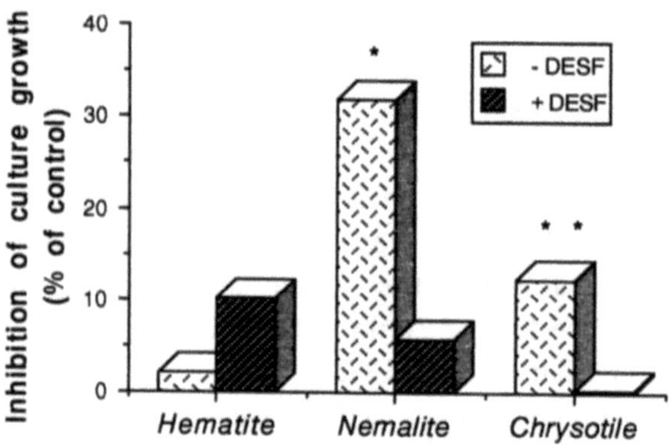

Fig. 3: Effect of mineral particles at 100 μg/cm² pretreated or not with desferrioxamine on culture growth after 24 hours of treatment. Results are expressed as a percentage of control values. Each point represents the mean of 6 values. Significative difference between desferrioxamine pretreated particle and untreated particle at (*) $P<0.08$ for nemalite and (**) $P<0.23$ for chrysotile (Student's *t*-test)

Considering the culture growth assay, it appears that a desferrioxamine pretreatment of nemalite significantly cancels its inhibitory effect on outgrowth extension (Fig. 3). The protective effect of the iron chelator with chrysotile is not so obvious since the difference of effect on culture growth between desferrioxamine pretreated chrysotile and untreated chrysotile is only statistically significant at $P<0.23$. In other respects an absence of protection of chrysotile with desferrioxamine was observed on the proliferation of Chinese hamster lung cells (V79-4) (Brown et al.,1987). Fontecave and al. (1987) have also shown that this chelator has no effect on chrysotile-induced peroxidation but was a very good inhibitor of lipid peroxidation catalyzed by nemalite.

We have shown that SOD and catalase never provide a protection against particles effect on culture growth (Fig. 4). It was not studied for hematite since it has not a marked effect on culture growth after 24 hours of treatment. The inefficiency of SOD and catalase in reducing the activity of nemalite and chrysotile was also shown for the same particles on lipid peroxidation (Fontecave et al., 1990). Moreover on tracheal epithelial cells, short (≤ 2 μm) chrysotile fibers were found to have not a reduced cytotoxic effect with enzyme cotreatment (Mossman et al., 1986).

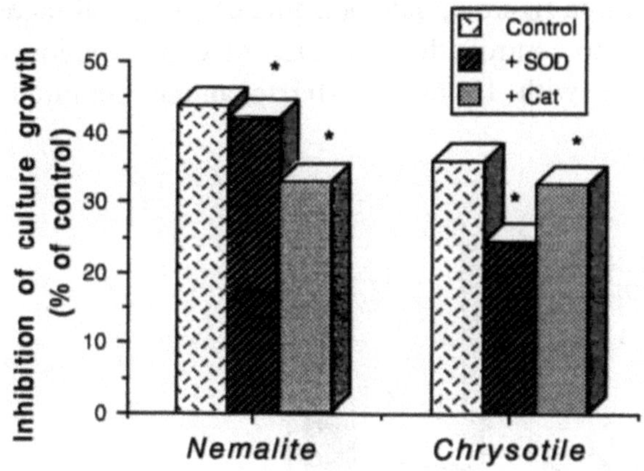

Fig. 4: Effect of mineral particles at 100 μg/cm2 cotreated or not with antioxidant enzymes SOD and catalase on culture growth after 24 hours of treatment. Results are expressed as a percentage of control values. Each point represents the mean of 6 values.* For both studies, values obtained for the enzyme-treated particle are not significantly different from the untreated particle $P<0.05$ (Student's *t*-test)

Squamous metaplasia is a transformation of the mucociliary epithelium in a stratified squamous epithelium which occurs under various of environmental stress. It is generally considered as a repair mechanism but is suspected in some cases to evolve to neoplasia. Among the cellular and molecular events leading to squamous metaplasia are the expression of the keratin 13 (Jetten et al., 1989- Baeza-Squiban et al., 1994) and the appearance of cells with cross-linked envelopes (CLE). They were the two criterion chosen to study the induction of squamous metaplasia in tracheal epithelium in primary culture under iron-containing particle treatment. Indeed *in vivo* and *in vitro* studies of different authors have demonstrated by morphological studies the ability of mineral fibers to induce such pathological differentiation (Woodworth et al., 1983b- Cameron et al., 1989). Based on the 2 criterion previously defined, some particles are able to induce the squamous differentiation: némalite increases the expression of keratin 13 and the number of cells with CLE whereas chrysotile only increases the keratin 13 expression. In the present study the counting of CLE was the unique criteria used to determine if the induction of squamous metaplasia by iron-containing particles can result from an oxidative stress. As shown in figure 5, only nemalite at 10 μg/cm^2 increases significantly (226% relative to control) the formation of CLE. Moreover a 4 hr-pretreatment of nemalite with 10 mM desferrioxamine inhibited this induction.

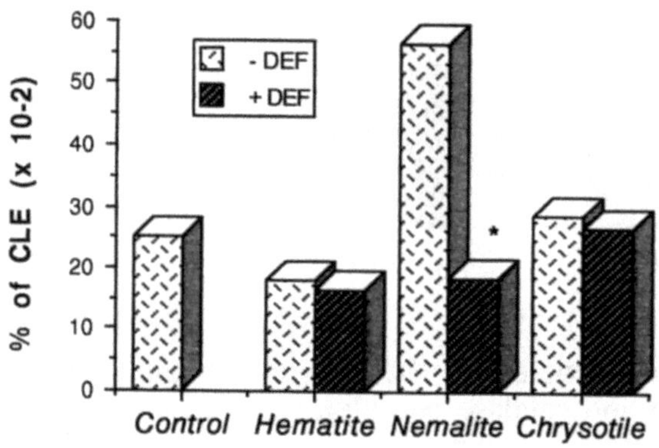

Fig. 5: Percentage of CLE produced in RTE cells treated at day 3 of culture for 7 days with mineral particles at 10 μg/cm^2 compared to control cultures. * $P < 0.05$: statistically significant decrease with desferrioxamine-pretreated particle in comparison to untreated particle (Student's *t*-test)

The present study reveals that Fe^{2+}-containing particles (nemalite and chrysotile) have an early effect on RTE cells inducing lipid peroxidation of the

plasmic membrane whereas the Fe^{3+}-containing particle (hematite) is without effect. When considering phagocytosis, it appears that all 3 particles are involved and this may lead to later effects on RTE cells. Some ultrastructural lesions seen in treated cells will need further investigations to determine their real specificity and their possible correlation with particle nature. Moreover an assessment of the quantity of endocyted particles will be interesting to be done to determine if particles are more or less endocyted according to their iron II content.
This study also clearly establish a correlation between Fe^{2+} content and cellular effects. Moreover if nemalite is pretreated with desferrioxamine its toxicity is canceled. Such observation supports the concept that iron II appearing during the leaching process, at the interface of nemalite particles with the medium, triggers a dioxygen reduction with formation of very electrophilic species (ferryl and/or perferryl, OH•), at the origin of the toxic effects (Pezerat, this volume). The nemalite content of the chrysotile being weak, it is normal that its toxic effects are less marked. The other localization of Fe^{2+} in chrysotile does not allow an AOS formation so efficient than with nemalite. Indeed Fe^{2+} in magnetite (an another contaminant) is not really a Fe^{2+}, but an iron mixed-valence state (Chouchane et al., this volume); and Fe^{2+} in the chrysotile network is less easily available to react with O_2, the leaching process of the chrysotile network being more difficult than the one of the nemalite network.
As a result of this study it appears that RTE cells in primary culture seems a suitable model to study the direct effects of mineral particles on respiratory cells. Among these effects we are now investigating the molecular events leading to squamous metaplasia in relation to particle treatment and more generally in relation to an oxidative stress using some other AOS-generating system.

We thank M.J. Falxa for technical assistance. This work was supported by grants from DRED, Action spécifique toxicologie de l'environnement, and MRE 91C0543 and 91739.

References

Baeza-Squiban A, Romet S, Moreau A, Marano F (1991) Progress in outgrowth culture from rabbit tracheal explants: balance between proliferation and maintenance of differentiated state in epithelial cells. In Vitro Cell Dev Biol 27A:453-460

Baeza-Squiban A, Boisvieux-Ulrich E, Houcine O, Geraud G, Guennou C, Marano F (1994) Extracellular matrix-dependent differentiation of rabbit tracheal epithelial cells in primary culture. In Vitro Cell Dev Biol in press

Banks-Schlegel SP, Gazdar AF, Harris CC (1985) Intermediate filament and cross-linked envelope expression in human lung tumor cell lines Cancer Res 45:1187-1197

Blanquart C, Romet S, Baeza A, Guennou C, Marano F (1991) Primary cultures of tracheal epithelial cells for the evaluation of respiratory toxicity. Toxic In Vitro 5:449-502

Brown RC, Poole A, Turver CJ, Vann C (1987) Role of iron-mediated free-radical generation in asbestos-induced cytotoxicity. Toxic. In Vitro 1:67-70

Cameron G, Woodworth CD, Edmondson S, Mossman BT (1989) Mechanisms of asbestos-induced squamous metaplasia in tracheobronchial epithelial cells. Environ Health Perspec 80:101-108

Chouchane S, Guignard J, Pezerat H (1994) Appearance of very electrophilic species generated by some iron oxides: effect of chelators and reducing agents. This volume

Fontecave M, Mansuy P, Jaouen M, Pezerat H (1987) The stimulatory effects of asbestos on NADPH-dependent lipid peroxidation in rat liver microsomes. Biochem J 241:561-565

Fontecave M, Jaouen M, Mansuy P, Costa D, Zalma R, Pezerat H (1990) Microsomal lipid peroxidation and oxy-radicals formation are induced by insoluble iron-containing minerals Biochem Biophys Res Comm 173:912-918

Guilianelli C, Baeza-Squiban A, Boisvieux-Ulrich E, Houcine O, Zalma R, Guennou C, Pezerat H, Marano F (1993) Effect of mineral particles containing iron on primary cultures of rabbit tracheal epithelial cells: possible implication of oxidative stress. Environ Health Perspec 101:436-442

Jetten AM, George MA, Smits HL, Vollberg TM (1989) Keratin 13 is linked to squamous differentiation in rabbit tracheal epithelial cells and down-regulated by retinoic acid. Exp Cell Res 182:622-634

Mossman BT, Marsh JP, Shatos MA (1986) Alteration of superoxide dismutase activity in tracheal epithelial cells by asbestos and inhibition of cytotoxicity by antioxidants. Lab Invest 54:204-212

Mossman BT, Marsh JP (1989) Evidence supporting a role for active oxygen species in asbestos-induced toxicity and lung disease. Environ. Health Perspec. 81:91-94

Mossman BT, Marsh JP, Dantona R, Bergeron M, Senoir A (1990) Involvement of active oxygen species (AOS) in injury, repair and proliferation of tracheobronchial epithelial cells after exposure to asbestos fibers. In "Biology, toxicology and carcinogenesis of respiratory

epithelium" (Thomassen DG, Nettesheim P, Eds), Hemisphere Publishing Corporation pp145-154
Ohkawa H, Ohishi N, Yagi K (1979) Assay for lipid peroxydes in animal tissues by thiobarbituric acid reaction. Anal Biochem 95:351-358
Pezerat H (1994) Role of transition metal compounds in the capacity of dust to generate electrophilic species. This volume.
Rieder CL, Sluder G, Brinkley BR (1991) Some possible routes for asbestos-induced aneuploidy during mitosis in vertebrate cells. In "Cellular and molecular fiber carcinogenesis" Brinkley BR, Lechner J, Harris C (Eds.) Cold Spring Harbor, NY pp1-26
Romet S, Schoevaert D, Marano F (1991) Dynamic image analysis applied to the study of ciliary beat on cultured ciliated epithelial cells from rabbit trachea. Biol Cell 71:183-190
Romet-Haddad S, Marano F, Blanquart C, Baeza-Squiban A (1992) Tracheal epithelium in culture: a model for toxicity testing of inhaled molecules. Cell Biol Toxicol 8:141-150
Woodworth CD, Mossman BT, Craighead JE (1983a) Squamous metaplasia of the respiratory tract. Possible pathogenic role in asbestos-associated bronchogenic carcinoma. Lab. Invest. 48:578-584
Woodworth CD, Mossman BT, Craighead JE (1983b) Induction of squamous metaplasia in organ cultures of hamster trachea by naturally occurring and synthetic fibers. Cancer Res 43:4906-4912
Zalma R, Bonneau L, Guignard J, Pezerat H, Jaurand MC (1987a) Production of hydroxyl radicals by iron solid compounds. Toxicol Environ Chem 13:171-187
Zalma R, Bonneau L, Jaurand MC, Guignard J, Pezerat H, (1987b) Formation of oxy-radicals by oxygen reduction arising from the surface activity of asbestos. Can J Chem 65:2338-2341

The Role of Iron in Asbestos-Induced Cancer

Ann E. Aust
Department of Chemistry and Biochemistry
Utah State University
Logan, UT 84322-0300
USA

Introduction

The mechanism by which asbestos and other mineral fibers cause cancer remains unknown despite intensive investigation. The most carcinogenic forms of asbestos contain iron and catalyze many of the same reactions that iron does. In fact, there is increasing evidence to suggest that iron is responsible for the biochemical reactivity of asbestos *in vitro*. Iron appears to be responsible for asbestos-dependent O_2 consumption [Aust and Lund (1991); Lund and Aust (1991)], ·OH formation [Weitzman and Graceffa (1984); Gulumian and Van Wyk (1987); Zalma *et al.* (1987); Kennedy *et al.* (1989); Aust and Lund (1991)], lipid peroxidation [Weitzman and Weitberg (1985); Turver and Brown (1987); Goodlick *et al.* (1989)], induction of deoxyribonuclease-S1 sensitive sites [Turver and Brwon (1987)], and induction of DNA single-strand breaks [Lund and Aust (1992)]. Iron may also play a role in one mechanism of phagocytosis of asbestos fibers [Hobson *et al.* (1990)]. Desferrioxamine B, an iron-specific chelator, reduced the cytotoxicity of asbestos to a variety of different cultured cell types, suggesting that iron may be involved in the cytotoxicity of asbestos [Goodlick and Kane (1986); Shatos *et al.* (1987); Garcia *et al.* (1988); Kamp *et al.* (1990); Goodlick and Kane (1990)].

Carcinogenicity of Iron

Iron has been shown to catalyze the formation of single-strand breaks [Henner *et al.* (1982)] and oxidized base products [Aruoma *et al.* (1989)] via the generation of oxygen radicals.

NATO ASI Series, Vol. H 85
Cellular and Molecular Effects
of Mineral and Synthetic Dusts and Fibres
Edited by J. M. G. Davis and M.-C. Jaurand

Thus, iron may cause cancer via damage to DNA. There is increasing evidence that iron from sources other than asbestos can increase the risk of cancer in humans [Pham *et al.* (1983); Niederau *et al.* (1985); Stevens and Kalkwarf (1990)]. In rodents, iron-nitrilotriacetate has been shown to induce renal adenocarcinoma [Ebina *et al.* (1986)], and iron-dextran to induce sarcomas at the site of injection [Richmond (1959)]. Perhaps more significantly, intraperitoneal injections of iron-saccharate in rats caused mesothelioma [Ebina *et al.* (1986)], the same type of tumor which is thought to be indicative of asbestos exposure in the pleural or peritoneal cavities. In each of these rodent experiments, iron was demonstrated to be required for the carcinogenic response.

Iron and the Generation of Oxygen Radicals

Since iron has the potential to catalyze the oxidation of DNA, lipid, and protein, living organisms have evolved proteins to transport and store iron in an unreactive form until it is needed to reconstitute the functional activity of certain proteins and enzymes [see Ponka *et al.* (1990) for review]. When iron is absorbed from the diet, it is bound to transferrin for transport in the blood. If iron is needed, transferrin receptors are synthesized and incorporated into the plasma membrane of cells for binding and endocytosis of transferrin. The iron is then passed from transferrin to ferritin, where it is stored until needed. However, there are conditions in which iron may be observed bound to compounds other than protein. Hemochromatosis is a genetic disease which results in the accumulation of iron to abnormally high levels in the body. Under these circumstances, citrate-chelated iron has been observed in the blood [Grootveld *et al.* (1989)]. It is thought that the presence of this abnormal chelate of iron is responsible for the symptoms in these patients, such as liver cirrhosis and cancer, arthritis, and heart disease. When iron is removed from these individuals, most of the symptoms are alleviated.

Uncontrolled iron is thought to damage biological systems

by catalyzing the formation of deleterious reactive oxygen species (e.g., $\cdot OH$) through a series of one-electron reductions of O_2 (see reactions 1-4 below).

$$Reductant^{(n)} + Fe(III) \longrightarrow Reductant^{(n-1)} + Fe(II) \quad (1)$$

$$Fe(II) + O_2 \longrightarrow Fe(III) + O_2^{\cdot -} \quad (2)$$

$$2O_2^{\cdot -} + 2H^+ \longrightarrow O_2 + H_2O_2 \quad (3)$$

$$Fe(II) + H_2O_2 \longrightarrow Fe(III) + OH^- + \cdot OH \quad (4)$$

The oxygen radicals $O_2^{\cdot -}$ and H_2O_2 do not induce detectable amounts of DNA damage unless transition metals, e.g. iron, are present to catalyze the formation of $\cdot OH$. Thus, it is the production of $\cdot OH$ or a similarly reactive species that is generally considered to be dangerous to DNA in the cell [Greenstock and Ruddock (1978)]. Therefore, the important reactions are those that generate Fe(II) and H_2O_2. When one considers the participation of asbestos or other iron-containing mineral fibers in this process, Fe(II) may already be present or may be generated, as in Reaction 1, by the reduction of Fe(III) by ascorbate or glutathione, present at millimolar concentrations in the cell, or under unusual circumstances, by $O_2^{\cdot -}$ produced by asbestos-activated macrophages. Hydrogen peroxide may come from an exogenous source, such as activated macrophages, or be generated by the spontaneous or enzyme-catalyzed dismutation of $O_2^{\cdot -}$ (Reaction 3). It is important to note that iron is a catalyst in Reactions 2 and 4 and will continue to redox cycle resulting in continued generation of $\cdot OH$ as long as there is sufficient reductant and either O_2 or H_2O_2. This is very different from other carcinogens where one molecule of carcinogen modifies one base in DNA. The continued redox cycling of iron means that one molecule of iron can generate many hydroxyl radicals leading to extensive damage.

It is important to consider the <u>reactivity of iron</u>. The electronic environment of iron created by the chelator can greatly affect its reactivity [Miller *et al.* (1990)]. For example, desferrioxamine B and ferrozine stabilize the Fe(III) or Fe(II) forms, respectively, rendering them virtually redox inactive in the presence of most reducing agents and O_2, by binding all of the coordination sites of iron. In contrast, citrate, EDTA, or nitrilotriacetate (NTA) enhance the redox activity of iron by leaving coordination sites open, resulting in reactions with O_2 to generate high levels of reactive O_2 species [Miller *et al.* (1990)]. Asbestos represents an insoluble complex of iron and appears to decrease the reactivity of iron when compared with other soluble chelators, such as citrate, EDTA, or NTA [Aust and Lund (1991); Lund and Aust (1991)]. This may be due to steric hindrance on the surface of the fiber which would slow reactions with reductants and O_2.

The location of iron within the cell is very important, since iron is required for the generation of $\cdot OH$. The $\cdot OH$ will travel only about 10 angstroms before reacting. Thus, iron must be present at the site of $\cdot OH$ damage. If the iron on the asbestos fibers is reactive, damage to biomolecules will only occur in the immediate vicinity of the fibers. However, if iron is mobilized by low-molecular-weight chelators, such as citrate or other chelators present in the cell, it could travel to sites distant from the fiber and damage a broader range of biomolecules. This would mean that the fibers would not need to be in the nucleus for DNA damage to occur. There is evidence to suggest that asbestos fibers do not have to make contact with the DNA of cells in order to cause DNA damage. Asbestos fibers were not observed in the nucleus of asbestos-treated rat pleural mesothelial cells in which unscheduled DNA synthesis (a measure of repair of DNA damage) occurred [Renier *et al.* (1990)]. In addition, asbestos-induced single-strand breaks were observed in all treated rat embryo cells in culture regardless of whether fibers were present in the cells or not [Libbus *et al.* (1989)]. These investigators proposed that DNA damage might be mediated by oxygen free radicals. If iron is responsible for this DNA damage, mobilization of iron from asbestos and the subsequent iron-catalyzed generation of reactive oxygen species in the vicinity of DNA may provide an explanation for these observations.

Mobilization of Iron from Asbestos

It seems clear from the work of many investigators that iron can be mobilized from asbestos *in vitro* [Morgan *et al.* (1973); Oberdorster *et al.* (1980); Spurny (1983); Lund and Aust (1990); Lund and Aust (1991)] and *in vivo* [Holmes and Morgan (1967); Hart *et al.* (1980)]. Mobilization may ultimately affect the reactivity and location of the iron. We have shown that chelators are required for the mobilization of iron *in vitro*, suggesting that iron mobilization *in vivo* was the result of chelation [Lund and Aust (1990)]. Both Fe(II) and Fe(III) were mobilized from crocidolite, amosite, or chrysotile. The rate of iron mobilization was greater for the iron-rich forms of asbestos (crocidolite or amosite) than for chrysotile of lower iron content, but was also dependent upon the pH, the chelator, and the surface area and crystalline structure of the asbestos. It should be noted that iron mobilization from crocidolite increased with decreasing pH [Lund and Aust (1990)]. This may be important since the pH within phagosomes, which often contain the internalized asbestos [Pinkerton *et al.* (1984)], can be significantly lower than the overall pH of the cell [Nyberg *et al.* (1991)]. Our data would suggest that this may lead to accelerated iron mobilization within the phagosome.

We have also shown that mobilization of iron from crocidolite by citrate, EDTA, or NTA increased its biochemical reactivity compared with its reactivity on the fibers *in vitro* resulting in increased O_2 consumption [Aust and Lund (1991); Lund and Aust (1991)], $\cdot OH$ formation [Aust and Lund (1991)], and induction of single-strand breaks in DNA [Lund and Aust (1992)]. If this occurs *in vivo*, it may lead to enhanced damage to biomolecules at intracellular sites distant from the phagocytized fibers.

Acquisition of Iron by Mineral Fibers

Asbestos, as well as other fibrous and nonfibrous particulates, can be coated with iron after inhalation to form asbestos or ferruginous bodies. It has been suggested that the iron in the coating material comes from ferritin or hemosiderin in macrophages [Davis (1964)]. However, the exact source of iron or the processes involved in the coating are not well understood. Although it has been proposed that this iron-coating of asbestos is used as a protective mechanism, the iron associated with the coating has been shown to be redox active and capable of catalyzing the formation of single-strand breaks in DNA [Lund *et al.* (in press)]. Thus, it may be possible that iron acquired within the body may be as important or more important in potentiating the biological effects of mineral fibers than the intrinsic iron.

Conclusions

The phagocytosis of iron-containing mineral fibers represents an uncontrolled entry of iron into cells. Within the cell, the fiber-associated iron, intrinsic and/or acquired, may generate ·OH at the fiber surface which would lead to damage of biomolecules, but only in the immediate vicinity of the fibers. In addition, the iron may be mobilized away from the fibers into a low-molecular-weight pool by chelators, such as citrate, which might enhance the redox activity of the iron and allow damage to occur at sites distant from the fibers. Entry of iron from mineral fibers into a low-molecular-weight pool may certainly contribute to the toxicity and carcinogenicity of the fibers. Reactions of iron at the fiber surface and after mobilization may be taking place simultaneously. However, one may be favored over the other, depending upon the chemistry of a particular type of mineral. This may contribute substantially to determining which minerals are more toxic and/or carcinogenic.

Because of the profound effect of the coordination environment of iron upon its reactivity, not all iron-containing minerals may be capable of catalyzing the formation of oxygen radicals on their surfaces and not all iron-containing minerals may be capable of releasing iron to soluble chelators. Thus, some iron-containing minerals may be relatively inert chemically and biologically. It may be possible to assess these properties by conducting biochemical assays, such as induction of DNA single-strand breaks *in vitro*, to predict which iron-containing minerals are biologically active.

References

Aruoma OI, Halliwell B, Dizdaroglu M (1989) Iron ion-dependent modification of bases in DNA by superoxide. J Biol Chem 264: 13024-13028

Aust AE, Lund LG (1991) Iron-catalyzed oxygen consumption and hydroxyl radical generation in the presence of cysteine. Brown, RC et al. (eds.), Mechanisms in Fibre Carcinogenesis, NATO ASI Series, Plenum Press New York, 397-405

Crichton RR (ed.) (1991) Inorganic Biochemistry of Iron Metabolism. Ellis Horwood Chichester

Davis JMG (1964) The ultrastructure of asbestos bodies from human lung. Br J Exp Path 45: 642-646

Ebina Y, Okada S, Hamazaki S, Toyokuni S, Midorikawa O (1989) Induction of mesothelioma by intraperitoneal injections of ferric saccharate in male Wistar rats. Br J Cancer 60: 708-711

Garcia JGN, Gray LD, Dodson RF, Callahan KS (1988) Asbestos-induced endothelial cell activation and injury. Demonstration of fiber phagocytosis and oxidant-dependent toxicity. Am Rev Respir Dis 138: 958-964

Goodglick LA, Kane AB (1986) Role of reactive oxygen metabolites in crocidolite asbestos toxicity to mouse macrophages. Cancer Res 46: 5558-5566

Goodglick LA, Kane AB (1990) Cytotoxicity of long and short crocidolite asbestos fibers *in vitro* and *in vivo*. Cancer Res 50: 5153-5163

Goodglick LA, Pietras LA, Kane AG (1989) Evaluation of the casual relationship between crocidolite asbestos-induced lipid peroxidation and toxicity to macrophages. Am Rev Respir Dis 139: 1265-1273

Greenstock CL, Ruddock GW (1978) Radiation activation of carcinogens and the role of $\cdot OH$ and $O_2^{\cdot -}$. Photochem Photobiol 28: 877-880

Grootveld M, Bell JD, Halliwell B, Aruoma OI, Bomford A, Sadler PJ (1989) Non-transferrin-bound iron in plasma or serum from patients with idiopathic hemochromatosis. J Biol Chem 264: 4417-4422

Gulumian M, Van Wyk JA (1987) Hydroxyl radical production in the presence of fibers by a Fenton-type reaction. Chem Biol Interact 62: 89-97

Hart RW, Kendig O, Blakeslee J, Mizuhira V (1980) Effect of cellular ingestion on the elemental ratio of asbestos, in The In Vitro Effects of Mineral Dusts. Brown RC, Chamberlain M, Davies R, Gormley IP (eds) Academic Press London: 191-199

Henner WD, Grunberg SM, Haseltine WA (1982) Sites and structure of radiation-induced DNA strand breaks. J Biol Chem 257: 11750-11754

Hobson J, Wright JL, Churg A (1990) Active oxygen species mediate asbestos fiber uptake by tracheal epithelial cells. FASEB J 4: 3135-3139

Holmes A, Morgan A (1967) Leaching and constituents of chrysotile asbestos *in vivo*. Nature 215: 441-442

Kamp DW, Dunne M, Anderson JA, Weitzman SA, Dunn MC (1990) Serum promotes asbestos-induced injury to human pulmonary epithelial cells. J Lab Clin Med 116: 289-297

Kennedy TP, Dodson RF, Rao NV, Ky H, Hopkins C, Baser M, Tolley E, Hoidal JR (1989) Dusts causing pneumoconiosis generate ·OH and produce hemolysis by acting as Fenton catalysts. Arch Biochem Biophys 269: 359-364

Libbus BL, Illenye SA, Craighead JE (1989) Induction of DNA strand breaks in cultured rat embryo cells by crocidolite asbestos as assessed by nick translation. Cancer Res 49: 5713-5718

Lund LG, Aust AE (1990) Iron mobilization from asbestos by chelators and ascorbic acid. Arch Biochem Biophys 278: 60-64

Lund LG, Aust AE (1991) Mobilization of iron from crocidolite asbestos by certain chelators results in enhanced crocidolite-dependent oxygen consumption. Arch Biochem Biophys 287: 91-96

Lund LG, Aust AE (1992) Iron mobilization from crocidolite asbestos greatly enhances crocidolite-dependent formation of DNA single-strand breaks in ϕX174 RFI DNA. Carcinogenesis 13: 637-642

Lund LG, Williams MG, Dodson RF, Aust AE (in press) Iron associated with asbestos bodies is responsible for the formation of single strand breaks in ϕX174 RFI DNA. Occupat and Environ Med

Miller DM, Buettner GR, Aust SD (1990) Transition metals as catalysts of "autoxidation" reactions. Free Rad Biol Med 8: 95-108

Morgan A, Lally AE, Holmes A (1973) Some observations on the distribution of trace metals in chrysotile asbestos. Ann Occup Hyg 16: 231-240

Niederau C, Fisher R, Sonnesberg A, Stremmel W, Trampish HJ, Strohmeyer G (1985) Survival and causes of death in cirrhotic and in noncirrhotic patients with primary hemochromatosis. N Engl J Med 313: 1256-1262

Nyberg K, Johansson U, Hohansson A, Camner P (1991) Phagolysosomal pH and location of particles in alveolar macrophages. Fund Applied Tox 16: 393-400

Oberdorster G, Boose CH, Pott F, Pfeiffer U (1980) *In vitro* dissolution rates of trace elements from mineral fibres, in The In Vitro Effects of Mineral Dusts. Brown RC, Chamberlain M, Davies R, Gormley IP (eds.) Academic Press London: 183-189

Pham QT, Gaertner M, Miur JM, Braun P., Gabiano M, Sadoul P (1983) Incidence of lung cancer among iron miners. Eur J Respir Dis 64: 534-539

Pinkerton KE, Pratt PC, Brody AR, Crapo JD (1984) Fiber localization and its relationship to lung reaction in rats after chronic inhalation of chrysotile asbestos. Am J Pathol 117: 484-498

Ponka P, Schulman HM, Woodworth RC (1990) Iron transport and storage. CRC Press Boca Raton

Renier A, Levy F, Pilliere F, Jaurand MC (1990) Unscheduled DNA synthesis in rat pleural mesothelial cells treated with mineral fibres. Mutat Res 241: 361-367

Richmond HG (1959) Br Med J i: 947-949

Shatos MA, Koherty JM, Marsh JP, Mossman BT (1987) Prevention of asbestos-induced cell death in rat lung fibroblasts and alveolar macrophages by scavengers of active oxygen species. Environ Res 44: 103-116

Spurny KR (1983) Measurement and analysis of chemically changed mineral fibers after experiments *in vitro* and *in vivo*. Environ Health Perspect 51: 343-355

Stevens RG, Kalkwarf DR (1990) Iron, radiation and cancer. Environ Health Prospect 87: 291-300

Turver CJ, Brown RC (1987) The role of catalytic iron in asbestos induced lipid peroxidation and DNA-strand breakage in C3H10T1/2 cells. Br J Canc 56: 133-136

Weitzman SA, Graceffa P (1984) Asbestos catalyzes hydroxyl and superoxide radical generation from hydrogen peroxide. Arch Biochem Biophys 228: 373-376

Weitzman S, Weitberg A (1985) Asbestos-catalyzed lipid peroxidation and its inhibition by desferrioxamine. Biochem J 225: 259-252

Zalma R, Bonneau L, Guignard J, Pezerat H, Jaurand MC (1987) Formation of oxy radicals by oxygen reduction arising from the surface activity of asbestos. Can J Chem 65: 2338-2341

TIME-LAPSE VIDEO LIGHT MICROSCOPIC AND ELECTRON MICROSCOPIC OBSERVATIONS OF VERTEBRATE EPITHELIAL CELLS EXPOSED TO CROCIDOLITE ASBESTOS

C. G. Jensen[1,2], L. C. W. Jensen[1,2], J. G. Ault[1], G. Osorio[1], R. Cole[1] and C. L. Rieder[1,3]

[1]Wadsworth Center for Laboratories and Research, New York State Department of Health, P.O. Box 509, Albany, N.Y. 12201-0509, USA
[2]University of Auckland, Private Bag 92019, Auckland, New Zealand
[3]Department of Biomedical Sciences, State University of New York, Albany, NY 12222, USA

INTRODUCTION

The most popular hypothesis for how asbestos transforms cells is based on its physical dimensions, not its chemistry. This view is derived from the finding that the cytotoxic effects and transforming ability of various mineral fibers are related primarily to fiber dimensions (i.e., the length/width aspect ratio -- See Hesterberg and Barrett, 1984; Stanton *et al*, 1981; reviewed in Harington, 1981). For this reason, a knowledge of how mineral fibers behave within interphase and dividing cells, and with what intracellular components they interact, becomes important in elucidating the mechanism(s) by which they cause cell injury.

Most data on mineral fibers/cell interactions have been collected on fixed material. Consequently, little is known regarding the behavior of mineral fibers in living cells. With this in mind, we have used video-enhanced light microscopy coupled with serial section electron microscopy to examine the behavioral and structural features of crocidolite fibers phagocytized by cultured vertebrate epithelial cells.

An important requirement for studying mineral fiber behavior within living cells is an appropriate cell model. As a rule, the cultured mammalian macrophages, fibroblasts and epithelial cells previously used for asbestos-related toxicity, mutagenicity, and carcinogenic studies are not suitable for high-resolution imaging of mineral fibers and cellular components because they are extremely small, and because they round up considerably during cell division and after incorporating asbestos fibers.

NATO ASI Series, Vol. H 85
Cellular and Molecular Effects
of Mineral and Synthetic Dusts and Fibres
Edited by J. M. G. Davis and M.-C. Jaurand

The two cell types, rhesus monkey kidney (LLC-MK_2)epithelial cells, and newt *(Taricha granulosa)* lung epithelial cells, which we have chosen for our studies, are especially suitable for high resolution observations on living cells. They remain flat after ingestion of asbestos and throughout the cell cycle, are large relative to fiber dimensions (250 μm diameter for newt and 50 μm diameter for LLC-MK_2), and possess excellent optical properties so that individual fibers and cellular components can be easily followed within the cell. The LLC-MK_2 cell line (American Type Culture Collection, passages 18-24) was cultured on coverslips in L-15 medium supplemented with 10% fetal calf serum and antibiotics. TIMA (Thermal Insulation Manufacturers' Association) crocidolite asbestos was added to the cultures at concentrations ranging between 0.3 and 1.0 mg/ml. For light microscopic observations the coverslip cultures were mounted on microperfusion chambers with fresh medium and placed on a heated (37°C) microscope stage. These chambers allowed cells to be examined with high resolution optics and periodically perfused with fresh medium. The newt lung primary cultures were prepared in Rose multi-purpose chambers as detailed by Rieder and Hard (1990). The coverslip cultures were mounted in microperfusion chambers for viewing with the light microscope (see Alexander and Rieder, 1991 for details of the video microscopic setup used). Using one or both of these two cell types, we have detailed the uptake of asbestos fibers by living cells, the behavior of fibers within interphase (non-dividing) cells, the relationship of fibers to intracellular membranes, the behavior of fibers during cell division, and their relationship to the production of nuclear abnormalities.

UPTAKE OF ASBESTOS BY MONKEY KIDNEY EPITHELIAL CELLS

The ingestion of asbestos has been shown to occur in a variety of cells, including pleural mesothelium (Broaddus *et al.*, 1993; Jaurand *et al.*, 1983), lung macrophages (Bruch, 1974; Davis, 1967; Suzuki *et al.*, 1972; reviewed in Miller, 1978), peritoneal macrophages (Johnson and Davies, 1980), Syrian hamster embryo cells (Hesterberg *et al.*, 1986), and newt lung cells (Cole *et al.*, 1991). Previous light and electron microscopic studies of fixed material have shown that the probable mechanism for incorporation of an asbestos fiber is similar to phagocytosis of other kinds of particles (reviewed in Alberts *et al.*, 1989; Holtzman, 1989; and Steinman *et al.*, 1983). During phagocytosis the fiber becomes encased in a portion of the plasma membrane and is then referred to as a *phagosome*. Subsequently, the phagocytic membrane may fuse with a lysosome, giving rise to a secondary lysosome, or *phagolysosome*. Cells appear to interact with additional fibers and continue to internalize them over periods of at least several days (Hesterberg *et al.*, 1986). A long fiber in the process of being incorporated may have one end outside the cell, the middle position in a phagosome, and the other end within a phagolysosome (Bruch, 1974; Johnson and Davies, 1980). In this case, the interior of the phagolysosome would be open to the outside of the cell and lysosomal enzymes would be released. Using LLC-MK_2 cells we have now confirmed and extended the previously described observations.

Earlier studies on fixed cells were not able to document the site of asbestos uptake, *i.e.*, whether fibers are incorporated at the edge of the cell, from the dorsal surface of the cell, or at both locations. We have used high-resolution time-lapse video-enhanced differential-interference-contrast (DIC) or phase-contrast light microscopy of LLC-MK_2 cells exposed to asbestos to resolve this question.

Incorporation from the edge of the cell. Although LLC-MK_2 cells are not migratory, when asbestos is added to a culture on a glass coverslip, the ruffled cell edge intermittently extends outward in different directions around the cell. The plasma membrane may come in contact with surrounding fibers which have settled down and become loosely attached to the coverslip; some of these fibers may then become phagocytized by the cell. Figure 1 documents this process. Individual fibers are progressively taken into the cell, and when completely incorporated, may begin to be actively transported within the cell (Fig. 1D). The edge of the cell membrane may also partially or completely cover a fiber without incorporating it (Fig. 1, fiber 3).

Incorporation from the dorsal surface of the cell. For these studies asbestos was added to LLC-MK_2 coverslip cultures in Rose chambers, which were placed on the heated stage of an inverted microscope equipped with phase-contrast optics. Confluent areas were immediately selected for filming and focus was maintained on the upper surface of a peripheral (non-nuclear) area of two or three cells. Filming was continued for up to 24 hours. Figure 2 shows a time-lapse sequence in which three fibers become incorporated and begin to be actively transported within the cell

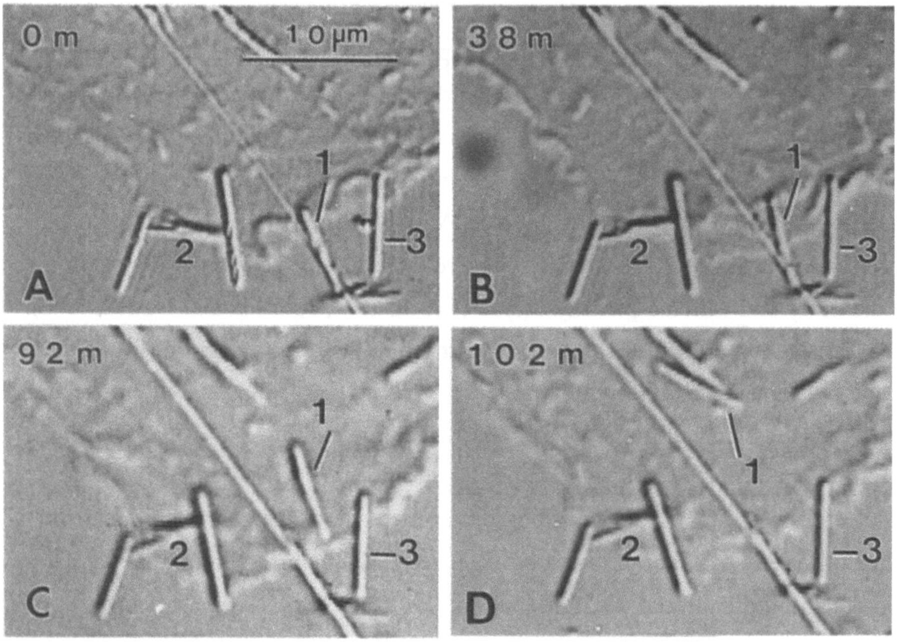

Fig. 1. Incorporation from the edge of an $LLCMK_2$ cell. Selected video frames viewed by DIC microscopy. Fiber 1 is progressively covered by the cell membrane in A and B. It is within the cell at C and has moved within the cell in D. A group of fibers (2) became incorporated and translocated within the cell later in the sequence. Another fiber (3) remained attached to the coverslip and was not incorporated during the 22 hours of filming.

during the ten hours of filming. Fibers begin to settle on the surface of the cells within minutes of their addition to the culture. They contact the cell surface either along their length or with their tip. Those which contact the cell surface with their tip may undergo a sequence of pivoting movements, giving the impression that the tip is attached to the cell surface (arrows in Fig. 2B). These pivoting fibers may subsequently become incorporated into the cell (Fig. 2D-H), or come to lie flat on the cell surface, or lose their association with the cell and float away as a result of currents in the surrounding medium. Fibers have also been seen to pivot on the glass coverslip. Fibers which lie on the cell surface can become incorporated along their length (Fig. 2B, fibers 1 and 2), or from the tip (Fig. 2D-F, fiber 3). Extensive membrane ruffling is visible along the fiber as it is incorporated (Fig 2F, fiber 3).

Many of the fibers which land on the surface of the cell do not become incorporated over a 24 hour period. They often move slowly about the cell surface in a non-specific fashion, suggesting they are riding along a moving plasma membrane (Bretscher, 1984). These unincorporated fibers progressively accumulate around the bulge produced by the nucleus.

Our observations thus demonstrate that cultured LLC-MK_2 cells may incorporate crocidolite fibers both at the edge, and from the dorsal surface, of the cell. Soon after a fiber becomes completely incorporated, it is often transported within the cell along cytoplasmic microtubules, providing indisputable evidence that the fiber is within the cell. Over a period of several days a cell will have almost completely cleared the immediate surrounding area of fibers (Fig. 3A).

BEHAVIOR OF ASBESTOS FIBERS IN LIVING CELLS

It has been well-documented that the majority of incorporated asbestos fibers progressively accumulate in the perinuclear region (Barrett *et al.*, 1989; Hesterberg *et al.*, 1986). However, studies on fixed cells were not able to document the mechanism(s) involved in fiber translocation. Our recent studies showed that in newt lung cells asbestos fibers became preferentially located near the nucleus (Cole *et al.*, 1991). This perinuclear accumulation was largely inhibited by disassembling the cytoplasmic microtubules with nocodazole. Fibers within the newt lung cells which were less than 5 μm in length were seen by video light microscopy to undergo saltatory transport at a maximum velocity of 1.18 μm/sec. By contrast, fibers over 5 μm in length rarely exhibited saltatory motion. Same cell-correlative light and electron microscopy revealed that fibers exhibiting saltatory behavior were enclosed within a membrane. From these observations we concluded that asbestos fibers undergo size-dependent active transport along cytoplasmic microtubules. Since the saltatory transport of cellular organelles is known to occur along the surface of microtubules (reviewed in Schroer and Sheetz, 1991 and Vallee and Shpetner, 1990), the membrane-bound asbestos fibers are being handled by the newt lung cells in the same manner as are other cellular components.

We have now confirmed these results in a primate cell line, LLC-MK_2, and

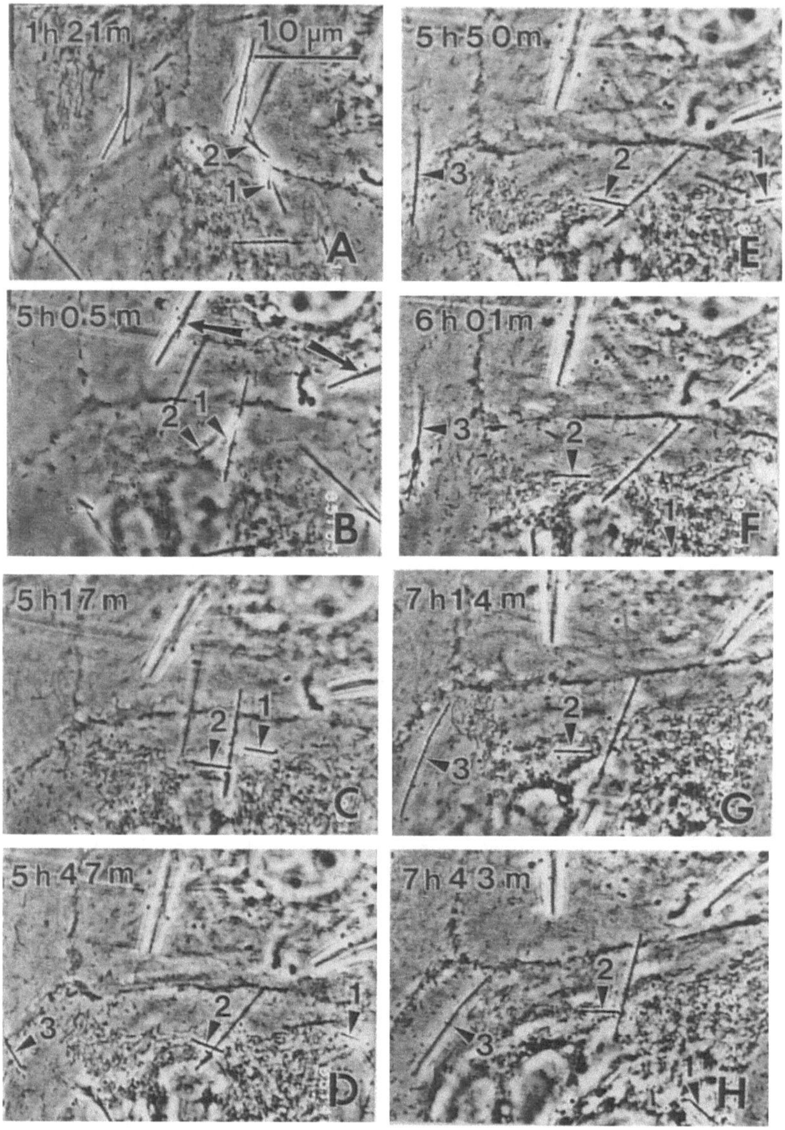

Fig. 2. Incorporation from the dorsal surface of an $LLCMK_2$ cell. Selected video frames of a cell viewed by phase-contrast optics. Three fibers (1,2,3) are incorporated and subsequently undergo translocation. 1 and 2 have just entered the cell in B, have started to move in C, and continue moving. 3 has landed on the cell surface in D, and dips into the cell in E and F, with increased membrane activity, especially visible in F (arrow). Fiber 3 has become completely incorporated in G and has translocated in H. Two fibers (arrows in B) pivot on the surface of the cell throughout the sequence.

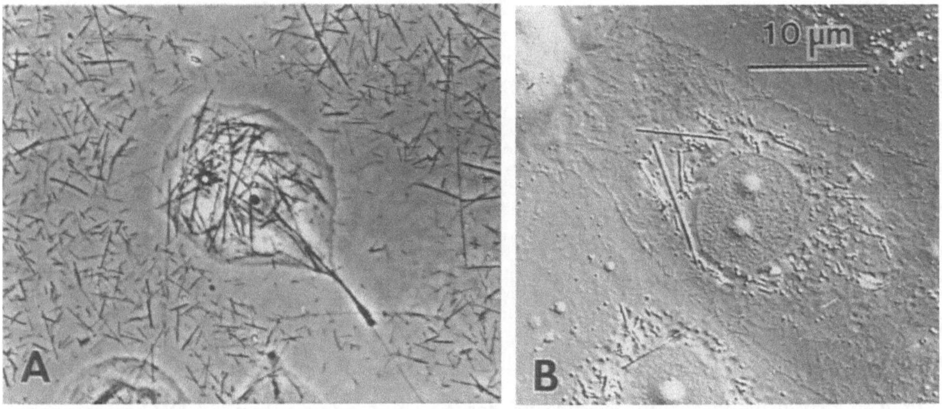

Fig. 3. Phase-contrast (A) and DIC (B) micrographs of $LLCMK_2$ cells exposed to asbestos for two days. The cell in A has cleared the surrounding area of all fibers. Many fibers have accumulated in the perinuclear region of the cell in B.

suggest that this feature is common to all vertebrate cells. Fibers in $LLC\text{-}MK_2$ cells, filmed with DIC or polarized light optics, are seen to undergo saltatory translocations at velocities up to 0.3 μm/sec (see Fig. 1C, D and Fig. 2C, D). These rates are similar to those measured for other microtubule-based movements in a variety of cells (reviewed in Vallee and Shpetner, 1990). Although the smaller fibers cover greater distances at faster velocities, fibers as long as 12 μm are seen to undergo translocation. The fibers progressively accumulate in the perinuclear region (Fig. 3B).

RELATIONSHIP OF ASBESTOS FIBERS TO INTRACELLULAR MEMBRANES

Demembranation of a phagocytosed mineral fiber could cause major complications for the cell. First, the active transport of fiber-containing phagosomes/phagolysosomes within the cell is likely to be mediated by microtubule motor proteins that are associated with the membrane surrounding the fiber (Cole *et al.*, 1991). Second, the lysis of a mineral fiber-containing phagolysosome would not only prohibit further fiber transport, but it would also liberate deleterious hydrolytic enzymes into the cytoplasm that may injure or kill the cell.

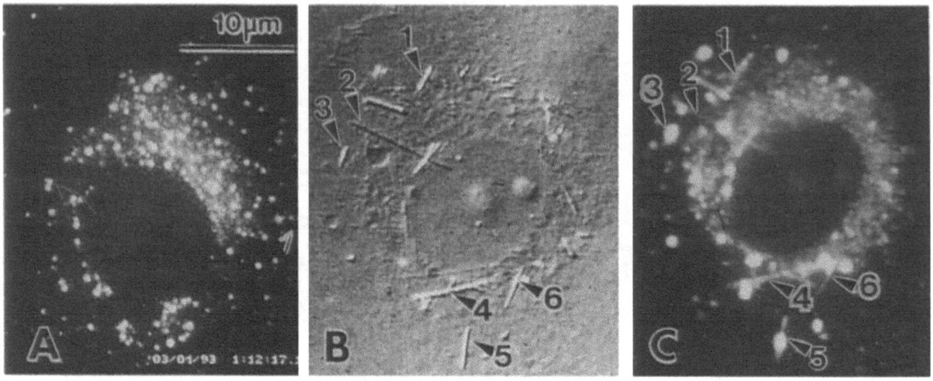

Fig. 4. Correlative vital fluorescence microscopy and video-enhanced DIC microscopy of LLCMK$_2$ cells using acridine orange to visualize lysosomes and phagolysosomes. (A) Lysosomes are visible as bright dots in the fluorescent image. Asbestos fibers within phagolysosomes (arrows 1-6) are visible in the DIC image (B) and the fluorescent image (C) of the same cell.

It would therefore be valuable to know whether or not asbestos fibers within the cell lose their surrounding membrane over time. There are previous reports of asbestos fibers within the cytoplasm of cells viewed with electron microscopy which do not show a surrounding membrane (Bruch, 1974; Johnson and Davies, 1980; reviewed in Miller, 1978). However, we have found that lysosomal/phagolysosomal membranes are often difficult to detect with certainty using standard fixation methods for electron microscopy. Therefore, to attempt to resolve the question of whether some fibers within the cell lose their surrounding membrane, we have applied two different techniques -- vital fluorescence microscopy using acridine orange to visualize lysosomal compartments of the living cell, and electron microscopy, using a newly-developed fixation protocol to enhance the contrast of membranes.

Fluorescence microscopy. Using vital fluorescence microscopy of LLC-MK$_2$ cells stained with acridine orange, a vital dye which binds to acidic cellular compartments (reviewed in Holtzman, 1989 and Steinman *et al.*, 1983), we confirm previous observations (reviewed in Holtzman, 1989) that the lysosomal population is located predominantly in the perinuclear region of the cell (Fig. 4A). Near-simultaneous DIC (Fig. 4B)/epi-fluorescence (Fig. 4C) microscopy of asbestos-containing LLC-MK$_2$ cells exposed to 0.6 μg/ml acridine orange for two minutes reveals that most of the fibers within the cell are surrounded by a lysosomal membrane. Some fluorescent images also show a bulge along the fiber, which is indicative of a large lysosomal component (Fig. 4C, arrows 2, 3 and 6). Some fibers visualized by DIC do not have a fluorescent counterpoint. These fibers may be either located outside the cell, may be surrounded by a primary phagosomal membrane, or may lack a membrane. This technique is unable to differentiate between these possibilities.

Electron microscopy. For optimal fixation and maximum visualization of membranes, LLC-MK_2 coverslip cultures were fixed for 5 minutes in a 0.1M phosphate-buffered 2% glutaraldehyde/1%osmium tetroxide mixture, followed by 1 minute in 0.1% aqueous ruthenium tetroxide. A more standard membrane fixation method of 5 minutes in buffered 2% glutaraldehyde, followed by 2 hours in 2% osmium tetroxide at 4°C was also used as a control. To facilitate thin sectioning and prevent tearing and distortion of areas containing asbestos, the asbestos fibers were removed from the Epon-Araldite embedment by a brief treatment with hydrofluoric acid at 4°C (Rieder and Bowser, 1987), leaving a hole equivalent to the fiber's dimensions (Cole *et al.*, 1991).

Figure 5A-C shows LLC-MK_2 cells cultured in the presence of asbestos for two days. An image seen frequently after exposure of LLC-MK_2 cells to asbestos is shown in Fig. 5A. The fibers penetrate between multiple lobes of the nucleus and are associated along their length with the outer surface of the nuclear envelope. The impression is that the nuclear envelope is surrounding the fiber in the same way that the plasma membrane surrounds a fiber during phagocytosis. Examination of serial sections reveals that such images represent single nuclei with multiple lobes rather than a multinucleate condition. It is possible that some, if not all, of the reported asbestos-induced multinucleation (Kenne *et al.*, 1986; Kodama *et al.*, 1993; Oshimura *et al.*, 1984) may actually represent such a multilobed condition. It is also possible that previous reports of fibers within the nucleus (Johnson and Davies, 1980) may in fact have been describing fibers between the lobes of a multilobed nucleus. The true morphology cannot be determined unless serial sections are examined.

In cells cultured in the presence of asbestos for two days, all the fibers are surrounded by membranes and are contained within either phagosomes or phagolysosomes (Fig. 5B, C). With our ruthenium tetroxide fixation method, the

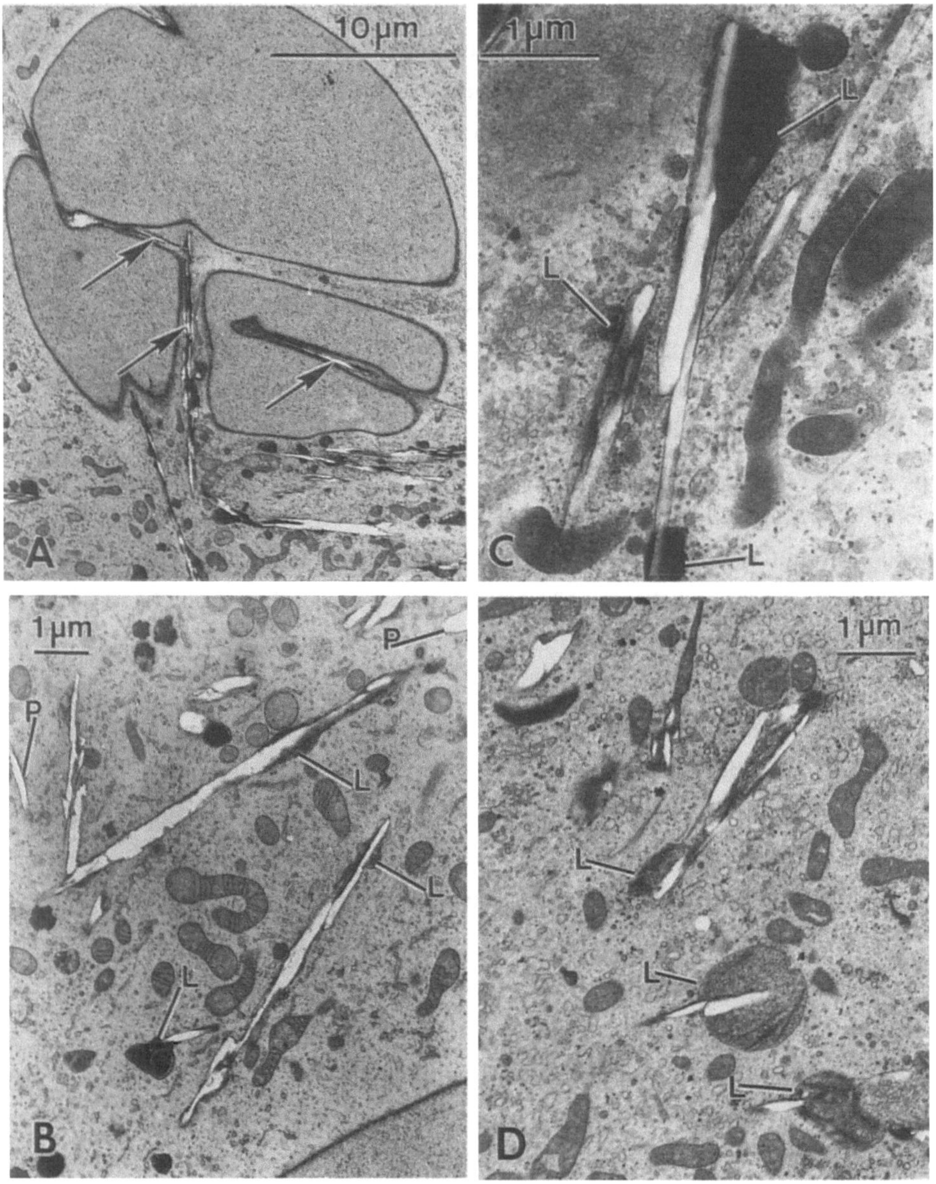

Fig. 5. LLCMK$_2$ cells exposed to asbestos. A-C. Two-day exposure. A. fibers (arrows) penetrate between nuclear lobes. B. several fibers are within phagosomes (P) or phagolysosomes (L). C. A ¼ μm thick section viewed with high voltage electron microscopy. Fibers are associated with lysosomes (L). D. Two-week exposure. All fibers are surrounded by membranes. Many are associated with lysosomes (L).

surrounding membranes are clearly visible as dark lines around the fibers.

To examine the effect of longer-term asbestos exposure, we cultured LLC-MK_2 cells in the presence of asbestos for two days, during which time the cells incorporated fibers. The cultures were then rinsed with fresh medium and cultured for an additional two weeks, with fresh medium being added every two days. The cells were not subcultured during this time to avoid possible disruption of cellular membranes which could result from rounding up of the cells following trypsinization. Because LLC-MK_2 cells do not show contact inhibition, they continued to multiply during this time. In cells containing asbestos for two weeks, virtually all incorporated fibers were still surrounded by membranes (Fig. 5D).

We compared the incidence of asbestos-containing phagosomes and asbestos-containing phagolysosomes in electron micrographs of cells which had been exposed to asbestos for two days or for two weeks, as described above. A control consisted of a two-week cell culture, with asbestos added for only the last two days. Fibers within 3 μm of the cell edge were compared with those within 3 μm of the nucleus. Fibers that lay within 3 μm of both the edge and the nucleus were scored for both areas. Asbestos-containing phagosomes were defined as fibers in which the surrounding membrane appeared tightly apposed to the fiber (P in Fig. 5B). Asbestos-containing phagolysosomes were identified by the presence of lysosomal "bulges" and/or cellular debris between the membrane and the fiber (L in Fig. 5B and C).

We found the highest incidence of phagolysosomes in the nuclear region of all three cultures (Table I). This correlates with the known distribution of lysosomes (see Fig. 4A). At the cell edge, where much of the phagocytosis of fibers takes place, the ratio of fiber-containing phagosomes to fiber-containing phagolysosomes was about equal. The distribution of fiber-containing phagosomes and phagolysosomes was similar in all three cultures (Table I), suggesting that an equilibrium in the distribution of fiber-containing phagosomes and phagolysosomes is reached within two days of exposure to asbestos and remains constant over at least two weeks.

CELL DIVISION IN THE PRESENCE OF ASBESTOS

As mentioned previously, the cytotoxic potency and transforming ability of various

TABLE I

The Incidence (%) of Asbestos-Containing Phagosomes and Phagolysosomes in LLC-MK_2 Cells after Two Days and Two Weeks of Culture

Treatment	Edge of Cell			Nucleus		
	Phago-somes	Phago-lysosomes	n	Phago-somes	Phago-lysosomes	n
2 da asbestos	55.0	45.0	40	28.4	71.6	95
2 wk asbestos	55.2	44.8	58	26.6	73.4	79
2 wk control	55.9	44.1	34	26.5	73.5	49

mineral fibers are related to fiber dimensions (Hesterberg and Barrett, 1984; Stanton *et al.*, 1981; reviewed in Harington, 1981). In general, long ($\geq 8\,\mu m$), thin ($\leq 0.25\,\mu m$) fibers are more toxic than shorter, thicker ones. At present, the mechanism(s) involved in this selective toxicity are unclear. To date, the most logical hypothesis is that long, thin fibers are more efficient at disrupting chromosome motion during cell division. The gain or loss of chromosomes, or chromosome parts, during cell division could then lead to the transformation of those cells that have genetic changes compatible with growth (Hesterberg *et al.*, 1986; Kenne *et al.*, 1986; Oshimura *et al.*, 1984; Rieder *et al.*, 1991). This hypothesis is compatible with the very long latency period of asbestos-induced mesotheliomas, and with the fact that asbestos-induced carcinomas are characterized by aneuploidy (Barrett, 1991; Lechner *et al.*, 1985, 1991; Oshimura *et al.*, 1984; Rieder *et al.*, 1991; Yegles *et al.*, 1993).

Nuclear division. To examine the behavior of asbestos in dividing cells, we have undertaken time-lapse studies of newt lung and LLC-MK_2 epithelial cells in the presence of asbestos. The results will be discussed in detail elsewhere. Briefly, we have found that, in both epithelial cell types, most fibers are prevented from entering the region of the forming spindle by the surrounding cage of intermediate filaments (Mandeville and Rieder, 1990). In contrast, mesothelial cells may be more sensitive to transformation by asbestos because of the plasticity of their cytoskeleton (Connell and Rheinwald, 1983; Lechner *et al*, 1985; Rieder *et al.*, 1991), which makes it easier for fibers to gain access to the spindle.

In newt lung and LLCMK_2 cells, thick fibers and other large membrane-bound cellular inclusions that are initially positioned near the chromosomes at the beginning of mitosis become laterally ejected from the forming spindle by "transverse equilibria" forces (Östergren, 1945). Many of these fibers subsequently attach to polar microtubules and exhibit rapid movements in and out of the spindle poles, but they seldom contact the chromosomes. It is noteworthy that fibers near the spindle that are not attached to microtubules are simply pushed aside by the moving chromosomes. This finding clearly demonstrates that asbestos fibers *in vivo* (which are presumably surrounded by a membrane - see above) do not have a particular affinity for chromatin.

Cleavage. We have found that LLC-MK_2 cells that have undergone repeated mitotic cycles in the presence of asbestos for two weeks possess a 28% higher incidence of nuclear abnormalities (*i.e.*, reconstituted nuclei, binucleate, and multinucleate cells) than controls. Binucleation, which has been reported previously (Jaurand *et al.*, 1983; Kenne *et al.*, 1986; Kodama *et al.*, 1993; Lechner *et al.*, 1991) as a feature of cultured cells exposed to asbestos, may result when an asbestos fiber becomes trapped between the separating anaphase chromosomes, thus inhibiting the completion of cleavage (Rieder *et a*l., 1991). However, this hypothesis has not been directly demonstrated.

We have found that, in LLC-MK_2 cells containing asbestos, cleavage is inhibited when one or more long fibers are present in the cleavage furrow. In summary, the data revealed that such cells proceeded in one of three ways: (1) In 18% of the cells the two potential daughter cells tugged back and forth on the fiber until one cell released it, resulting in two normal cells. (2) In 18% of the cells the potential daughter cells approached each other and eventually (8-10 hours later) formed a single binucleate cell containing the fiber within the cytoplasm. (3) In the remaining 64% of the cells containing an interzonal fiber, the potential daughter cells did not separate completely; instead they remained together throughout the time (up to 48 hours) of filming, with the fiber plus a persistent midbody extending between them.

Electron micrographs of cleaving cells containing asbestos fibers within the elongated interzone showed a persistent midbody surrounding the fibers. They were intimately associated with long, attenuated strands of nuclear envelope in both cells, indicating a close affinity of crocidolite fibers for the surface of the nuclear envelope.

CONCLUSIONS

We have used video light microscopy and serial section electron microscopy to examine behavioral and structural features of crocidolite fibers phagocytized by cultured vertebrate epithelial cells. We found that: (1) Asbestos fibers are incorporated at both the edge and the dorsal surface of vertebrate epithelial cells. (2) The phagosomal/phagolysosomalmembranes surrounding incorporated asbestos fibers persist for at least two weeks. (3) Most asbestos fibers are eliminated from the spindle in dividing cells. Fibers do not stick to chromosomes, as evidenced by the fact that they are passively pushed by the chromosome arms. (4) Fibers positioned in the interzone during cleavage can produce binucleation by inhibiting cytokinesis.

ACKNOWLEDGEMENTS

The work discussed in this chapter was partly supported by American Lung Association Grant #RG-018-N to J.G.A and by Grant #RR01219, awarded by the National Center for Research Resources (Department of Health and Human Services/PublicHealth Service), to support the Wadsworth Center's Biological Microscopy and Image Reconstruction Facility as a National Biotechnological Resource.

REFERENCES

Alberts B, Bray D, Lewis J, Raff M, Roberts K, and Watson JD (1989) Molecular Biology of the Cell. Garland Publ, New York, 2nd Ed pp 323-340

Alexander SP and Rieder CL (1991) Chromosome motion during attachment to the vertebrate spindle: Initial saltatory-like behavior of chromosomes and quantitative analysis of force production by nascent kinetochore fibers. J Cell Biol 113:805-815

Barrett JC, Lamb PW and Wiseman RW (1989) Multiple mechanisms for the carcinogenic effects of asbestos and other mineral fibers. Environ Health Perspect 81:81-89

Barrett JC (1991) Role of chromosomal mutations in asbestos-induced cell transformation. In CC Harris, JF Lechner and BR Brinkley (eds) Cellular and Molecular Aspects of Fiber Carcinogenesis. Cold Spring Harbor Laboratory Press. pp 27-39

Bretscher MS (1984) Endocytosis: relation to capping and cell locomotion. Science

224:681-686
Broaddus VC, Yung R, Sanan D, Sheppard D, Boylan AM (1993) Attachment of crocidolite asbestos fibers to pleural mesothelial cells via integrin $\alpha V\beta 5$. Mol Biol Cell Vol 4, Suppl p 284a
Bruch J (1974) Response of cell cultures to asbestos fibers. Environ. Health Perspect 9:253-254
Cole RW, Ault JG, Hayden JH, and Rieder CL (1991) Crocidolite asbestos fibers undergo size-dependent microtubule-mediated transport after endocytosis in vertebrate lung epithelial cells. Cancer Res 51:4942-4947
Connell ND and Rheinwald JG (1983) Regulation of the cytoskeleton in mesothelial cells: reversible loss of keratin and increase in vimentin during rapid growth in culture. Cell 34:245-253
Davis JMG (1967) The effects of chrysotile asbestos dust on lung macrophages maintained in organ culture: an electron-microscope study. Brit J Exp Pathol 48:379-385
Harington JS (1981) Fiber Carcinogenesis: Epidemiologic observations and the Stanton hypothesis. J Natl Cancer Inst 67:977-989
Hesterberg TW and Barrett JC (1984) Dependence of asbestos- and mineral dust-induced transformation of mammalian cells in culture on fiber dimension. Cancer Res 44:2170-2180
Hesterberg TW, Butterick CJ, Oshimura M, Brody AR and Barrett JC (1986) Role of phagocytosis in Syrian hamster cell transformation and cytogenetic effects induced by asbestos and short and long glass fibers. Cancer Res 46:5795-5802
Holtzman E (1989) Lysosomes. Plenum Press New York London
Jaurand MC, Bastie-Sigeac I, Bignon J. and Stoebner P (1983) Effect of chrysotile and crocidolite on the morphology and growth of rat pleural mesothelial cells. Environ Res 30:255-269
Johnson NF and Davies R (1980) The effect of crocidolite and chrysotile on peritoneal macrophages: a study by transmission and scanning electron microscopy. In RC Brown, M Chamberlain, R Davies and IP Gormley (eds) The *in vitro* effects of mineral dusts. Academic Press New York pp 97-103
Kenne K, Ljungquist S and Ringertz NR (1986) Effects of asbestos fibers on cell division, cell survival, and formation of thioguanine-resistant mutants in Chinese hamster ovary cells. Environ Res 39:448-464
Kodama Y, Boreiko CJ, Maness SC and Hesterberg TW (1993) Cytotoxic and cytogenetic effects of asbestos on human bronchial epithelial cells in culture. Carcinogenesis 14:691-697
Lechner JF, Tokiwa T, LaVeck M, Benedict WF, Banks-Schlegel S, Yeager Jr H, Banerjee A, and Harris CC (1985) Asbestos-associated chromosomal changes in human mesothelial cells. Proc Natl Acad Sci USA 82:3844-3888.
Lechner JF, Gerwin BI, Reddel RR, Gabrielson EW, Van der Meeren A, Linnainmaa K, Somers ANA and Harris CC (1991) Studies on human mesothelial cells: effects of growth factors and asbesti-form fibers. In CC Harris, JF Lechner, and BR Brinkley (eds) Cellular and Molecular Aspects of Fiber Carcinogenesis. Cold Spring Harbor Laboratory Press pp 115-129
Mandeville EC and Rieder CL (1990) Keratin filaments restrict organelle migration into the forming spindle of newt pneumocytes. Cell Motil Cytoskel 15:111-120
Miller K (1978) The effects of asbestos on macrophages. CRC Critical Rev Toxicol

5:319-354

Oshimura M, Hesterberg TW, Tsutsui T and Barrett JC (1984) Correlation of asbestos-induced cytogenetic effects with cell transformation of Syrian hamster embryo cells in culture. Cancer Res 44:5017-5022.

Östergren G (1945) Transverse equilibria on the spindle. Bot Not 4:467-468

Rieder CL and Bowser SS (1987) Correlative light and electron microscopy on the same epoxy section. In MA Hyatt (ed) Correlative Microscopy in Biology: Instrumentation and Methods. Academic Press New York pp 249-277

Rieder CL and Hard R (1990) Newt lung epithelial cells: cultivation, use and advantages for biomedical research. Int Rev Cytol 122:153-220

Rieder CL, Sluder G and Brinkley, BR (1991) Some possible routes for asbestos-induced aneuploidy during mitosis in vertebrate cells. In CC Harris, JF Lechner and BR Brinkley (eds) Cellular and Molecular Aspects of Fiber Carcinogenesis. Cold Spring Harbor Laboratory Press pp 1-26

Schroer TA and Sheetz MP (1991) Functions of microtubule-based motors. Ann Rev Physiol 53:629-652

Stanton MF, Layard M, Tegeris A, Miller E, May M, Morgan E and Smith A (1981) Relation of particle dimension to carcinogenicity in amphibole asbestoses and other fibrous minerals. J Natl Cancer Inst 67:965-975

Steinman RM, Mellman IS, Muller WA and Cohn ZA (1983) Endocytosis and the cycling of plasma membrane. J Cell Biol 96:1-27

Suzuki Y, Churg J and Ono T (1972) Phagocytic activity of the alveolar epithelial cells in pulmonary asbestosis. Am J Pathol 69:373-388

Vallee RB and Shpetner HS (1990) Motor proteins of cytoplasmic microtubules. Ann Rev Biochem 59:909-932

Yegles M, Saint-Etienne L, Renier A, Janson X and Jaurand MC (1993) Induction of metaphase and anaphase/telophase abnormalities by asbestos fibers in rat pleural mesothelial cells *in vitro*. Am J Respir Cell Mol Biol 9:186-191

The effect of MMMF on the kinetics of cytokine expression in murine lung populations : Role in fibre-induced disease

K. Miller, B.N. Hudspith and C. Meredith

Department of Immunotoxicology
BIBRA Toxicology International
Woodmansterne Road
Carshalton
Surrey SM5 4DS
UK

Introduction

Our knowledge concerning mineral fibre induced disease has increased over the last few decades due to the intensive research effort in this area. Based upon expert clinical observations, extensive epidemiological studies and the use of experimental systems, there is now clear evidence that indicates the development of fibrosis or carcinogenesis is related to differences in overall dimensions, chemical composition and durability of the fibre in lung tissue (Liddell and Miller, 1991). However, the precise cellular and molecular events leading to the observed pathological changes remain to be elucidated. One of the most important areas that is now being investigated is the effect of man made mineral fibres (MMMF) on the development of pulmonary pathology. This is because these products are being introduced as replacements for asbestos based products and little human data is available.

Recent advances in our knowledge of cellular and molecular biology has led to a better understanding of the interaction of cytokines and other factors responsible for the development of lung disease (Kelley, 1990; Zhang *et al.*, 1993). It has been demonstrated that alveolar macrophages exposed to MMMF *in vivo* and *in vitro* release a range of humoral factors, including cytokines and prostaglandins (Elias *et al.*, 1990). Inhalation of certain fibres activate airway macrophages which release a range of these factors including interleukin-1 alpha and beta (IL-α and IL-ß), tumour necrosis factor-alpha (TNF-α) and platelet derived growth factor (PDGF). These mediators then activate other pulmonary cells to

NATO ASI Series, Vol. H 85
Cellular and Molecular Effects
of Mineral and Synthetic Dusts and Fibres
Edited by J. M. G. Davis and M.-C. Jaurand

release more factors and this cytokine cascade can then modulate the activity of pleural cells which could have a role in the development of the observed fibre induced pathological changes.

We have previously reported on studies in which groups of PVG rats were exposed to different man-made mineral fibres (MMMF's) via inhalation 6 hours per day for 5 days, and alveolar and pleural cell populations examined for fibre-induced alterations and population characteristics (Miller, Hudspith and Meredith, 1991). Rat pleural macrophages were found to differ in their phenotypic expression with respect to all other macrophage populations and also showed low levels of esterase activity. Elevated levels of fibronectin gene expression after fibre exposure (as determined by RNA dot blot analysis) was demonstrated. An important finding was the high numbers of inflammatory cells (neutrophils, eosinophils and mast cells) present in the pleural cavity. High numbers of inflammatory cells had also previously been reported by other laboratories using a rat model (Lehnert *et al.*, 1985; Oberdorster *et al.*, 1983) and these findings led us to question to what extent potentially pathological changes in the pleura might be influenced by factors released by inflammatory cells already present in the cavity. Serosal mast cells are known to express high levels of cytokines and adhesion factors i.e. TNF-α, MIP-1α, MIP-ß (Gordon, 1991; Miller and Krangel, 1992) and neutrophils to express and release growth factors such as M-GSF and GM-CSF (Ralph and Sampson-Johannes, 1990). Of even greater concern to us was the knowledge that increased levels of inflammatory cells, eosinophils in particular, are frequently found in pleural effusions from asbestos-exposed and asbestotic individuals (Herbert, 1989; Martensson *et al.*, 1987). We therefore thought it important to investigate the use of an animal model where pleural cavity cells would not contain populations that could augment fibrotic and/or other pathological processes.

Although the rat has been the most widely used in studies of the effect of fibres in the lung (probably because it is the standard animal in toxicology work) a murine inhalation model of fibre-induced pulmonary disease has been established for some years (Bozelka *et al.*, 1983). The preferential deposition pattern of fibres has also been shown to be at the alveolar duct byfurcation located nearest the bronchiolar-alveolar junction i.e. at the alveolar level, in the lungs of both rats and mice (Donaldson and Brown, 1993). Furthermore, most cytokine probes are of human or mouse origin and the use of a murine model, as described by Piquet and co-workers (1990), would therefore have a number of advantages in studying the mechanism of fibre-induced pulmonary disease.

We now report on the population characteristics of mouse pleural cavity and alveolar cells, and on optimum culture conditions of murine alveolar and pleural macrophages needed for measuring cytokine mRNA expression.

Materials and Methods

Animals. In preliminary studies alveolar and pleural cell populations were studied using 8-10 week old male BALB/c, CBA and C3H mice. In subsequent experiments the C3H strain was used as this is the one that the majority of similar studies have been carried out in.

Collection of cells from pleural and alveolar compartments. Pleural lavage was performed by opening the abdomen of the mouse and inserting a 1 mm gauge polythene cannula through the diaphragm and then instilling 1 ml of normal saline into the thoracic space, the mouse was then gently massaged. The fluid was removed back through the cannula using a 1 ml syringe. This process was then repeated once more and the fluid from both lavages was pooled. Alveolar lavage was then carried out by inserting a 1 mm gauge polythene cannula into the lungs via the trachea. Then 0.5 ml of normal saline was slowly instilled into the lung which was then gently massaged and then left for 5 minutes. The fluid was then removed by gentle suction and the process repeated 3 times without the 5 minute delay. All four lavage fluids were pooled. It was found that the cell return from individual mice was not sufficient to study the cytokine profile of these cells, therefore several lavages from different mice had to be pooled. These lavage samples were centrifuged for 10 minutes at 900 g and resuspended in 2 ml RPMI culture medium and counted.

Cell culture. The pleural and alveolar cell populations were adjusted to 1 x 10 ml in RPMI 1640 containing 10% FCS, and 1 ml of these cell suspensions were plated in 35 mm diameter plastic dishes for 1 hour at 37°C in an atmosphere of 5% CO_2 in air. After this time the non-adherent cells were removed and the plates washed with fresh culture medium. These adherent cells were cultured overnight in fresh culture medium to enable the cells to become quiescent with respect to the expression of the immunoregulatory genes of interest, as it has been demonstrated that the adherence process induces a strong but transient signal which temporarily enhances the expression of these genes (Meredith *et al.*, 1990). After this culture period the cells were stimulated with one of the MMMFs at a dose of 10 ug/ml in the presence or absence of 10 ug/ml lipopolysaccharide (LPS) and mRNA isolated 3 hours later.

Cytokine specific mRNA analysis. Total RNA was obtained from cells immediately following lavage and cultured cells, some of which having been activated with LPS and treated with MMMF following the method described previously (Meredith *et al.*, 1990). Cytokine specific mRNA was then measured by Northern and dot-blot techniques using GeneScreen membranes (NEN research products) and following the manufacturers instructions. The membranes were then exposed to X-rat film (Fuji) for 3-7 days at -70° and autoradiographs quantified by scanning densitometry.

Monoclonal antibody staining. Cytocentrifuge cell preparations using 5×10^4 cells/slide were fixed in acetone for 10 minutes at -30°C. The slides were then air dried. This preparation was treated with 3% hydrogen peroxide for 5 minutes to block endogenous peroxidase. After rinsing in PBS the monoclonal antibody was applied for 1 hour at a dilution of 1 in 500 and the slides were then rinsed in PBS. A biotinylated goat anti-rat secondary antibody was applied for 45 minutes, after washing in PBS an avidin-conjugated peroxidase reagent was applied for 20 minutes. Peroxidase activity was visualised by a 10 minute incubation in a solution of 0.2 mg/ml 3-amino-9-ethylcarbozole (ACE) in 0.05 acetate buffer at pH 5.0 containing 0.01% H_2O_2. The slides were washed in PBS and counterstained with Meyers haematoxylin.

Results and Discussion

Preliminary investigations using several different mouse strains revealed that pleural cavity cells recovered by lavage consisted mainly of macrophages and lymphocytes with the proportions of one to the other population differing somewhat according to strain. C3H mice were selected for all subsequent studies.

Little difference was found in cell populations recovered by alveolar lavage from either mouse or rat with well over 90% of cells consisting of alveolar macrophages (Table 1). In sharp contrast, great differences were found in populations obtained by pleural lavage (Table 2), in that mouse cells consisted of more than 50% macrophages (compared to 35% in the rat) and contained very few neutrophils and mast cells. No eosinophils were present in any of the preparations.

Only two monoclonal antibodies are available to delineate subsets of macrophages in the mouse, MAC 1 that recognised mature macrophages only and MOMA 2 which labels all cells of the macrophage/monocyte lineage. Little difference was found in the ability of pleural and alveolar macrophages to bind to the two antibodies (Table 3). Differences between the two populations, however, were found when using an antibody associated with complement function, the CR3b receptor, and in MHC Class II antigen expression. Only 11% of the alveolar macrophages expressed the complement receptor whereas 35% of pleural cell populations (i.e. approximately 70% of pleural macrophages) recognised the CR3b antibody. The expression of MHC Class II molecules on the surface membranes, said to correlate with the ability to present antigen and stimulate autologous lymphocyte proliferation also differed markedly between mice pleural and alveolar cells. Approximately 16% of pleural cell populations were positive for Class II whereas only 2% of alveolar cells expressed the antigen. These differences between alveolar and pleural macrophage phenotype which are related to phagocytosis and antigen presentation are similar to those found earlier when using the rat as an animal model.

In both species pleural macrophages appear more able to interact with foreign material than alveolar cells. Indeed, in studies where murine pleural and alveolar adherent cells were incubated *in vitro* with MMMF at 10 μg/ml for as long as 24 hr the phagocytic index of alveolar cells was only 25.3 compared to 46.3 for the pleural cells (Table 4). Although pre-incubation with the bacterial antigen LPS activated both populations, these essential differences in phagocytes remained, with alveolar cells phagocytosis fibres increased up to 32% but the pleural cell phagocytic index increasing by 50% or more. This is a somewhat unexpected finding as alveolar macrophages *in vivo*, at least in theory, would be far more likely to meet up with foreign particulate matter than pleural cells.

The kinetics of cytokine expression also differed between the two macrophage populations in the mouse although in a more orthodox fashion. We have established optimum conditions for culture of murine alveolar and pleural macrophages and baseline levels of mRNA expression of cytokine IL-α, IL-1ß, IL-6, TNF-α and others (Meredith *et al.*, 1990) in order to gain information useful in establishing the kinetics of cytokine expression to inhalation studies. As expected the interleukins and TNF-α were constitutively expressed at very low levels provided the macrophage populations had been cultured overnight to minimise interference from the pulse of cytokine mRNA synthesis which accompanies adherence of the macrophage to tissue culture ware. Stimulation of both alveolar and pleural cells with 10 μg/ml lipolysaccharide (LPS) for 3 hr, induced marked increases in the level of mRNA expression of IL1-α, IL-1ß and IL-6 but had no effect on the steady-state levels of mRNA transcripts for actin. Increased expression of TNF-α mRNA levels were much higher and longer-lasting than those previously demonstrated in peritoneal macrophage populations (Meredith *et al.*, 1994). However levels of TNF-α in pleural cells had returned to those observed in unstimulated cells at 24 hr, whereas levels were still increased five-fold at 24 hr in alveolar cells. Thus the kinetics of cytokine expression during stimulation differ between macrophage populations, indicating a maximum capacity to synthesize TNF by alveolar cells.

Our intention was to study the LPS-augmented response of the two cell populations after fibre exposure to mimic the *in vivo* situation, but we were unable to clearly differentiate the influence of fibres from that of LPS *in vitro*. We therefore added fibres at concentrations of 1, 10 or 100 μg/ml alone or simultaneously with LPS. Northern blot analysis clearly showed that all fibre types when added alone to the cultures increased the expression of mRNA for IL-1ß and TNF-α in both alveolar and pleural cell populations although the increase was higher in pleural cells. When fibres were added together with LPS, expression was further increased (Tables 5 & 6), although a slight but consistent reduction of the LPS-induced mRNA was observed.

One explanation for this diminution of cytokine synthesis may well be that the fibres caused macrophages to release prostaglandins which would down-regulate the synthesis of cytokines such as IL-1ß in *in vitro* systems. Chrysotile asbestos fibres have previously been reported to induce primary macrophage cultures to release membrane arachidonic acid and PGE2 (Sirois, Rola-Pleszczynski and Begin, 1980).

Summary

These studies have provided useful time- and dose-response relationship prior to the planned inhalation exposures. However, whereas *in vitro* exposure of macrophage populations demonstrated that glass and refractory ceramic fibres all induced synthesis of the cytokines measured, preliminary *in vivo* investigations indicate that only refractory ceramic fibres (in particular RCF-1) induce a lasting increased expression of IL-1ß and TNF-α. These findings are now being further investigated and analysed in relation to fibre characteristics. We would conclude that *in vitro* studies followed by short-term inhalation can be used for investigating important early changes in the lung; also that molecular biological analysis is an important tool for determining such changes. However, in spite of familiarity with the rat, the use of a murine model would be more productive.

Acknowledgements

These investigations have been supported by the UK Health & Safety Executive to whome are thanks are due.

References

Bozelka BA, Sestini P, Gaumer HR, Hammad Y, Heather CJ, Salvaggio JE (1983) A murine model of asbestosis. Am. J. Path. 112:326-337

Donaldson K, Brown GM (1993) Bronchalveolar lavage in the assessment of the cellular response to fiber exposure. In: Fiber Toxicology Warheit DB (ed) Academic Press New York 117-139.

Elias JA, Freundlich B, Kern JA, Rosenbloom (1990) Cytokine networks in the regulation of inflammatory and fibrosis in the lung. Chest 97:1439-1445.

Gordon JR, Burd DR, Galli SJ (1990) Mast cells as a source of multifunctional cytokines. Immunol. Today 11:458-460.

Herbert A (1989) Pathology of pleurisy, pleural fibres and mesothelial proliferation. Thorax 41:176-189.

Kelley J (1990) Cytokines of the lung. Am. Rev. Respir. Dis. 141:765-788.

Lehnert BE, Valdez YE, Bomalaski SH (1985) Lung and pleural cell response to the intrapulmonary deposition of particles in the rat. J. Toxicology and Environmental Health 16:823-839.

Liddell D, Miller K (eds) (1991) Mineral fibres and health. CRC Press Boca Raton, Florida.

Martensson G, Hagberg S, Pettersen K, Thiringer G (1987) Asbestos pleural effusion. Thorax 42:646-651.

Meredith C, Scott MP, Pekelharing H, Miller K (1990) The effect of Biostim on the expression of cytokine mRNA in murine peritoneal macrophages *in vitro*. Toxicology Letters 53:327-337.

Miller MD, Krangel MS (1992) Biology and Biochemistry of the chemokines. Crit. Rev. Immunol. 12:17-46.

Miller K, Hudspith BN, Meredith C (1991) Rat pleural cell populations effects of MMMF inhalation on cytokine mRNA expression and population characteristics In: Mechanisms in fibre carcinogenesis Brown RC, Hoskins JA, Johnson NF (eds). Plenum Press New York 367-376.

Oberdoerster G, Ferin J, Marcello NL, Meinhold SH (1983) Effects of intrabronchially instilled amosite on lavagable lung and pleural cells. Environmental Health Perspectives 51:41-48.

Piquet PF, Collart MA, Grau GE, Sappino AP, Vasselli P (1990) Requirement of tumour necrosis factor for development of silica-induced pulmonary fibrosis. Nature 34:245-247.

Ralph P, Sampson-Johannes A (1990) Macrophage growth and stimulation factor, M-CSF. In: Hematopoietic growth factors in transfusion medicine. Wiley Liss New York 42-63.

Sirois P, Rola-Pleczcynski M, Begin R (1980) Phospholipase A activity and prostaglandin synthesis from alveolar macrophages exposed to asbestos. Prostaglandins Med. 5:31-37.

Zhang Y, Lee TC, Guillemin B, Yu M-C (1993) Enhanced IL-1ß and tumour necrosis factor-α release and messenger RNA expression in macrophages from idiopathic pulmonary fibrosis or after asbestos exposure. J. Immunol. 150:4188-4196.

Table 1. Alveolar cell counts

	BALB/c	CBA	C3H
Total cell recovery (x 10^6 mouse)	0.56	0.56	0.59
Differential cell counts (%)			
Macrophages	93	97	96
Lymphocytes	1	1	2
Neutrophils	4	1	1
Ciliated epithelial	2	1	1

Table 2. Pleural cell counts

	BALB/c	CBA	C3H
Total cell recovery (x 10^6 mouse)	0.70	0.78	0.81
Differential cell counts (%)			
Macrophages	64	47	54
Lymphocytes	27	53	36
Neutrophils	6		2
Eosinophils			
Mast cells	3		8

Table 3. Monoclonal antibody staining % of total cells

	Pleural cells	Alveolar cells
MAC 1	43	83
MOMA 2	46	91
CR3b	35	11
Class II	16	2

Table 4. Phagocyte index of pleural and alveolar macrophages after 24 hr exposure to the MMMF, RCF 1 (mean ± SD)

	Pleural macrophages	Alveolar macrophages
Control cells	46.2 ± 8.6	25.3 ± 6.0
Cells + LPS (10 μg/ml)	73.4 ± 10.8	36.2 ± 5.3

Table 5. Assessment of cytokine specific mRNA following *in vitro* exposure to MMMF

A. Interleukin-1 beta		
	–LPA	+LPS
1. Pleural cells		
Control	+	+++++
Glass fibre	++	+++
RCF 1	++	+++
RCF 2	++	+++
2. Alveolar cells		
Control	–	++++
Glass Fibre	+	++
RCF 1	+	++
RCF2	+	++

Table 6. <u>Assessment of cytokine specific mRNA following *in vitro* exposure to MMMF</u>

A. Tumour necrosis factor-alpha		
	–LPA	+LPS
1. Pleural cells		
Control	–	+++++
Glass fibre	++	++++
RCF 1	++	++++
RCF 2	++	++++
2. Alveolar cells		
Control	–	++++
Glass Fibre	+	++
RCF 1	+	++
RCF2	+	++

The Second Messenger System in Diseases Caused by Mineral Dusts

J A Hoskins, R C Brown, C E Houghton and G M Brown[1].
MRC Toxicology Unit,
Hodgkin Building,
University of Leicester,
P O Box 138, Lancaster Road,
Leicester LE1 9HN,
U K.

INTRODUCTION

Evidence is accumulating that promoters of carcinogenesis, although themselves neither mutagenic nor carcinogenic, enhance the cellular division and proliferation necessary for neoplastic transformation. *In vitro* studies with a number of cell types indicate that certain mineral fibres, such as asbestos, may function as tumour promoters. Several *in vivo* experimental studies have demonstrated synergy between asbestos and a chemical carcinogen (e.g. Harrison & Heath, 1988) which could support this proposition. There is also epidemiological evidence for its tumour-promoting activities including the synergism between asbestos exposure and smoking in the aetiology of lung cancer in humans (NRC, 1984). However, since fibres are not gene mutagens in bacterial and mammalian cell systems, they are not classified as initiators for bronchogenic tumours although it is known that *in vitro* fibres can enhance the effect of chemical or physical carcinogens causing cell transformation (Brown *et al.*, 1983; Hei *et al.*, 1985).

In bronchogenic cancer asbestos appears to act solely as a promoter but in the case of mesothelioma fibres seem to have both initiating and promoting activities and act as a complete carcinogen (Brown *et al.*, 1990a). Unlike most carcinogens asbestos does not cause base substitution and frameshift mutations in bacterial mutation assays (Chamberlain & Tarmy, 1977) or in mammalian cell systems (Shelby, 1988). It has been reported to be weakly

---

[1]Institute of Occupational Medicine, Edinburgh

NATO ASI Series, Vol. H 85
Cellular and Molecular Effects
of Mineral and Synthetic Dusts and Fibres
Edited by J. M. G. Davis and M.-C. Jaurand

mutagenic in Chinese hamster lung cells (Huang, 1979) but not in liver epithelial cells (Reiss *et al.*, 1983) or in Syrian hamster embryo fibroblasts (Oshimura *et al.*, 1984).
In spite of the evidence for tumour promoting activity the mechanism of promotion is unknown. Asbestos minerals seem to act as if they were agonists. It appears, however, that indirect activation of protein kinase C (PKC), a membrane-associated calcium and phospholipid-dependent enzyme activated by diacylglycerols (DAGs), may be involved. Studies have shown that PKC and early response genes including c-*fos* and c-*jun* are activated by asbestos (Heintz *et al.*, 1993).

Stimulation of certain cell-surface receptors can, through their coupling via G proteins, activate phospholipase-C which cleaves phosphatidylinositol 4,5-bisphosphate to inositol triphosphate and DAG. The DAG then activates PKC by decreasing the Km of the enzyme for calcium ion (Huang *et al.*, 1988). This same enzyme is also activated directly by the classic tumour promoter 12-*O*-tetradecanoylphorbol-13-acetate (TPA) which appears to be a DAG analogue. This is not a simple analogy since under some circumstances TPA stimulation of cells can induce release of DAGs. For example, stimulation of receptors for the phosphatidylinositol second messenger system by TPA produces a burst of DAGs (Berridge & Irvine, 1989). A quantitatively similar release of diacylglycerol occurs following addition of chrysotile, or amosite, to P388D1 or C3H10T1/2 cells but over a time scale two orders of magnitude greater. A phenomenon possibly due in part to the asynchrony of cell-fibre interaction (Brown *et al.*, 1989).

Continuous stimulation of cells over several hours with TPA produces a down-regulation of the pathway, in contrast cell/fibre interaction can produce DAG release which persists for many hours and takes several hours to reach a maximum. DAGs are a link in a second messenger chain between receptors which are G-protein coupled to phospholipase and protein kinase C. Recent work (for example, Lefkowitz, 1993; Parma *et al.*, 1993; Shenker *et al.*, 1993) has shown that if a second messenger system is switched on permanently it can have serious consequences for the host such as hyperplasia, over-production of hormones and cancer.

In the present work in which three cell types were treated with a number of different fibre types under various schedules of treatment increased levels of DAGs the results differ slightly from each other but confirm the interference with some aspects of the second messenger pathway. Surprisingly the effect can also be demonstrated *in vivo* in the rat following intratracheal instillation and inhalation of fibres and particulates.

MATERIALS AND METHODS

Minerals

UICC samples of chrysotile B, amosite and crocidolite were used. Long amosite fibres were a gift from the Institute of Occupational Medicine, Edinburgh. Samples of UICC amosite and crocidolite were milled in an agate ball mill to produce "short" fibres. Other fibres used were samples of the Thermal Insulation Manufacturer's Association (TIMA) MMVF21 (rock wool) and MMVF22 (slag wool) and a respirable fraction of Johns Manville code 100/475 glass fibre.

Coated and modified fibres

Amosite was surface-modified by reaction in boiling toluene with octyldimethylchlorosilane or octadecyldimethylchlorosilane as described previously (Brown *et al.*, 1990b). Two modified materials were produced designated C_8-amosite and C_{18}-amosite.

Fibres were coated with poly-D-lysine by sonicating 2 mg/ml suspensions of fibres in a solution of poly-lysine (0.1 mg/ml) for 2 min and centrifuging at 18,000 g for 12 min. Fibres were treated with water alone to prepare control material. After treatment the fibres were washed in water, respun and the supernatant solution removed by pipette. The fibres were resuspended in medium and sonicated immediately before use.

Fibres were coated with fibronectin by adding a solution of fibronectin (1 mg/ml) to a 2 mg/ml suspension of fibres, shaking vigorously and allowing to stand at room temperature for 10-30 min. The mixture was used as such and added to cells to give a final concentration of fibronectin of 30 μg/ml.

For all experiments with these fibres samples were autoclaved, suspended in medium and sonicated for at least 10 min prior to addition to the cells.

Cells and their culture

P388D1 macrophage-type cells were grown in Dulbeccos Modified Eagles Medium (DMEM) and C3H10T1/2 fibroblasts were grown in low glucose DMEM, both media supplemented with 10% foetal calf serum and glutamine. The cells were incubated at 37 °C in an atmosphere of 8% CO_2. V79-4 cells were grown in Minimum Essential Medium (MEM) supplemented with 15% heat inactivated serum and glutamine. They were incubated at 37 °C in an atmosphere of 5% CO_2.

Cells were initially grown in 9.40 cm^2 tissue culture plates but in later studies the larger 148 cm^2 dishes were used to provide a greater yield of cells and consequently of DAGs for analysis.

***In vitro* experiments**

Fibres were added to the cultures when the cells were 80% confluent and incubated for 18 h. DAGs were then extracted from the cells using a modified Bligh and Dyer (1959) technique, acetylated with ^{14}C-acetic anhydride and the mono-, di- and tri-glycerides separated by tlc on silica gel in chloroform/acetone (96:4) as described previously (Hoskins and Evans, 1989).

Coated fibre studies

At confluence the cells were suspended in medium, transferred to silanised bottles and incubated in a 37 °C water bath for 30 min. The fibres were added to the cells, 2 ml samples taken and centrifuged at 2000 g for 5 min. 300 µl perchloric acid (10%) and 2 ml PBS were added to the pellet which was shaken and recentrifuged. After removing the supernatant solution 2 ml of chloroform/methanol (2:1) was added followed by 300 µl of EDTA in water (5 mM). The organic layer was evaporated to dryness and acetylated as described above.

Exposure of cells to fibres

Autoclaved fibres were made up in medium to give a suspension of 2 mg/ml, sonicated for 5 min and added to cells to give the required concentration.

Either: The fibre suspension was added to near confluent cell cultures and, after incubation for 18 h, the cells were washed twice with complete medium. The medium was then removed and the cells killed by adding 150 µl of a 10% TCA solution followed by suspension of the mixture in 1 ml of PBS.

Or: With cells in suspension 1 ml of the fibre suspension was added to a 9 ml volume of cells and medium in a shaking water bath at 37 °C to give a final concentration of fibres of 200 μg/ml. Two ml aliquots were taken at desired time points, spun at 2000 g for 5 min, the medium removed, 300 μl of 10% perchloric acid added and the mixture resuspended in 2 ml PBS.

The mixtures were spun at 2000 g for 5 min, the supernatant solution discarded and the pellet extracted with 2 ml of $CHCl_3$/MeOH (2:1) (Bligh & Dyer, 1959) for 1 h. 300 μl of 5 mM EDTA solution was added and the bottom layer removed and evaporated to dryness.

Cytidinediphosphate-DAG (CDP-DAG) extraction

At confluence cells were suspended in complete medium containing 120 kBq/ml ^{3}H-cytidine. The cells were divided between silanised 100 ml bottles and shaken in a water bath at 37 °C for 60 min. The cells were incubated with LiCl (10 mM, final concentration) for 10 min. before addition of the fibres. Incubations were continued for 50 min then 2 ml samples centrifuged at 2000 g for 5 min. The supernatant solutions were removed and the reactions stopped by the addition of 0.94 ml chloroform/methanol (2:1). The CDP-DAGs were extracted using the method of Godfrey (1989).

***In vivo* studies**

Male Hans Wistar rats were either exposed to fibre in whole body exposure chambers or had mineral intratracheally instilled under general anaesthesia. At selected time-points after exposure the animals were killed, the lungs ex-sanguinated, removed and frozen immediately in liquid nitrogen. For analysis the lung tissues were homogenised and ground to a powder under liquid nitrogen. After addition of 1 ml TCA (10%) and 5 ml PBS the samples were centrifuged at 4000 g for 5 min. Five ml of chloroform/methanol (2:1) were added and the samples kept at -20 °C overnight. EDTA in water (1.5 ml; 5 mM) was added, the tubes centrifuged at 4000 g for 2 min. and then the organic layer was evaporated to dryness and acetylated with 740 kBq ^{14}C-acetic anhydride and 80 μl pyridine (Hoskins and Evans, 1989).

Levels of DNA were estimated from optical density measurements at 260 and 280 nm (Brown *et al.*, 1994). To determine protein levels samples of the lungs were homogenised in 5 ml NaOH (2M) and incubated at 37 °C overnight. The samples were neutralised by the addition of 5 ml hydrochloric acid (2M), mixed by inversion and diluted 10-fold before estimating protein levels using the Biorad protein microassay (Bio-Rad Laboratories Ltd., Hemel Hempstead, Herts.).

RESULTS

Cell studies

Two cell types, P388D1 and C3H10T1/2 were treated with chrysotile and the two amphiboles amosite and crocidolite. A further cell type, V79-4, was used in the studies of cytidinediphosphate-DAGs. Preliminary experiments were also carried out with TIMA rock wool (MMMF21) and slag wool (MMMF22) to compare them with amosite as a positive control. The experiments are best described with reference to the minerals examined.

Most of the asbestos experiments were carried out with amosite since experience has shown that this material is the most homogeneous of the UICC samples. UICC samples generally are only claimed to be homogeneous to 100 mg, at least an order of magnitude greater in amount than required for any *in vitro* experiment.

Chrysotile

The brucite surface of chrysotile is positively charged and its fibres are attracted to negatively charged cell surfaces but there is still a delay of some hours before there is a measurable increase in intracellular DAG levels in P388D1 cells. When fibres were added to near confluent layers of cells at loadings of 4, 8 and 16 $\mu g/cm^2$ no increase was seen after 2 h but levels had increased by 8 h although the increase was only statistically significant at a fibre loading of 8 $\mu g/cm^2$. Earlier experiments (Hoskins *et al,* 1990) had shown that DAG release was at a maximum by 18 h. Measurements made at this time show an apparent dose related increase in DAG levels up to the top loading of 16 $\mu g/cm^2$. However, at high loading levels the probability of cell death, because of the sheer number of fibres on the cells, distorts the results (figure 1).

Amosite

We had shown previously that the effect of amosite on DAG levels in C3H10T1\2 after 18 h exposure is to produce significant increases with a range of loadings from 12.5 $\mu g/cm^2$ to 50 $\mu g/cm^2$. (Hoskins *et al,* 1990).

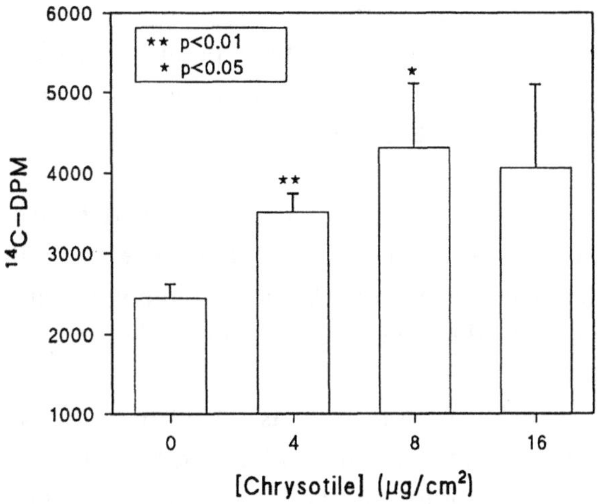

Figure 1. DAG levels in P388D1 cells exposed to chrysotile for 18 h.

These experiments have been repeated using P388D1 cells which have been shown to react in the same manner.

The role of fibre length in DAG release was explored using prepared samples of amosite with different length distributions. When amosite was milled the resultant mix of isomorphous particles produced no rise in DAG levels at any loading tested from 0 to 50 μg/cm^2. At these high loadings the number of particles is very much larger than for a similar weight of fibres.

A series of experiments was carried out using C3H10T1/2 cells with size selected amosite fibres. Three amosite samples were used: short amosite, which contains virtually no fibres greater than 5 μm in length, long amosite which contains a high proportion of long fibres (Davis, 1989; Davis *et al.*, 1986) and the older UICC standard which contains fibres of intermediate length to the other groups. Although, due to inhomogeneity, there were exceptions, generally after 18 h exposure DAG levels were found to have increased above control levels in the order: control <short amosite <UICC amosite < long amosite.
The response time of cell/fibre interaction is reduced when fibres are coated with protein. This effect is simulated with the polypeptide poly-D-lysine[2].

[2]Poly-D-lysine rather than poly-L-lysine is used because of its greater resistance to proteases. This is important in other *in vivo* experiments which are in progress.

Using fibres pre-coated with peptide there was an immediate increase in DAG levels. Although sustained at time points of 10 and 65 min these increases did not, by these times, reach statistical significance but were greater than the responses at 8 h using uncoated fibres.

When these experiments were repeated using milled fibre, with or without peptide coating there was no increase in DAG levels. Similarly, using fibre samples with a surface made hydrophobic by binding dimethylsilylalkyl (R = 8 or 18) ligands produced no increase in DAG levels.

Poly-D-lysine coated UICC amosite was added to cultures of both V79-4 and C3H10T1/2 cells both with and without added lithium chloride and the mixture incubated for 60 min. The cells were then analysed for their CDP-DAG content. There was no difference in the CDP-DAG levels in the four groups of V79-4 cells: controls and fibre-treated, with and without lithium chloride. Using C3H10T1/2 cells there was an apparent increase in CDP-DAG levels only for the poly-D-lysine coated amosite group but this was not significant.

Crocidolite

Very similar results to those obtained with amosite were found using crocidolite samples. When P388D1 or C3H10T1/2 cells are exposed for 18 h to UICC crocidolite, a long fibre crocidolite prepared by TIMA or a milled sample of crocidolite at a loading of 14 $\mu g/cm^2$ the resultant DAG levels were found to increase significantly above control levels in the order control = or < milled < UICC < TIMA. Microscopic examination of the milled sample showed that it was still mainly fibrous, which accounts for this result (figure 2).

MMMF

Raised levels of DAGs may also be produced by man-made vitreous fibres. A few experiments were carried out using 'standard' samples prepared by TIMA. A sample of rock wool (MMMF21) produced no response in studies with P388D1 cells while slag wool (MMVF22), which had been found in other cell studies to be toxic, produced quantitatively the same effect as amosite used as a positive control.

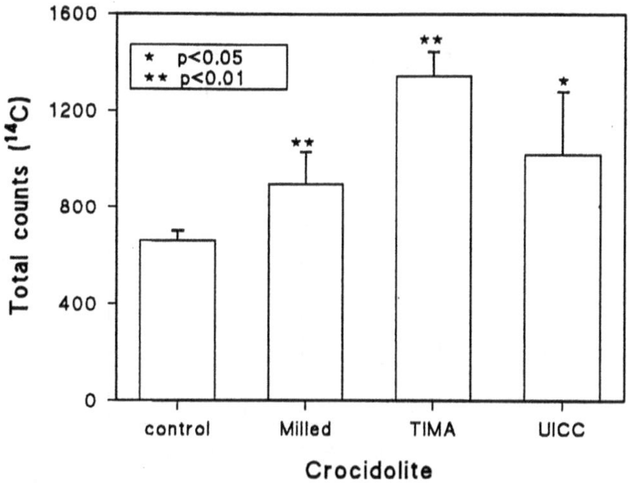

Figure 2. DAG levels in C3H10T1/2 cells exposed to crocidolite.

other cell studies to be toxic, produced quantitatively the same effect as amosite used as a positive control.

In vivo

Animals were exposed to fibres either by intratracheal instillation or by inhalation. At certain times after the exposure the animals were killed and their lungs removed and immediately frozen in liquid nitrogen. Changes in DAG levels were measured in samples prepared from whole lungs.

Crocidolite

UICC crocidolite or a powdered quartz sample, DQ12 was intratracheally instilled into Hans Wistar rats and the animals sacrificed one and three days later. One day after instillation there was no difference in lung DAG levels between crocidolite treated animals and control animals treated with the PBS vehicle alone. After three days both minerals had produced apparent increases in DAG levels.

The presence of the minerals in the lung can produce inflammation and/or recruitment of cells. The degree of inflammation can be assessed by protein assay and cell recruitment by DNA assay. Using these techniques it was shown that DNA levels were about the same in controls and animals treated with DQ12 and higher, though not significantly so, in those treated with crocidolite. Protein assays showed that DAG levels in control animals were

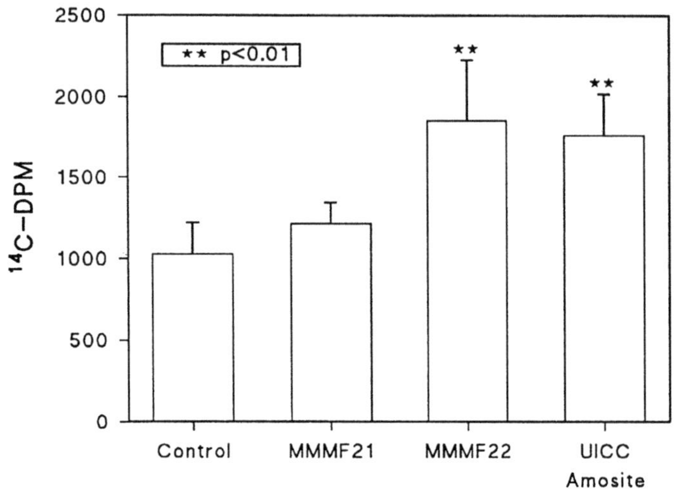

Figure 3. DAG levels in P388D1 cells exposed to MMVF and amosite.

less than in those treated with DQ12 or crocidolite but the difference was only significant ($P<0.05$) for the crocidolite sample.

Amosite

An inhalation experiment in a whole body exposure chamber comparing long fibre amosite with code 100/475 glass fibre was carried out. DAG levels were raised after one day but the levels in all groups were not significantly different after 14 days exposure. None of the increases were statistically significant. There was a similar picture when DAG levels were compared to protein content but when compared to DNA levels there was an apparent, but not statistically significant, increase.

DISCUSSION

The reaction of a soluble agonist with cells in culture produces a transient increase in intracellular DAGs (Huang *et al.*, 1988). Mineral fibres behave similarly but the reaction is less consistent probably because of the heterogeneity of the fibre samples at the concentrations used though it is likely that there are other contributory factors responsible for this variability. The initial cell-fibre interaction which produces this effect can seem to be then the fibres can become attached to cells through integrin receptors. These receptors are normally used to anchor cells to a substrate and recognise key tri-peptide sequences in substrate proteins, notably RGD (arg-gly-asp) which is found in fibronectin and LDV (leu-asp-val) (Newell, 1992). Non-specific binding of fibres to cells occurs when other proteins carrying a net positive charge are bound to fibres, such as amphiboles, which have a net negative charge and facilitate the movement of the fibre through the electrostatic barrier which normally acts to inhibit fibre-cell interaction. The DAG released activates membrane bound PKC. This enzyme is also activated directly by the classic tumour promoter 12-*O*-tetradecanoylphorbol-13-acetate (TPA) which appears to be a DAG analogue. The analogy, though, is not simple since under some circumstances TPA stimulation of cells can induce release of DAGs. For example, stimulation of receptors for the phosphatidylinositol second messenger system by TPA produces a burst of DAGs.

While a soluble agonist produces a DAG response within seconds the response to fibres is slow. We have treated a variety of cell types in culture with asbestos and shown a sustained increase in intracellular DAGs which occurs over a period of hours. For example a release of diacylglycerol occurs following addition of UICC chrysotile, or amosite, to P388D1 cells - which are macrophage-like cells - or C3H10T1\2 fibroblasts and this occurs over a time scale two orders of magnitude greater than for a soluble agonist such as TPA. These experiments were carried out using cell monolayers or cells in suspension, used to eliminate any part of the signal due to attachment to a plastic surface. We have reported previously (Hoskins *et al.*, 1990) that a near confluent layer of P388D1 cells treated for 18 h with amosite showed a dose-dependent increase in DAG levels, significant ($p<0.05$) at a fibre load of 23.4 $\mu g/cm^2$. When P388D1 cells were treated with chrysotile there was no increase in DAG levels after two hours but some increase after 8 h which was significant at a loading of 8 $\mu g/cm^2$. The increased DAG release only became significant after 18 h.

The slow response to the effects of fibres is in marked contrast to the action of TPA but there is another difference between the two systems. Continuous stimulation of cells over several hours with TPA produces a down-regulation of the phosphatidyl inositol pathway. This phenomenon does not seem to occur with fibres. If it did it is unlikely that one would see any significant or sustained release of DAGs after fibre treatment.

It has been shown (Godfrey, 1989) that treatment of cerebral cortex tissue with thrombin and carbachol causes an increase in the levels of CDP-DAGs particularly in the presence of lithium. Lithium blocks the conversion of phosphatidylinositol to myo-inositol and so stops any re-cycling of the phosphoinositides. Phosphatidic acid reacts with CTP to form CDP-DAGs which further react with myo-inositol to form phosphatidylinositol and CMP. The measurement of the CDP-DAGs therefore gives a measure of the level of phospholipid metabolism. Since fibres also turn on the phosphatidylinositol pathway the levels of CDP-DAGs following the reaction of poly-D-lysine coated amosite with cells was measured. No significant increase in the levels of these metabolites was found.

In most of the experiments described in this paper there is an inconsistent dose dependence. This is to be expected bearing in mind the heterogeneous nature of the systems. A suspension of fibres settling onto a monolayer of cells which only present at most half of their surface for interaction will, in time, cover the exposed area completely. The duration of our experiments is not long enough for settlement to be complete. UICC samples are only homogeneous down to 100 mg and it is unlikely that other mineral samples used were any better. Therefore when cells are treated with fibres there is asynchrony in any response because of the time taken for cells and fibres to meet and the heterogeneity at that level of the fibre sample. These factors can obscure any dose relationship especially at low concentrations. In the case of amphibole fibres there is a charge-barrier to be overcome which further complicates cell/fibre interaction. Although from the results a dose response relationship seems to be demonstrated this is almost certainly not linear. A certain number of fibres have to land on cells to elicit a measurable response, while at a very high load the response falls off possibly due to the sheer number of fibres killing cells.

Unlike some other measures of the effect of minerals on cells DAG release is quite definitely related to the morphology. To examine the importance of this phenomenon cells were treated with amosite, milled using an agate ball mill to avoid metal contamination. After milling the particles have a similar diameter to the original fibres but few have an aspect ratio greater than 3:1, which is usually taken to define a fibre. The particles, which have identical chemistry to the amosite fibres, do not cause DAG release in the cell systems studied. Treatment of C3H10T1/2 cells for 18 h with milled amosite at doses of 12.5 to 50 $\mu g/cm^2$ produced no increase in intracellular DAG levels. The effect of amosite on DAG levels in C3H10T1\2 after 18 h exposure is to produce significant increases in levels from 12.5 μg/cm^2 to 50 μg/cm^2. Conversely a size selected sample containing mostly long amosite fibres could be shown to have a greater effect on DAG release than UICC amosite which contains a greater spread of fibre lengths.

Shielding the anionic polysilicate of the amosite surface by effectively bonding hydrocarbon chains to it (C_8- and C_{18}-amosite) produces a product which does not stick to cells (Brown *et al.*, 1990b). Reaction of these surface modified materials with C3H10T1/2 cells for 18 h did not produce an increase in DAG levels.

Since an increase in intracellular DAG levels appears to signal an interaction with a surface, part of the measured DAGs in a cell monolayer must result from the attachment of the cell to the plastic surface of the dish. This provides a constant background in any measurements of DAG release. It can be removed by using cells in suspension. All of the subsequent work described here has been carried out using cells in this way.

To reduce the asynchrony of fibre-cell interaction mediated by charge fibres were coated with poly-lysine to render them strongly positively charged. Fibres so treated reacted almost immediately with cells, which have a negatively charged surface. For example, when amosite is coated with poly-D-lysine and added to C3H10T1/2 cells the protein-fibre complex sticks more rapidly to cells than native amosite and produces a faster response and a greater release of DAGs. There is about a 50% increase in DAG levels in C3H10T1/2 cells, interaction was immediate and levels were sustained for at least 75 mins. When milled amosite coated with poly-lysine was used there was no increase in DAG levels. Poly-D-lysine is equivalent to poly-L-lysine and was used because of its use in concomitant *in vivo* experiments not reported here. Fibres were also coated with fibronectin, a cell adhesion molecule which binds to cells through integrin receptors. This protein is involved in the attachment of cells to plastic culture dishes. It is believed that the first step in cell fibre interaction is the coating of the fibre with protein from the medium. When a cell culture medium is depleted of fibronectin it supports less cell/fibre interaction: reported elsewhere (Brown *et al.*, 1991). Also, fibronectin binding can be blocked by peptides containing the RGD sequence. Fibronectin, like any other protein, will change the surface charge of the fibre which may assist in its interaction with a cell.

Fibronectin was added to a suspension of C3H10T1/2 cells in medium immediately before addition of an amosite suspension. The final concentration of fibronectin was 50 μg/ml. Following addition there was generally a slight increase in DAG levels after 75 min but this was not statistically significant in all experiments. Other end-points of cell/fibre interaction had given a positive indication by this time. In some experiments fibronectin alone produced an increase in DAG levels. Pre-coating the fibre with fibronectin before adding it to the cells gave the same result. The degree of cell/fibre reaction is very dependent upon the concentration of fibronectin which may explain these negative results.

MMMF

Because of the current interest in man-made fibres we have begun some preliminary work with these materials using two of the samples prepared by TIMA: a rock wool (MMVF21) and a slag wool (MMVF22). We found that MMVF21 had no effect on the DAG levels in C3H10T1/2 cells while MMVF22 was as active as amosite used as a positive control. This result mirrors certain other cell toxicity studies in which rockwool has been shown to be inactive and slagwool active. There is no obvious reason that we have discovered so far to explain this.

In vivo

The problems of relating experimental studies using cells in culture to those in whole animals has been much discussed. We decided to try and measure changes in DAG levels in whole lungs removed from animals treated with fibre by instillation or inhalation. If the target cells are those which form the inner lining of the lung then these must be but a small part of the whole organ. Consequently, it was felt that significant DAG increases would not be seen. In the event this was so, but increases in DAG levels were demonstrated in relation to other biological parameters.

The presence of the minerals in the lung can produce inflammation and/or recruitment of cells. The degree of inflammation can be assessed by protein assay and that of cell recruitment by DNA assay. In both the instillation and the inhalation experiments significant increases in DAG levels could only be demonstrated in relation to one of these parameters. When a mineral was instilled, a more severe insult than inhaling fibres, absolute rises in DAG levels were seen after three days with both DQ12 and crocidolite but these never quite reached significance. After only 1 day there was no difference in DAG levels. When the DAG levels are related to DNA levels in the lung, as a measure of the total number of cells, they appear to be higher in the crocidolite treated animals though not significantly so. When the DAG levels are related to protein levels as a measure of inflammation then the crocidolite treated animals do have significantly increased levels. Exposing animals to long fibre amosite by inhalation produced an apparent effect after one day but this had decreased after 14 days to an insignificant level. The increase in DAG levels was significant ($p<0.05$) only for exposure to long fibre amosite and in relation to lung protein levels as for the exposure to crocidolite above.

DAGs are a link in a second messenger chain between receptors which are G-protein coupled to phospholipase and protein kinase C. Recent work (Lefkowitz, 1993; Parma *et al.*, 1993; Shenker *et al.*, 1993) has shown that if a second messenger system is switched on permanently it can have serious consequences for the host such as hyperplasia, over-production of hormones and cancer. Fibres turn on the inositol phosphate second messenger pathway for a period several orders of magnitude greater than in normal operation. This may be sufficient for there to be serious sequelae and is evidence that the disruption of the second messenger pathway by the interaction of fibres with cell-surface receptors has a role in fibre carcinogenesis.

CONCLUSIONS

Overall our results show that the phosphatidylinositol second messenger pathway, which when activated by a soluble agonist is a fast response and short duration system, is switched on by fibres for hours rather than seconds. This is probably because cell/fibre interaction is irreversible while soluble agonists have a finite residence time on a receptor. Amosite, crocidolite and chrysotile fibres, as well as certain man-made fibres, but not non-fibrous particles turn on the phosphatidyl inositol pathway and cause an increase in DAG levels in cells in culture. The degree of response that occurs depends in some way the surface chemistry of the fibre. When the surface of the fibre is hydrophobic through coating with co-valently bound hydrocarbon chains, cells and fibres do not interact and there is no increase in intracellular DAGs. Fibres coated with a synthetic protein, poly-D-lysine, have a positively charged surface, are attracted to cells and do cause an immediate rise in DAG levels. Coating the fibre with the cell attachment protein fibronectin should increase the speed of cell/fibre interaction and the amount of intracellular DAGs. However, we were unable to demonstrate such an increase probably because the concentration of fibronectin in the medium is critical for cell growth. Destroying the fibre morphology by milling produced a material which was inactive in all our experiments. Finally we have presented some evidence that the *in vitro* results can, to a degree, be demonstrated *in vivo*.

REFERENCES

Berridge,M.J. and Irvine,R.F. (1989) Inositol phosphates and cell signalling. *Nature* **341**:197- 205.

Bligh,E.G. and Dyer,W.J. (1959) A rapid method of total lipid extraction and purification. *Canad. J. Biochem. Physiol.* **37**:911-917.

Brown,R.C., Poole,A. and Fleming,G.T.A. (1983) The influence of asbestos dust on the oncogenic transformation of C3H10T1/2 cells. *Cancer Letters* **18**:221-227.

Brown,R.C., Hoskins,J.A., Cole,K.J., Evans,C.E. and Sara,E. (1989) Effects of Asbestos on some aspects of the Cell Second Messenger System. In "Effects of Mineral Dusts on Cells", Mossman,B.T. and Bégin,R.O., eds. NATO ASI Series Vol **H30**, pp. 375-381.

Brown,R.C., Hoskins,J.A., Miller,K. and Mossman,B.T. (1990a) Pathogenetic Mechanisms of Asbestos and other Mineral Fibres. *Molec. Aspects Med.* **11**:325-349.

Brown,R.C., Carthew,P., Hoskins,J.A., Sara,E. and Simpson,C.F. (1990b) Surface modification can affect the carcinogenicity of asbestos. *Carcinogenesis* **11**:1883-1885.

Brown,R.C., Hoskins,J.A. and Evans,C.E. (1991) Factors affecting the interrelationship of asbestos fibres with mammalian cells: a study using cells in suspension. *Ann. Occup. Hyg.* **35**:25-34.

Brown,R.C., Pickering,P. and Hoskins,J.A. (1994) Oxidative damage to 2-deoxyguanosine in solution and in DNA after treatment with mineral fibres. (submitted to *Carcinogenesis*).

Chamberlain,M. and Tarmy,E.M. (1977) Asbestos and glass fibres in bacterial mutation tests. *Mutation Res.* **43**:159-164.

Davis,J.M.G., Addison,J., Bolton,R.E., Donaldson,K., Jones,A.D. and Smith,T. (1986) The pathogenicity of long versus short fibre samples of amosite asbestos administered to rats by inhalation and intraperitoneal injection. *Br. J. Exp. Path,* **67**: 415-430.

Davis,J.M.G. (1989) Mineral fibre carcinogenesis: experimental data relating to the importance of fibre type, size, deposition, dissolution and migration. In: "Non-occupational exposure to mineral fibres." (Bignon,J., Peto,J. and Saracci,R. eds.) IARC publication no. 90. Published by the International Agency for Cancer Research, pp 33-46.

Godfrey, P.P. (1989) Potentiation by lithium of CMP-phosphatidate formation in carbachol-stimulated rat cerebral-cortical slices and its reversal by myo-inositol. *Biochem. J.* **258**:621-624.

Harrison,P.T.C. and Heath,J.C. (1988) Apparent synergy between chrysotile asbestos and *N*-nitrosoheptamethyleneimine in the induction of pulmonary tumours in rats. *Carcinogenesis* **9**: 2165-2171.

Hei,T.K., Geard,D.E., Osmak,R.S. and Travisano,M. (1985) Correlation of *in vitro* genotoxicity and oncogenicity induced by radiation and asbestos fibres. *Br. J. Cancer* **52**:591-597.

Heintz,N.H., Janssen,Y.M. and Mossman,B.T. (1993) Persistent induction of c-*fos* and c-*jun* proto-oncogene expression by asbestos. *Proc. Natl. Acad. Sci.* **90**:*3299-3303.*

Hoskins,J.A. and Evans,C.E. (1989) Measurement of intracellular diacylglycerols by acetylation and thin-layer chromatography. *J. Chromatog. Biomed. Appl.* **490**:439-443.

Hoskins,J.A., Brown,R.C. and Evans,C.E. (1990) Promoting effects of fibres, Fibres and the second messenger pathways. In, "Mechanisms in Fibre Carcinogenesis" (Brown,R.C., Hoskins, J.A. and Johnson, N.F. eds.) NATO ASI series No 223, Plenum Press, New York, pp 459-467.

Huang,S.J., Monk,P.N., Downes,C.P. and Whetton,A.D. (1988) Platelet-activating factor-induced hydrolysis of phosphatidylinositol-4,5-bisphosphate stimulates the production of reactive oxygen intermediates in macrophages, *Biochem. J.* **249**:839-845.

Huang,S.L. (1979) Amosite, chrysotile and crocidolite asbestos are mutagenic in Chinese hamster lung cells. *Mutation Res.* **68**:265-274.

Lefkowitz,R.J. (1993) Turned on to ill effect. *Nature* **365**:603-604.

Newell,J. (1992) A sticky end for disease. *New Scientist,* 5 December pp 29-31.

NRC, National Research Council (1984) "Asbestiform Fibers: Non-occupational Health Risks" National Academy Press, Washington DC.

Oshimura,M., Hesterberg,T.W., Tsutsui,T. and Barrett,J.C. (1984) Correlation of asbestos-induced cytogenetic effects with cell transformation of Syrian hamster embryo cells in culture, *Cancer Res.* **44**:5017-5022.

Parma,J., Duprez,L., Van Sande,J., Cochaux,P., Gervy,C. Mockel,J., Dumont,J. and Vassart,G. (1993) Somatic mutations in the thyrotropin receptor gene cause hyperfunctioning thyroid adenomas. *Nature* **365**:649-651.

Reiss,B., Tong,C., Telany,S. and Williams,G.M. (1983) Enhancement of Benzo[a]pyrene mutagenicity in rat liver epithelial cells. *Environ. Res.* **31**:100-104.

Shelby,M.D. (1988) The genetic toxicity of human carcinogens and its implications. *Mutation Res.* **204**:3-15.

Shenker,A., Laue,L., Kosugi,S., Merendino,J.J. Minegishi,T. and Cutler,G.B. (1993) A constitutively activating mutation of the luteinizing hormone receptor in familial male precocious puberty. *Nature* **365**:652-654.

Transcriptional Regulation of IL-1β and IL-6 by TNF-α in Asbestosis is Mediated by NF-IL6-like Motifs

Yihong Zhang, MD, PhD and William N. Rom, MD, MPH
Division of Pulmonary and Critical Care Medicine
Departments of Medicine and Environmental Medicine
and Chest Service, Bellevue Hospital Center
New York University Medical Center
550 First Avenue
New York, NY 10016

ABSTRACT

Asbestosis is an inflammatory and fibrotic interstitial lung disease characterized by a lower respiratory tract alveolitis and bronchiolitis with increased macrophages, lymphocytes, and neutrophils. The alveolar macrophages have a morphological appearance of activation and spontaneously secrete exaggerated quantities of growth factors, matrix proteins, and cytokines. We performed bronchoalveolar lavage of nine asbestos insulators (category $\geq$ 1/0 parenchymal opacities) and five normal controls finding increased release into 18 hr supernatants of tumor necrosis factor-α (TNF-α asbestos-exposed 953 $\pm$ 192 pg/ml vs normals 290 $\pm$ 38 pg/ml, $p < 0.005$) and interleukin-1β (IL-1β asbestos-exposed 760 $\pm$ 120 pg/ml vs normals 138 $\pm$ 24 pg/ml, $p < 0.001$). Since TNF-α is also a potent mediator of fibrosis in animal models of silicosis, we evaluated its ability to stimulate other cytokines. TNF-α maximally stimulated a -130/+15 IL-1β genomic fragment and a -158/-49 IL-6 fragment using the CAT assay. We found two NF-IL6 motifs in these fragments that mediated transcription in an independent manner. Interestingly, the IL-6 gene had a NF-κB motif downstream of the NF-IL6 sites that appeared to be even more potent. We also identified NF-IL6 sites near the transcription start site of IL-8 and in the promoter region of TNF-α. The NF-IL6 motifs belong to a family of nuclear binding proteins characterized by a leucine zipper and may play a role in inflammation by stimulating cytokine production contributing to an acute phase response localized to the lower respiratory tract.

NATO ASI Series, Vol. H 85
Cellular and Molecular Effects
of Mineral and Synthetic Dusts and Fibres
Edited by J. M. G. Davis and M.-C. Jaurand

INTRODUCTION

Asbestosis represents an interstitial lung disease with both fibrosis and inflammation (1,2). The inflammatory cells in the lower respiratory tract can be sampled by bronchoalveolar lavage (BAL) and consist of activated alveolar macrophages with increased percent of lymphocytes and neutrophils in some patients (1-5). A variety of mediators have been measured in BAL fluid or spontaneously released into culture medium by BAL cells. These include oxidants (superoxide anion, hydrogen peroxide), growth factors (alveolar macrophage derived growth factor, platelet derived growth factor), matrix proteins (fibronectin) and enzymes (2,6-8). Since tumor necrosis factor-α (TNF-α) is a predominant cytokine identified in silicotic lesions in animal models of silicosis, we hypothesized that cytokines such as TNF-α and interleukin-1β (IL-1β) would also be released by alveolar macrophages (9). Cytokine networks have also been described where TNF-α, for example, can not only amplify its own release, but stimulate the release of other cytokines such as IL-1β, IL-8, IL-6, etc. (10). We evaluated the IL-1β and IL-6 genes identifying a series of 9 bp motifs with high homology to the NF-IL6 family of nuclear binding proteins that function as promoters. The NF-IL6 motif on genes that mediate inflammation may be a common pathway for mononuclear phagocytes to mediate inflammation in the lower respiratory tract.

METHODS

We recruited nine asbestos-exposed individuals who had been exposed to asbestos for 31 $\pm$ 2 yr as insulators. All were non-smokers or ex-smokers > 5 yr. Their mean age was 56 $\pm$ 2 yr. All had irregular opacities on their chest radiographs and bilateral circumscribed pleural plaques or diffuse pleural fibrosis. Pulmonary function tests performed according to American Thoracic Society recommendations revealed mean VC of 81 ± 5%, FEV_1 of 85 ± 6%, and diffusing capacity 87 ± 7%. BAL recovered a mean of 302 ± 30 x 10^3 cells/ml fluid returned with a cell differential of 87 ± 5% macrophages, 11 ± 5% lymphocytes, 1 ± 0.4 neutrophils, and 0.7 ± 0.6% eosinophils.

We included in this study five normal controls (all males) with normal physical examinations, chest radiographs, and pulmonary function tests. BAL recovered a mean of 208 ± 15 x 10^3 cells/ml with a cell differential of 83 ± 2% macrophages, 15 ± 2% lymphocytes, 2 ± 2% neutrophils, and < 1% eosinophils.

Isolation of RNA and Northern blot analysis

AM were lysed by addition of 5.5 M guanidinium isothiocyanate buffer and cytoplasmic RNA was isolated through $CsCl_2$ gradient ultra-centrifugation. Total RNA was fractionated by electrophoresis through a 1% agarose-6% formaldehyde denaturing gel, and transferred onto a nitrocellulose filter (BA 85, Schleicher and Schuell, Keene, NH). The baked filter was incubated in 40 ml of prehybridization solution (50% formamide, 0.5% SDS, 10X Denhardt's solution, 2.5% herring sperm DNA, and 4X SSPE) at 42°C for 6 to 12 h. IL-1β cDNA (provided by S. Gillis, Immunex, Seattle, WA) and TNF-α cDNA (Genentech, So. San Francisco, CA), were nick-translated with [α-^{32}P]dCTP (sp. act. 3000 Ci/mmol, New England Nuclear, Boston, MA) with a nick translation kit (Boehringer Mannheim, Indianapolis, IN). The hybridizations were carried out at 42°C for 10 h. The filter was then washed in a solution containing 2X SSC and 0.5% SDS at room temperature for 20 min followed by 0.1X SSC plus 0.5% SDS at 65°C for 30 min. To ascertain that equal amounts of RNA were loaded in individual lanes, the filters were reprobed with ^{32}P-labeled pHe7 cDNA (housekeeping gene) whose mRNA level is non-cell-cycle dependent. Autoradiography was performed at -70°C.

The TNF-α ELISA kit was purchased from Biokine (Minneapolis, MN); the IL-1β ELISA kit was kindly supplied by Dr. R. Dondero, Cistron Biotech, Pine Brook, NJ.

Screening of human genomic library and plasmid construction

A human genomic library (Clontech) was screened by using human IL-1β cDNA and the synthetic oligonucleotide sequence of IL-1β as probes. Eight positive clones were picked up, and one of the eight was used for subcloning. An IL-1β DNA fragment from positions -1130 to +15 (-1130/+15 fragment) was cut out with restriction enzymes *Xba*I and *Taq*I, and subcloned into a polylinker site in front of the chloramphenicol acetyltransferase (CAT) structural gene on the plasmid vector pTK(-).CAT (from Herbert Samuels, New York University Medical Center). A series of deletion mutant fragments were created by restriction enzyme cutting from the 5' end, namely, *Sau*3 (-680), *Sty*I (-490), *Dra*I (-313), *Hin*dIII (-131), and *Dde*I (-69) fragments.

Transient transfection of human suspension cultures and assay of CAT activity

THP-1 cells were transfected with serially constructed plasmids by the DEAE-dextran method. Briefly, 10^7 cells were washed in STBS solution (25 mM Tris [pH 7.4], 137 mM NaC1, 5 mM KC1, 0.6 mM Na_2HPO_4, 0.7 mM $CaC1_2$, 0.5 mM $MgC1_2$) and were transfected with 10 μg of $CsC1_2$-purified plasmids in 1 ml of STBS solution containing 500 μg of DEAE-dextran per ml at 37^oC for 90 min. The transfected cells were then washed with STBS solution and incubated in complete medium for 24 h in the absence or presence of inducing agents. The transfected cells were lysed by three cycles of freeze-thawing, and equal amounts of protein from different cell extracts were assayed for CAT activity. The protein concentration was determined with Bio-Rad reagents. For the CAT assay, 100 μg of protein was incubated with 0.1 μCi of [^{14}C] chloramphenicol-250 mM Tris (pH 7.5)-360 μg of acetyl coenzyme A per ml in a total volume of 170 ml at 37°C for 5 h. The reactions were stopped by addition of 0.5 ml of cold ethyl acetate. After extraction with ethyl acetate, the upper layer was saved, dried, and spotted onto a thin-layer chromatography plate. After being developed in organic solution (95% chloroform and 5% methanol), the plate was air-dried and exposed to X-ray film. Site-directed mutagenesis was carried out according to directions by the Bio-Rad site-directed mutagenesis kit.

Statistics

Data were assessed using one-way Kruskal-Wallis Analysis of Variance (ANOVA). For both IL-1β and TNF-α the ANOVA was significant leading to Dunnett's test. Dunnett's test follows a significant ANOVA when one wants to compare disease or exposure groups with normal subjects.

RESULTS

BAL recovered significantly more cells in asbestos-exposed ($302 \pm 30 \times 10^3$ cells/ml) than in normal controls ($208 \pm 15 \times 10^3$ cells/ml, $p < 0.05$). The spontaneous release of IL-1β and TNF-α into culture medium over 18 h was higher in asbestos-exposed individuals than in normal controls (TNF-α: asbestos-exposed 953 ± 192 pg/ml; normal controls 290 ± 38 pg/ml $p < 0.005$, Figure 1).

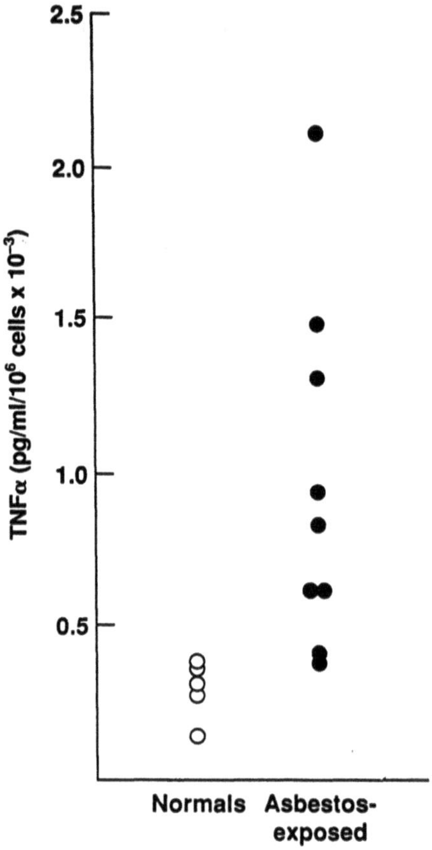

Figure 1. Spontaneous release of TNF-α from alveolar macrophages from normal controls and asbestos-exposed individuals as measured by ELISA in supernatants collected over 18 hr. Reprinted by permission from *J Immunol* 1993; 150:4188-96.

The asbestos-exposed released significantly more IL-1β than normal subjects (IL-1β: asbestos-exposed 760 ± 120 pg/ml, normal subjects 138 ± 24 pg/ml $p < 0.001$, Figure 2).

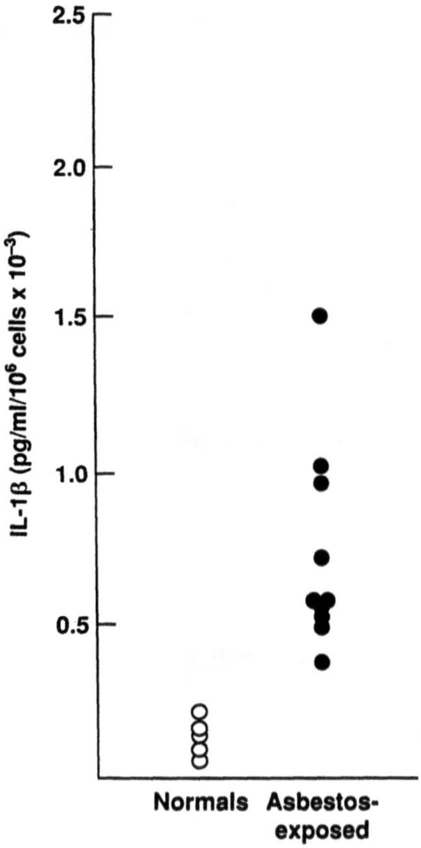

Figure 2. Spontaneous release of IL-1β from alveolar macrophages from normal controls and asbestos-exposed individuals as measured by ELISA in supernatants collected over 18 hr. Reprinted by permission from *J Immunol* 1993; 150:4188-96.

In addition to spontaneous protein release, fresh AM lavaged from individuals studied with asbestos exposure were evaluated for IL-1β and TNF-α gene expression by Northern analysis. Compared with a normal control, AM from asbestos-exposed individuals expressed increased mRNA levels for both IL-1β and TNF-α, whereas levels of the housekeeping gene pHe 7 revealed similar amounts of RNA in each lane (Figure 3).

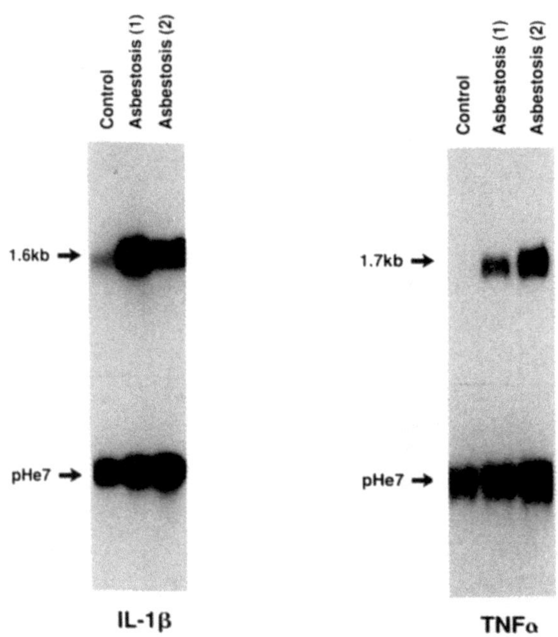

Figure 3. Northern analysis of IL-1β and TNF-α mRNA in AM from a control and two individuals with asbestosis. Equal amounts of total RNA were loaded into each lane as evaluated by hybridization with the housekeeping gene pHe 7 cDNA probe.

<u>Analysis of IL-1β promoter region by deletion analysis</u>

In order to investigate the molecular mechanism involved in the activation of IL-1β gene expression, we initiated experiments aimed at the identification of DNA sequences responsible for the regulation of the IL-1β gene by LPS or TNF-α. A 1,145-bp IL-1β DNA fragment (-1130/+15) was isolated from a human genomic library and cloned into the pTK(-).CAT plasmid. Plasmids containing the 5' IL-1β fragment linked to a CAT reporter gene and transfected into THP-1 cells could be stimulated up to 16-fold by TNF-α or 13-fold by IL-1β as evaluated by CAT assay (Figure 4).

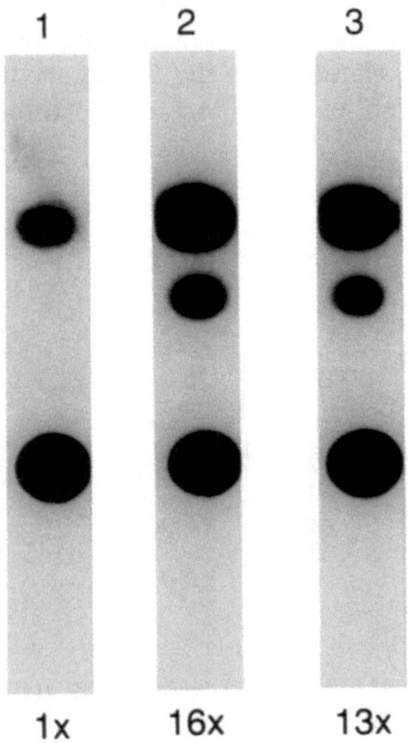

Figure 4. Transient transfection and CAT assay. THP-1 cells were transfected with the plasmid pTK (-) CAT.IL-1β by the DEAE-dextran method and the CAT activity was determined by thin layer chromatography. The transfected cells were cultured for 24 hr with the following stimuli: lane 1, control; lane 2 TNF-α (10 ng/ml); lane 3 IL-1β (10 ng/ml). The number on the bottom indicates the relative increase (fold induction) in CAT activities calculated by dividing the percent conversion into the acetylated form of chloramphenical in extracts from cultures treated with the respective stimuli by that in control extracts.

This DNA fragment contains several sequences resembling known regulatory elements, as schematically illustrated in Figure 5. To identify *cis*-acting elements responsive to LPS or TNF-α, four IL-1β DNA fragments with deletions from the 5' end were fused to the CAT reporter gene. Activation of the IL-1β promoter region was determined by measuring CAT activities in THP-1 cells transfected with these constructs after stimulation with the different reagents. Our results showed that CAT activity gradually increased as deletion advanced from positions -1130 to -313 and to -131 (data not shown). The data suggested that major positive regulating elements on the IL-1β gene in response to stimulation by LPS or TNF-α were located between positions -131 and +15.

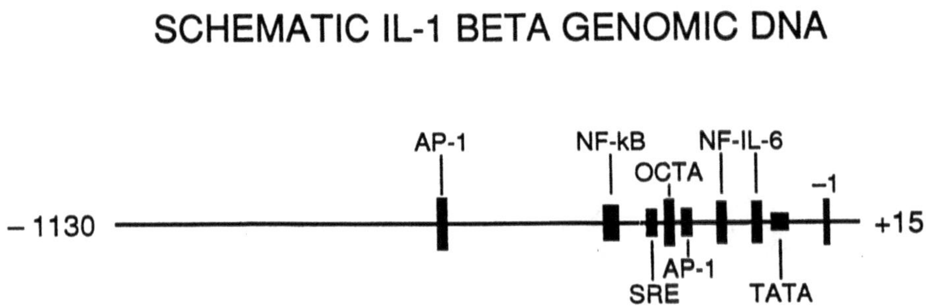

Figure 5. Schema of the 5' flanking region of the human IL-1β gene. Reprinted by permission from *Mol Cell Biol* 1993; 13:3831-37.

There are two sequences on IL-1β that share the NF-IL-6 consensus sequence T(T/G)NNGNAA(T/G). They are 5'-TTGTGAAAT-3' (-90/-82) and 5'-TTTTGAAAG-3' (-40/-32). To test whether the two sequences are the positive regulatory element(s), we performed site-directed mutagenesis of the IL-1β DNA fragment (Figure 6).

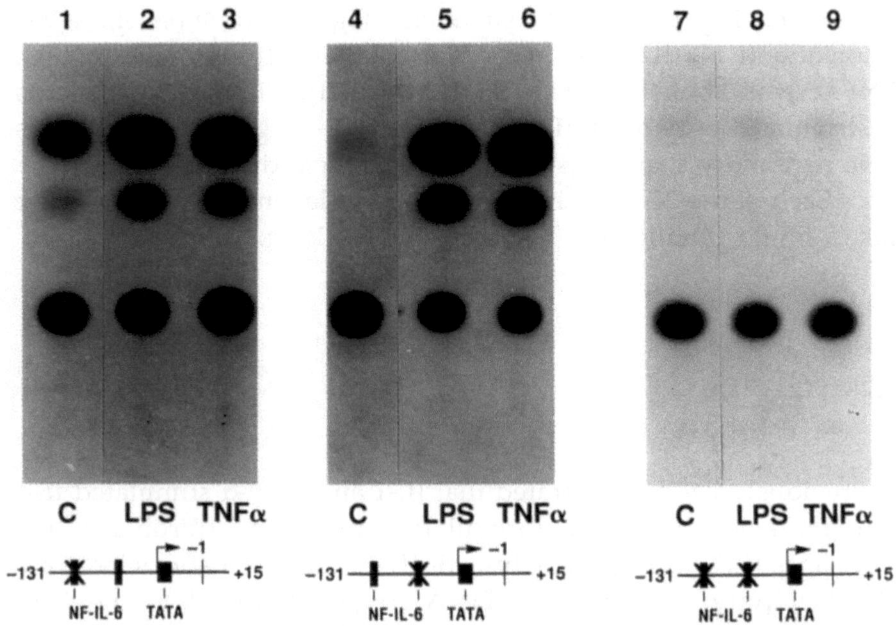

Figure 6. Analysis of the IL-1β promoter region by site-directed mutagenesis. Three plasmids constructed with one NF-IL6 mutation or double NF-IL6 mutations were tested by transient transfection and CAT assays as shown on the bottom. The boxes labeled X represent the mutated NF-IL6 motifs and the stimuli are labeled below each lane with C = control, LPS = lipopolysaccharide (100 ng/ml), and TNF-α = tumor necrosis factor α (10 ng/ml). Reprinted by permission from *Mol Cell Biol* 1993; 13:3831-37.

The mutations were made as follows: (i) the NF-IL6 sequence (-90/-82) was mutated to 5'-CCGTAAAAG-3'; (ii) the second NF-IL6 sequence (-40/-32) was mutated to 5'-CCTTAAAAG-3'. Three kinds of mutations were carried out by mutating either one of the NF-IL6-like sites and leaving the other one intact and mutating both NF-IL6-like sites. The mutants were sequenced to confirm the correct mutation. Induction of CAT activity by LPS or TNF-α was striking, with mutation of the 5' NF-IL6 site leaving the 3' site intact (Fig. 6, left panel) or mutation of the 3'

NF-IL6 site leaving the 5' NF-IL6 site intact (Fig. 6, middle panel). After mutation of both NF-IL6 sites, inducibility of the IL-1β promoter by LPS, or TNF-α was abolished (Fig. 6, right panel). These experiments suggested that the two NF-IL6 sites on the IL-1β gene functioned as positive regulatory sequences in response to the induction with LPS or TNF-α. Each of the NF-IL6 sites functioned independently of the other because a single mutation did not affect CAT activity.

Analysis of IL-6 gene

Previously we demonstrated that IL-1 and TNF-α stimulated IL-6 protein release and mRNA upregulation in human fibroblasts and macrophages (LPS) (11,12). Analysis of IL-6 promoter regions localized maximal activity to a -158/-49 DNA fragment linked to the CAT reporter gene following stimulation with LPS in THP-1 cells. There were two NF-IL6 and one NF-κB site located 3' to the NF-IL-6 sites within this fragment. Following stimulation, these sites demonstrated transcriptional activation and nuclear protein binding similar to our results with IL-1β.

DISCUSSION

Alveolar macrophages recovered by BAL spontaneously release increased quantities of TNF-α and IL-1β that can be detected in 18 hr supernatants (11). The mRNAs for these two cytokines are up-regulated as demonstrated in Northern blots in two patients. We have also demonstrated that TNF-α can stimulate the release of IL-6 from human fibroblasts as well as macrophages (LPS) (12). IL-6 activates B cells to synthesize immunoglobulins and is a mediator of the acute phase response (13). TNF-α also stimulates the release of IL-8, a potent chemoattractant for neutrophils and lymphocytes; both of these cell types have been reported to be increased in BAL (1,10,14).

We have also shown that IL-1β transcription is mediated by two NF-IL6-like motifs (15). These 9 base motifs were stimulated by TNF-α and LPS and acted independent of each other to enhance transcription. Previously we have used the gel mobility shift assay to demonstrate that TNF-α and LPS stimulate nuclear binding proteins that bind to these motifs in THP-1 cells, a macrophage-monocyte like cell line (15). These NF-IL6 motifs thus have the two features of binding to DNA, and enhancement of transcription by nuclear factors.

The NF-IL6 nuclear factor is a member of a family of nuclear factors termed C/EBP for CCAAT/Enhancer Binding Proteins primarily represented by acute phase proteins from the liver (16). Activation of the NF-IL6 proteins is an early event in the inflammation cascade resulting in the release of IL-1β, IL-6, and IL-8 inflammatory cytokines (17). TNF-α also has a NF-IL6 sequence, but it and IL-6 contain NF-κB sequences 3' to NF-IL6 closer to the transcription start site (17,18). Since TNF-α also activates NF-κB and NF-κB can act independently of NF-IL6, this may provide a further augmentation of signals enhancing an inflammatory response focused in the lower respiratory tract (19-21). Many of these inflammatory signals are involved in fibrosis: both TNF-α and IL-1β upregulate collagen I and III mRNAs, increase fibronectin mRNA, and promote fibroblast replication (10,11). IL-1β stimulates mRNA of PDGFA chain over 6-8 hr and TNF-α has been reported to upregulate IGF-I mRNA (22,23). PDGF and IGF-I act synergistically to stimulate fibroblast replication and increased numbers of fibroblasts and their products result in fibrosis (24).

Table 1 illustrates the NF-IL6 consensus and NF-IL6-like sequences in cytokines IL-1β, IL-6, IL-8, TNF-α, and IL-1α (17,25). The C/EBP rat liver nuclear factor has a leucine zipper structure followed by the DNA binding domain that is highly homologous to NF-IL6 and differs in the N-terminal domains (17,26). The leucine zipper is characterized by five leucines in every seventh position, and leucines have a symmetrical end that could interlock with adjoining leucines forming an α-helix three-dimensional structure (17,26). NF-IL6 has been cloned and is expressed in lung tissue after exposure to LPS, IL-1, TNF-α, or IL-6 (17). Leucine zipper regions bind to dimers and a recently described NF-IL6β forms a heterodimer with NF-IL6 enhancing transcriptional activation of cytokine genes (27). In addition to lung tissue, NF-IL6 has been shown to be in macrophage cell lines, e.g.

macrophage differentiation in HL-60 cells and especially in maturation of U937 cells (28). NF-IL6 mRNA was inducible in peripheral blood monocytes following stimulation with LPS (28). The NF-IL6 factor may also dimerize or interact with NF-κB since NF-IL6 and NF-κB are immediately adjacent to each other on IL-8 and both elements are required for an IL-8 response to stimuli. In this regard, the NF-κB-p50 and NF-IL6 proteins interact in the leucine zipper domain (29). The NF-IL6 promoter can also be upregulated by p53 mutants (val-135 and phe-132) demonstrating a link between cytokines and tumor cells which can alter responsiveness to cytokines (30). In the presence of IL-6, the time course of NF-IL6 and C/EBP are opposite to each other as demonstrated in livers of IL-6 transgenic mice (16,17). We speculate that IL-6 may alternatively participate in a generalized acute phase response, or a localized inflammatory response specifically the lower respiratory tract (18). In this regard, the NF-IL6 nuclear factors may play a central role in macrophages orchestrating an inflammatory/fibrotic response.

TABLE 1

NF-IL6 CONSENSUS SEQUENCE		5' T(T/G)NNGNAA(T/G) 3'		
IL-1β	(-90/-82)	TT	GTGAAA	T
	(-40/-32)	TT	TTGAAA	G
IL-6	(-153/-145)	TT	GTGCAA	T
	(-83/-75)	TT	ATGAAA	T
TNF-α	(-193/-185)	TT	TGGAAA	G
IL-8	(-64/-56)	TA	ATGAAA	A
IL-1α	(-146/-138)	TT	TAGCCA	G
	(-199/-191)	TT	CTGGAA	T

REFERENCES

Akira S and Kishimoto T (1992) IL-6 and NF-IL6 in Acute-phase response and viral infection. Immunol Rev 127:25-50

Akira S, Isshiki H, Sugita T, Tanabe O, Kinoshita S, Nishio Y, Nakajima T, Hirano T, Kishimoto T (1990) A nuclear factor for IL-6 expression (NF-IL6) is a member of a C/EBP family. EMBO 1897-1906

Cantin A, Allard C, Begin R (1989) Increased alveolar plasminogen activator in early asbestosis. Am Rev Respir Dis 139:604-609

Carre PC, Mortenson RL, King TE Jr, Noble PW, Sable CL, Riches DWH (1991) Increased expression of the interleukin-8 gene by alveolar macrophages in idiopathic pulmonary fibrosis. J Clin Invest 88:1802-1810

Economou JS, Rhoades K, Essner R, McBride WH, Gasson JC (1989) Genetic analysis of the human tumor necrosis factor α/cachectin promoter region in a macrophage cell line. J Exp Med 170:321-326

Furutani Y, Notake M, Fukui T, Ohue M, Nomura H, Yamada M, Nakamura S (1986) Complete nucleotide sequence of the gene for human interleukin 1 alpha. Nucleic Acids Res 14:3167-3179

Garcia JGN, Griffith DE, Cohen AB, Callahan KS (1989) Alveolar macrophages from patients with asbestos exposure release increased levels of leukotriene B_4. Am Rev Respir Dis 139:1494-501

Kinoshita S, Akira S, Kishimoto T (1992) A member of the C/EBP family, NF-IL6β, forms a heterodimer and transcriptionally synergizes with NF-IL6. Proc Natl Acad Sci USA 89:1473-1476

Kishimoto T (1989) The biology of interleukin-6. Blood 74:1-10

Landschulz WH, Johnson PF, McKnight SL (1988) The leucine zipper: a hypothetical structure common to a new class of DNA binding proteins. Science 240:1759-1764

LeClair KP, Blanar MA, Sharp PA (1992) The p50 subunit of NF-κB associates with the NF-IL6 transcription factor. Proc Natl Acad Sci USA 89:8145-8149

Margulies L and Sehgal PB (1993) Modulation of the human interleukin-6 promoter (IL-6) and transcription factor C/EBPβ (NF-IL6) activity by p53 species. J Biol Chem 268:15096-15100

Natsuka S, Akira S, Nish Y, Hashimoto S, Sugita T, Isshiki H, Kishimoto T (1992) Macrophage differentiation-specific expression of NF-IL6, a transcription factor for interleukin-6. Blood 79:460-466

Noble PW, Lake FR, Henson PM, Riches DWH (1993) Hyaluronate activation of CD44 induces insulin-like growth factor-1 expression by a tumor necrosis factor-α-dependent mechanism in murine macrophages. J Clin Invest 91:2368-2377

Piguet PF, Collart MA, Grau GE, Sappino A-P, Vassalli P (1990) Requirement of tumour necrosis factor for development of silica-induced pulmonary fibrosis. Nature 344:245-247

Raines EW, Dower S, Ross R (1989) Interleukin-1 mitogenic activity for fibroblasts and smooth muscle cells is due to PDGF-AA. Science 243:393

Rhoades KL, Golub SH, Economou JS (1992) The regulation of the human tumor necrosis factor α promoter region in macrophage, T cell, and B cell lines. J Biol Chem 267:22102-22107

Robinson BWS, Rose AH, James A, Whitaker D, Musk AW (1986) Alveolitis of pulmonary asbestosis. Bronchoalveolar lavage studies in crocidolite-and chrysotile-exposed individuals. Chest 90:396-402

Rom WN (1991) Activated alveolar macrophages from individuals with asbestosis release peptide growth factors. In Cellular and Molecular Aspects of Fiber Carcinogenesis. Cold Spring Harbor; NY: Cold Spring Harbor Laboratory Press, pp. 103-114

Rom WN and Travis WD (1992) Lymphocyte-macrophage alveolitis in nonsmoking individuals occupationally exposed to asbestos. Chest 101:779-786

Rom WN, Basset P, Fells G, Nukiwa T, Trapnell BC, Crystal RG (1988) Alveolar macrophages release an insulin-like growth factor I-type molecule. J Clin Invest 87:1685-1693

Rom WN, Bitterman PB, Rennard SI, Cantin A, Crystal RG (1987) Characterization of the lower respiratory tract inflammation of nonsmoking individuals with interstitial lung disease associated with chronic inhalation of inorganic dusts. Am Rev Respir Dis 136:1429-1434

Rom WN, Travis WD, Brody AR (1991) Cellular and molecular basis of the asbestos-related diseases. Am Rev Respir Dis 143:408-422

Shakhov AN, Collart MA, Vassalli P, Nedospasov SA, Jongeneel CV (1990) κB-type enhancers are involved in lipopolysaccharide-mediated transcriptional activation of the tumor necrosis factor α gene in primary macrophages. J Exp Med 35-47

Vilcek and Lee TH (1991) Tumor necrosis factor: new insights into the molecular mechanisms of its multiple actions. J Biol Chem 266:7313-7316

Wallace JM, Oishi JS, Barbers RG, Batra P, Aberle DR (1989) Bronchoalveolar lavage cell and lymphocyte phenotype profiles in healthy asbestos-exposed shipyard workers. Am Rev Respir Dis 139:33-38

Zhang Y and Rom WN (1993) Regulation of the interleukin-1β (IL-1β) gene by mycobacterial components and lipopolysaccharide is mediated by two nuclear factor-IL6 motifs. Mol Cell Biol 13:3831-3837

Zhang Y, Broser M, Rom WN (in press) Activation of the interleukin-6 gene by *Mycobacterium tuberculosis* or lipopolysaccharide is mediated by NF-IL6 and NF-κB. Proc Natl Acad Sci, USA
Zhang Y, Lee TC, Guillemin B, Yu Ming-Chih, Rom WN (1993) Enhanced IL-1β and tumor necrosis factor-α release and messenger RNA expression in macrophages from idiopathic pulmonary fibrosis or after asbestos exposure. J Immunol 150:4188-4196
Zhang Y, Lin J-X, Vilcek J (1990) Interleukin-6 induction by tumor necrosis factor and interleukin-1 in human fibroblasts involves activation of a nuclear factor binding to a κB-like sequence. Mol Cell Biol 10:3818-3823

TYPE II EPITHELIAL CELL PROLIFERATION AND SURFACTANT LIPID TURNOVER IN SILICOSIS

Raymond Bégin, Olivier Lesur, and Boris Melloni
Unité de recherche pulmonaire et Service de pneumologie,
Université de Sherbrooke, Sherbrooke, Québec, Canada

INTRODUCTION

Recent advances in the isolation procedures and culture conditions maintaining the type II pneumocyte in a differentiated state in vitro (Mason et al, 1977) have documented that the type II epithelial cell, in addition to producing surfactant, acts as a stem cell to repair injured type I epithelium. Type II cells can also synthesize a large variety of factors implicated in inflammatory and fibrogenic reactions.

In experimental silicosis, type II epithelial cell hyperplasia and hypertrophy and the excessive accumulation of phospholipids in the lung are well established (Adamson and Bowden, 1984; Bégin et al, 1989; Reiser et al,1982), but the biologic regulation and the origin of these enhanced activities are incompletely understood.

In recent years, our laboratory has explored the direct effect of silica particles on type II epithelial cells (Lesur et al, 1992 a), looked at the alveolar fluids from normal and silicotic sheep to alter the proliferation of type II epithelial cells in culture, characterized the growth factors (Lesur et al 1992b, Melloni et al, 1993b) and identified one origin of the factors (Melloni et al, 1993 a). Lastly, we explored the capability of silica dusts and alveolar fluids from normal and silicotic sheep to alter the lipid metabolism of quiescent type II cells in culture and to alter the surfactant lipid clearance ability of alveolar macrophages (Lesur et al, 1993).

In this paper, we summarize selected contributions of these investigations on the regulation of epithelial type II cells in normal and silicotic conditions, in the sheep model.

MATERIALS AND METHODS

Dusts and alveolar fluids

Silicon dioxide [SiO_2] [Minusil-5™, PGS Pennsylvania Glass Sand Corp, Pittsburg, PA]

NATO ASI Series, Vol. H 85
Cellular and Molecular Effects
of Mineral and Synthetic Dusts and Fibres
Edited by J. M. G. Davis and M.-C. Jaurand

was chosen for dust-related assays, prepared in an initial suspension of 500 µg / ml and vigorously vortexed before using in successive dilutions as previously described (Lesur et al, 1992 a). Ranges from 500 to 0.6 µg / ml were tested for the experiments.

Alveolar fluids were obtained from bronchoalveolar lavage [BAL] on sheep of ongoing experiments (Bégin et al, 1989, 1993). The two groups were a control group exposed to phosphate buffered saline [PBS] and a group of sheep chronically exposed to silica, having simple ILO profusion of radiographic opacities of category 2 silicosis [grade 2 silicosis] on the basis of the International Labor Organization [ILO] classification system (anonymous, 1980), after 36 month of exposure. A third group of sheep with an earlier stage of silicosis [10 month of exposure, ILO category 1 silicosis] [grade 1 silicosis] was also used for the experiments with alveolar macrophages. In these sheep, exposure were carried out through bronchoscopic canulation and multiple 10-day-interval slow infusion of PBS [100 ml] or silica suspended in 100 ml PBS into the trachea. Randomly selected sheep were studied in each group. Alveolar fluids were processed before adding on type II cells as previously described (Lesur et al, 1992 b).

Isolation and culture of alveolar macrophage

Alveolar macrophages were obtained by bronchoalveolar lavage [BAL] from normal sheep and silicotic sheep as previously reported (Bégin et al, 1989). The cells were resuspended at a concentration of 2×10^6 cells / ml in MEM with 10 % of heat-inactived fetal bovine serum [FBS HI] containing 50 mg / ml gentamycin and 2.5 mg / ml amphotericin B. Then, aliquots of 0.5 ml were seeded into 24-well tissue culture plates [Falcon Inc, Lincoln Park, NJ]. The cells were allowed to attach in 5 % CO_2 / air for one hour and then the plates were extensively washed 3 times with warm PBS to remove non adherent cells [ie: lymphocytes, neutrophils]. After the last wash, 0.5 ml of fresh serum-free MEM, supplemented by glutamine [2mM] with or without the mineral particles of this study, was added in each well. The adherent cells were maintained for 24 h at 37°C in 5 % CO_2 / air. Cell differentials were performed on cytocentrifuge smears of adherent cells and stained by Wright-Giemsa. In all cases, the resulting cells were > 92 % alveolar macrophages.

Production of macrophage conditioned media (CM)

Macrophages from control sheep were maintained for 24 h under the following conditions: in medium alone, in medium plus 10, 25, 50 or 100 µg / ml of either silica, or titanium [Kronos Inc.,

Paris, France] or aluminum-treated quartz. Silica, silica- aluminum and titanium dioxyde particles were no larger than 5 µm at 95 %. Alveolar macrophages from silicotic sheep were maintained in medium alone, without added mineral particles, under the same gas conditions. After 24 hour incubation, the supernatants were removed, centrifuged, filtered through a 0.22 µm filter [Millex-GS, Millipore Corp., Bedford, MA] and stored in aliquots at -70°C until use.

In vitro dust cytotoxicity for alveolar macrophages

The release of lactate dehydrogenase enzyme [LDH] into the supernatants assessed the cytotoxic effects of mineral dusts after exposure on control sheep alveolar macrophages for 24 hour. The LDH activity was determined by spectrophotometric method. Sheep control alveolar macrophages were incubated for 24 h in MEM alone [negative control], MEM with 10, 25, 50, 100, 500, 1000 µg / ml quartz, MEM with 50 µg / ml titanium or aluminum-treated quartz. Positive control was assessed by incubating alveolar macrophages in MEM with 1 % triton X-100 [J.T. Baker Chemical Co., Pittsburgh, PA]. The cytotoxic effects of mineral dusts on alveolar macrophages were expressed as means ± SEM of four separate experiments.

Isolation, culture and identification of type II pneumocytes

Fetal rat lung type II pneumocytes were isolated and prepared as previously described (Lesur et al, 1992). These cells have acquired, and maintained a differentiated phenotype, in addition to grow in culture; morphological characteristics were identified by tannic acid polychrome staining, transmission electronic microscopy and cytokeratin-19 expression (Lesur et al, 1992 a & b).

Proliferation of epithelial type II cells

The effects of BALF of control and silicotic sheep and alveolar macrophage CM on type II epithelial cell growth were assessed by measuring [^{3}H] thymidine incorporation into DNA and direct cell counting (Lesur et al, 1992 a). Data were analyzed using the means and SEM of cell counts from each well in duplicate and from three separate experiments. The results are expressed in percentage of control value.

Surfactant lipid synthesis activity: labeled precursor incorporation

The synthesis activity of surfactant lipids by type II cells was studied by the analysis of [1-^{14}C] acetic acid incorporation [specific activity 54 mCi / mmol] [Amersham Can Ltd, Oakville, Ont] into subconfluent cells [50 to 75 % of well covering], as previously described (Lesur et al, 1992 a).

Surfactant lipid secretion activity: release of labeled lipids

The [1-^{14}C] acetic acid [2.5 µCi / well] [Amersham] was previously incorporated into subconfluent type II cells for 24 hour at 37°C, 1% O_2 / 5 % CO_2. Then, silica dust suspension (range of 2.5 to 50 µg / ml) or alveolar fluid dilutions [at 100 %, 20 % / 50 %, 2 %] in MEM D-Valine 2 % FBS HI were applied on type II cells in triplicate wells for 3 days. Supernatants plus two washes PBS, and monolayers harvested with 0.5 ml trypsin 0.05 %: EDTA 0.02 % plus two washes PBS were extracted and analyzed according the method described above. Data were expressed in relative percentage of total lipid released in dust related-assays. Unlabeled cells seeded in similar conditions and in the same plates were numerized as described above for the experiments with alveolar fluids, and a secretion index established as: dpm release in supernatants and normalized to cell count. Isoproterenol [Sigma, St Louis, MO] at 10 µg / ml was chosen as positive control.

Surfactant lipid reuptake activity by type II cells: uptake of labelled dipalmitoyl phosphatidylcholine [DPPC]

Subconfluent type II cells were exposed to processed alveolar fluids [as above] in addition to a standard concentration of [L-3-Phosphatidylcholine, 1, 2-Di (1-^{14}C) palmitoyl] [specific activity 112 mCi / mmol] [Amersham]. An emulsion of the toluene: ethanol solution of labelled PC was made with MEM and sonicated 30 min at 37°C before adding into each well. This allowed to homogeneize the lipid and aquous components into liposome-like structures of similar size. Incubations of 1 and 3 hour were performed before discarding the supernatants, washing vigorously three times and harvesting type II cells with 0.5 ml trypsin 0.05 %: EDTA 0.02 % plus two washes PBS. The cells were then extracted and analyzed according the method described above. Data were expressed as percentage of PC reuptake compared to negative control. Terbutaline [Sigma] at 0.1mM was chosen as positive control.

Surfactant lipid uptake by alveolar macrophages: uptake of DPPC

Alveolar macrophages from the three groups of sheep were used: control normal sheep, sheep with grade 1 silicosis and sheep with grade 2 silicosis. Alveolar fluids were processed as described above and added on alveolar macrophages at 2 % dilutions. Three different animals randomly selected from each group were lavaged. Alveolar cells were pelletized, counted, and seeded at 1.5×10^6 cells per wells in MEM 10 % FBS HI. An incubation time of 3 hours was selected on the basis of pilot assays. ^{14}C-PC was added to each well as for reuptake experiments. Thereafter, macrophages were harvested with 0.5 ml trypsin 0.05 %: EDTA 0.02 % plus two washes PBS following discarding supernatants and washing vigorously twice the cells with PBS. Cells recovered were extracted and analyzed as described above. Data were expressed as percentage in dpm of PC uptake per well.

Cytotoxicity tests: labelled adenine release by type II cells

To differentiate a nonspecific release of lipids by cell injury and an active secretion of lipids by type II cells in culture, experiments were performed with [8-^{14}C] adenine [specific activity 55 mCi / mmol] [Amersham] according to the method previously described (Lesur et al, 1992 a).

Statistical analysis

Statistical analysis was done using a variance analysis (ANOVA) with an α value of 0.05 and a β value of 0.20. A Dunnett t test for comparison with control data was used for each assay.

RESULTS

The mineral dust proliferation and cytotoxicity assays on type II cells

Exposure of type II cells to the silica mineral dust produced a significant 50-60 % stimulation of ^3H thymidine incorporation up to 20 µg / ml of SiO_2, paralleled by cell number increases. The inert dusts, SiO_2Al and TiO_2 did not affect the rate of the radionucleotide integration. Higher doses of the 3 tested dusts reduced ^3H thymidine incorporation [fig 1].

The cytotoxicity index demonstrated the toxic activity of silica on type II cells in a dose dependent manner, with a maximum of 45 ± 2.5 % [$p < 0.0001$] at 250 µg / ml of SiO_2. Low doses [< 50 µg / ml] of silica induced a slight cytotoxic effect: 9 ± 1 % for 10 µg / ml of SiO_2 [$p < 0.0002$]. SiO_2 Al produced little release of adenine at any dose, with a maximum at 500 µg / ml [4 ± 0.1 %]. Similarly, TiO_2 was not cytotoxic at any level tested.

Proliferation of type II cells WITH sheep BALF exposure

The effects of BALF from control and silicotic sheep documented a dose dependent stimulatory effect [by ^{3}H thymidine incorporation and cell counting] for the two alveolar fluids, with a peak of activity at 50 % of dilution for the normal lung derived materials [80 % of rise vs negative control, $p < 0.025$], and at 20 % of dilution for the silicotic lung derived-materials [110 % of rise vs negative control, $p < 0.005$]. Thus, the stimulatory effect of silicotic sheep BALF on type II cells was higher than control sheep BALF [$p < 0.05$] [fig 2].

Effect of macrophage conditioned media on proliferation of type II cells

Supernatants from unstimulated alveolar macrophages increased [^{3}H] thymidine incorporation and cell count in type II cells [n = 3]. This effect was dose dependent with a maximal activity at high concentrations. At 1:1 dilution, macrophage CM were found to double basal type II cell DNA synthesis [202 ± 10 % versus medium control, $p < 0.05$] [fig 3].

Supernatants from *in vitro* silica exposed alveolar macrophages increased the thymidine incorporation on type II cell in a dust concentration-dependent manner [n = 3]. When supernatants were serially diluted, maximal activity was at dilution 1:1 [fig 3].

For the *in vivo* silica exposed macrophages, as from our silicotic sheep, stimulation was also increased, maximally at dilution 1:1 [322 ± 13 % versus control type II cells and 149 ± 9 % versus unstimulated alveolar macrophages, $p < 0.05$ for each] [fig 3].

Surfactant lipid synthesis activity

The kinetic study of acetate incorporation demonstrated a steeper slope for the incorporation of acetate than that of number of cells per well , thus documenting active incorporation over time.

Silica exposure within the minimally cytotoxic concentrations [< 50 µg / ml] did not induce any significant change in surfactant lipid synthesis. The more toxic concentrations [eg: 500 µg / ml] reduced by 40 % the acetate incorporation. [$p < 0.05$ vs control].

Exposing monolayers to alveolar fluids and indexing the data per cell, we found that most of the stimulation of lipid synthesis was at the low concentration of 2 % alveolar fluids, which does not support a proliferative effect on type II cells. Total alveolar fluids from silicotic sheep were more stimulatory than control fluids [1.5-fold increase, $p < 0.05$ vs control] [figure 4a].

Surfactant lipid secretion/ release activity of type II cells

A 3 day silica exposure at growing concentrations resulted in gradually enhanced amounts of surfactant lipids in type II cell supernatants. These amounts were significantly higher when monolayers were exposed to silica from the concentration of 10 µg / ml [$p < 0.05$ vs control]. However, the lipid release in supernatants was parallelled and correlated with the adenine release [$r^2 = 0.876$, $p < 0.01$].

Applying alveolar fluids to the monolayers produced the followings: 100 % fluids from silicotic sheep increased the amount of surfactant lipids in cell supernatant [$p < 0.05$ vs control], at least by a nonspecific mechanism as suggested by the adenine release; 2 % total fluids from silicotic sheep shared an inhibitory effect on type II cell lipid secretion [$p < 0.05$ vs control] [fig 4b]. The positive control [isoproterenol 10 µg / ml] produced a 2.5-fold increase of type II cell lipid secretion .

DPPC reuptake by type II cells

Adding alveolar fluids from the 3 groups of sheep to DDPC for 1 hour resulted in a decreased uptake of the phospholipid by type II cells [$p < 0.05$ vs control], significantly reversed at 3 hour for total fluids and reaching a stimulatory level of reuptake [$p < 0.05$ vs control] [fig 4c]. The positive control [Terbutaline 0.1 mMol] produced a 2-fold increase of type II cell DPPC reuptake.

DPPC uptake by alveolar macrophages

DPPC uptake by sheep alveolar macrophages at 3 hour incubation demonstrated a decreased ability of in vivo silica-exposed macrophages to ingest the phospholipid [$p < 0.05$ vs control macrophages] [fig 4d].

Adding 2 % total alveolar fluids to the monolayers induced an important inhibition of DPPC uptake [$p < 0.05$ vs control] [fig 4e].

DISCUSSION

In these studies, we have documented several direct effects of silica on lung cells: 1- silica is directly cytotoxic for type II pneumocytes and that the degree of cytotoxicity is dependent upon the concentration of silica; 2- exposure of the type II epithelial cells to the weakly/non-cytotoxic concentrations of silica increases proliferation of these cells but the proliferative response was absent or suppressed at high cytotoxic silica doses; 3- the direct exposure of type II cells to silica dusts does not influence surfactant lipid synthesis activity of type II cells and the lipid released in supernatant of cell culture is proportional to the cytotoxicity; 4- silica acts on the phagocytic macrophages, beyond the established direct cytotoxic effect (Larivée et al, 1991), decreasing the ability of macrophages to ingest phospholipids, the clearance pathway for surfactant lipids. These in vitro observations of direct effects of silica on lung cells are clear and relevant to the in vivo condition, as already suggested by a prior human silicosis study (Schuyler et al, 1980).

The second type of alterations are indirect and relates to the influence of the molecules in the alveolar fluids of silicotic sheep on the type II epithelial cells: 1- the alveolar fluids of silicotic sheep have an excess of molecules that up-regulates the proliferation of epithelial type II cells, molecules that are PDGF-, IGF- and FGF-like cytokines (Lesur et al, 1992 b); 2- these molecules are at least in part coming from the alveolar macrophage which produces more of PDGF-like cytokine after a recent exposure and more IGF- and FGF-like cytokines in the chronic state of disease (Melloni et al, 1993 a & b).

Finally, regarding the surfactant lipid turnover, alveolar fluids at dilution in the proliferative area [2 %] (Lesur et al, 1992 a & b) produced the followings: 1- an increase in lipid synthesis, 150 % that of normal alveolar fluids; 2- a decreased in lipid secretion by type II cells, 150 % that of normal alveolar fluids; 3- an 50 % increased reuptake of lipids by the type II cells; and 4- a 50 % decreased lipid uptake by alveolar macrophages of silicotic sheep (Lesur et al, 1993)

Thus, silica has several activities on lung cells beyond its cytotoxic effect. The disease process it initiates is also influenced by at least 3 cytokines produced by lung cells that alter the proliferation of epithelial type II cells. In the disease process, the surfactant lipid turnover is altered directly by the silica dust and indirectly via molecules in the lung fluids. Further characterization of these molecules and their actions are on going in our laboratory.

Supported by MRC Canada.

REFERENCES

Adamson IYR, Bowden DH. Role of polymorphonuclear leucocytes in silica-induced pulmonary fibrosis. Am J Pathol 117: 37-43, 1984.

Anonymous. International Labor Organization/ University of Cincinnatti. International classification of radiographs of pneumoconioses 1980, no 22 revised, Occupational safety and health series. Geneva: International Labour Office, 1980.

Bégin R, Possmayer F, Ormseth MA, Martel M, Cantin A, Massé S. Effect of aluminum inhalation on alveolar phospholipid profiles in experimental silicosis. Lung 167: 107-115, 1989.

Bégin R, Dufresne A, Cantin A, Possmayer F, Sébastien P: Quartz exposure, retention and early silicosis in sheep. Exp Lung Res 15: 409-428, 1989.

Bégin R, Cantin A, Massé S, Dufresne A, Perreault G, Sébastien P. Aluminum inhalation in sheep silicosis. Int J Exp Pathol 74; 299-307, 1993.

Larivée P, Cantin A, Dufresne A, Bégin R: Enzyme activities of lung lavage in silicosis. Lung 168 : 151-158,1990.

Lesur O,Melloni B, Cantin A, Tanswell AK, Beaulieu J-F, Bégin R: Silica exposure induces cytotoxicity and proliferation activity of type II pneumocytes. Exp Lung Res, 18 : 173-190, 1992 a.

Lesur O, Melloni B, Cantin A, Bégin R: Silica-exposed lung fuids have a proliferative activity for type II epithelial cells. A study on human and sheep alveolar fluids. Exp Lung Res 18; 633-54, 1992 b.

Lesur O, Bouhadiba T, Melloni B, Cantin A, Bégin R. Alterations of surfactant lipid turnover in silicosis. Submitted for publication 1993.

Mason RJ, Williams MC, Greenleaf RD, Clements JA. Isolation and properties of type II cells from rat lung. Am Rev Respir Dis 115: 1015-1026, 1977.

Melloni B, Lesur O, Cantin A, Bégin R: Silica-exposed macrophages release a growth-promoting activity for type II pneumocytes. J Leuc Biol 53; 327-335, 1993 a.

Melloni B, Lesur O, Bouhadiba T, Cantin A, Bégin R. Partial characterization of the proliferative activity for type II cells produced by silica-exposed alveolar macrophages. submitted for publication 1993 b.

Reiser KM, Hesterberg TW, Haschek WM, Last JA. Experimental silicosis: 1. Acute effects of intratracheal instilled quartz on collagen metabolism and morphologic characteristics of rats lungs. Am J Pathol 107: 176-185, 1982.

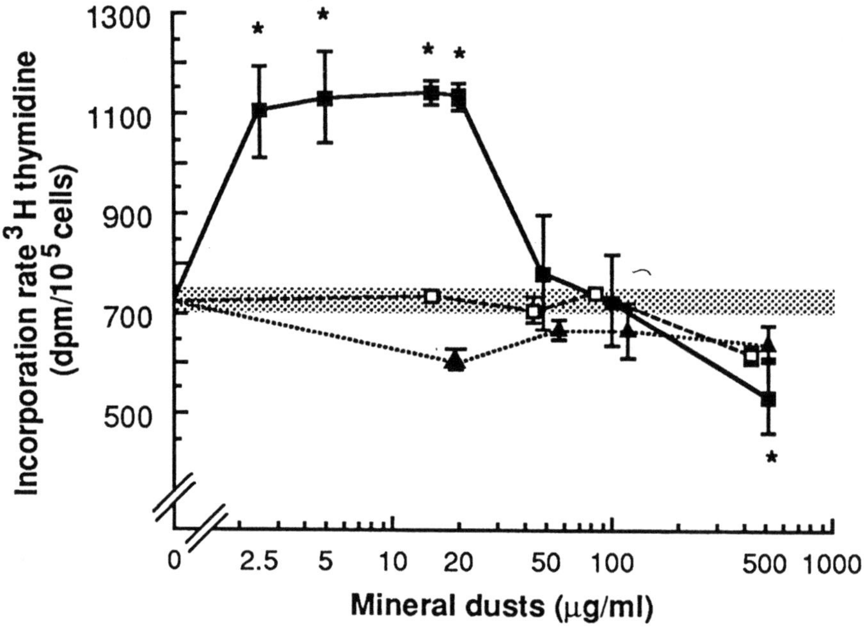

Figure 1

Effects of mineral dusts on [^3H] thymidine incorporation of type II cells. Cells were incubated in the presence of either SiO_2 [dark square], SiO_2-Aluminum [dark triangle], TiO_2 [open square], or media without dust [shaded area]. The data are means ± standard error. * =p < o.o5 compared to control. Reprinted with permission of editor, Experimental Lung Research 18: 173-180, 1992.

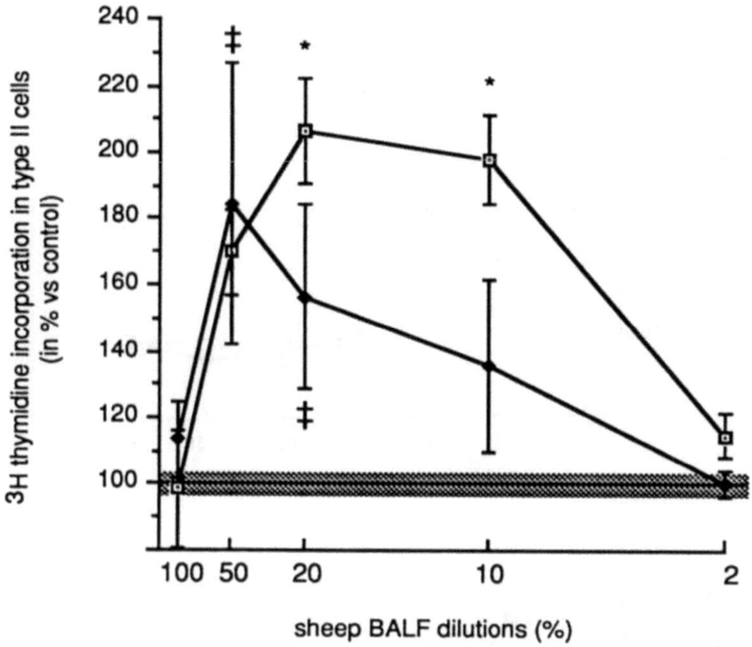

Figure 2

Effects of BALF on [^{3}H] thymidine incorporation of type II cells. The BALF were obtained at month 24 of exposure to either saline [control sheep, n = 10, dark square] or silica [silicotic sheep, n = 13, open square]. Negative control [shaded area] consisted of MEM D Valine with 2 % FBS. * = $p < 0.005$ vs control , ‡ = $p < 0.025$ vs control.

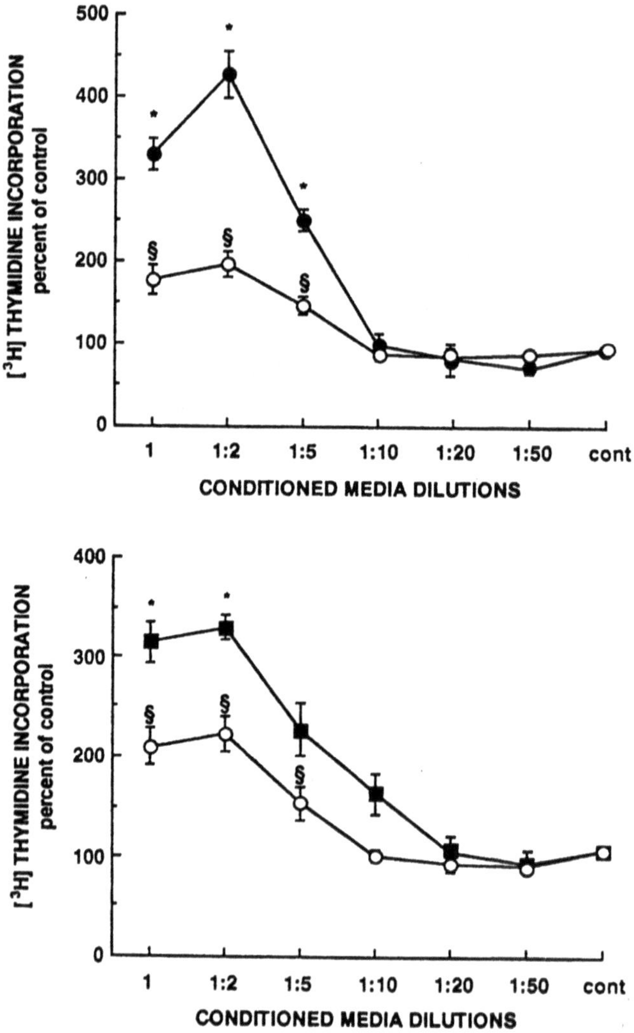

Figure 3

In the upper panel, the effects of supernatants from unstimulated [open circle] and in vitro silica-exposed alveolar macrophages [dark square] on type II cell [^{3}H] thymidine incorporation. In the lower panel, the effects of supernatants from unstimulated [open circle] and in vivo silica-exposed alveolar macrophages [dark square] on type II cell [^{3}H] thymidine incorporation. The thymidine incorporation data are expressed as percentage of control, MEM D-Valine with 2 % FBS. The results represent the mean ± SEM of four independent experiments, each carried out in triplicate. § = $p < 0.05$ vs control, * = $p < 0.05$ vs unstimulated macrophage supernatants. Reprinted with permission of editor, J Leukocyte Biology 53: 327-335, 1993.

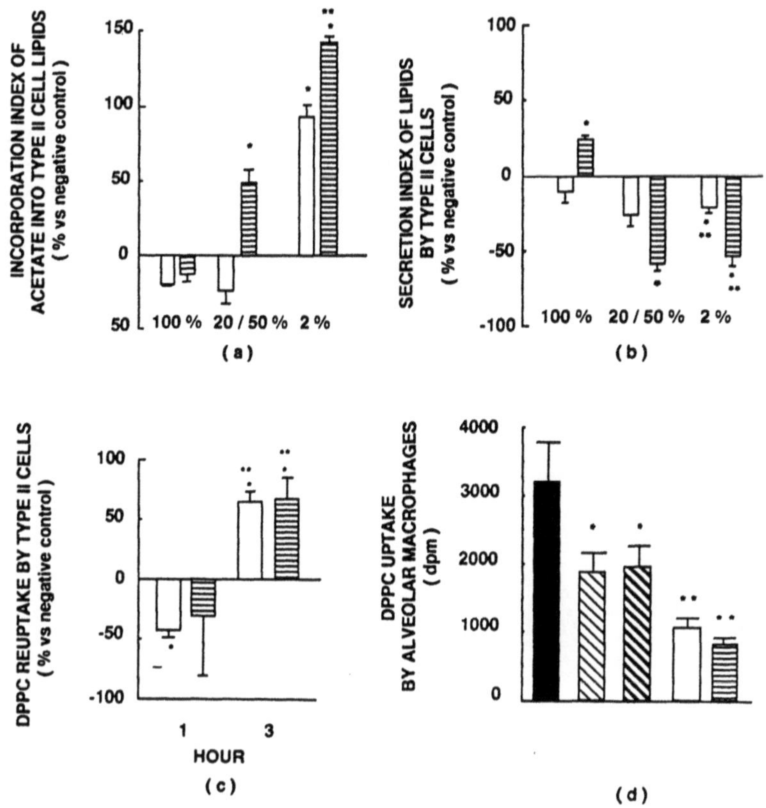

Figure 4

Effects of alveolar fluids from normal and silicotic sheep on surfactant metabolism cycle as determined in 4 assays: synthesis in (a), secretion (b), recycling by type II cells (c) and clearance-uptake by alveolar macrophages (d). In (a), (b) and (c), open bars are for control sheep and horizontally hatched bars for silicotic sheep. Each bar is the mean±SD of six independent experiments in triplicate, expressed in percentage of negative control. *: $p < 0.05$ vs negative control, **: $p < 0.05$ vs control sheep in (a) and (b) and vs hour 1 in (c). In (d), the dark bar is negative control [control macrophages], the white hatched bar is for macrophages from 10 month's silica exposed sheep, the black hatched bar for macrophages from 36 month's silica exposed sheep, the open bar for 2 % alveolar fluids from normal sheep and the horizontally hatched bar for 2 % alveolar fluids from silicotic sheep.* = $p < 0.05$ vs control for the in vivo exposed sheep, ** = $p < 0.05$ vs control for the effect of alveolar fluid vs control.

Role of Nitric Oxide Radicals in Asbestos-Induced Injury.

George Thomas[2], Tasuke Ando[1], Kiran Verma[1] and Elliott Kagan[1]
Departments of Pathology[1] and Physiology & Biophysics[2]
Georgetown University School of Medicine
3900 Reservoir Road, N.W.
Washington, D.C. 20007, USA

Introduction

It is well recognized that inhalation of asbestos fibers is linked to the causation of a variety of clinical disorders including interstitial pulmonary fibrosis, parietal pleural plaques, bronchogenic carcinoma and diffuse malignant mesothelioma of the pleura and peritoneum (Kagan, 1985; Mossman and Gee, 1989). Although the pathogenesis of these diseases has not been elucidated fully, there is evidence that alveolar macrophages (AM) play a pivotal role in mediating asbestos-related injury. Thus, studies in asbestos-exposed rodents and in asbestos workers have shown that AM are recruited to the sites of deposition of inhaled asbestos fibers (Kagan, 1988; Rom *et al.*, 1991). It has also been demonstrated that asbestos fibres can activate AM to secrete a diverse group of inflammatory mediators including cytokine growth factors, chemoattractants and arachidonic acid metabolites (Kagan, 1988; Rom, *et al.*, 1991). Several studies also have demonstrated that the *in vitro* phagocytic uptake of asbestos fibers can generate a variety of reactive oxygen species (ROS) such as the superoxide anion (O_2^{*-}), the hydroxyl radical (OH*) and hydrogen peroxide (Shull, *et al.*, 1992; Mossman, *et al.*, 1986). It also has been reported that scavengers of oxygen free radicals, such as superoxide dismutase (SOD) and catalase, may ameliorate the injurious effects of asbestos exposure (Mossman, *et al.* 1986; Mossman, *et al.* 1990). These studies are predicated on O_2^{*-}-driven, iron-catalyzed, Fenton-Haber-Weiss reactions which

NATO ASI Series, Vol. H 85
Cellular and Molecular Effects
of Mineral and Synthetic Dusts and Fibres
Edited by J. M. G. Davis and M.-C. Jaurand

generate the OH∗ radical (Grisham, 1992), and suggest that ROS may have an important role in the pathobiology of asbestos-related disease.

Recently, considerable attention has been focused on the nitric oxide (NO∗) free radical (Ignarro, 1990; Grisham, 1992). Nitric oxide, which is synthesized by the enzyme, NO∗ synthase, is a labile molecule which has important biologic properties. Thus, NO∗ has been shown to be a potent vasodilator as well as an inhibitor of platelet aggregation and smooth muscle proliferation (Ignarro, 1990; Moncada and Higgs, 1991). There is considerable evidence that many cells (including endothelial cells, smooth muscle cells, neutrophils and macrophages) can generate NO∗ or an NO∗-like molecule. A recent study also has shown that NO∗ can be synthesized by rat AM (Jorens, *et al.*, 1991). Moreover, the induction of NO∗ synthase activity within macrophages is greatly augmented by the cytokine, interferon-γ (Jorens, *et al.*, 1991; Nussler and Billiar, 1993). Not all the effects of NO∗ are, however, considered to be beneficial, since this free radical has been implicated in inflammation and immune complex-mediated cell injury (Mulligan, *et al.*, 1991; Seekamp, *et al.*, 1993). It has been suggested that many of these effects are mediated by the generation of secondary radicals as a consequence of the interaction of O_2^{*-} and NO∗, which yields the peroxynitrite anion ($ONOO^-$). The latter, when protonated, can undergo homolytic cleavage, generating OH∗ and NO_2^* radicals (Beckman, *et al.*, 1990) which, in turn, can produce oxidation of tissue sulfhydryl groups.

The present study was undertaken to investigate the effect of *in vitro* asbestos exposure on NO∗ production by rat AM. For this purpose, comparisons were made between two NIEHS samples of asbestos, chrysotile (a serpentine form) and crocidolite (an amphibole). The possible interaction of asbestos fibers and interferon-γ in the induction of NO∗ synthesis also was studied, since it has been shown previously that asbestos workers' bronchoalveolar cells spontaneously secrete excessive amounts of interferon-γ in comparison with unexposed individuals (Robinson, *et al.*, 1988; Rom and Travis, 1992). To determine whether any observed effects were unique for asbestos, comparisons also were made with the effects of the non-fibrogenic particulate, carbonyl iron.

Materials and Methods

The AM were obtained by bronchoalveolar lavage from adult male Sprague-Dawley rats (Miller and Kagan, 1977). Subsequently, the AM were suspended in RPMI 1640 medium (without phenol red) containing 25 mM HEPES buffer, and supplemented with L-glutamine, 10% fetal bovine serum, penicillin (100 U/ml), streptomycin (100 μg/ml) and fungizone (0.25 μg/ml). Thereafter, the AM (5×10^5 cells/well) were allowed to attach for 2 h in 24-well culture plates and then cultured for an additional 4-48 h in the presence or absence of NIEHS asbestos fibers (2.5-10 μg/ml). Some experiments were performed in the presence or absence of recombinant rat interferon-γ (250-500 IU/ml). Parallel experiments also were performed with cultures containing 10 μg/ml of carbonyl iron spheres (Sigma). All the cultures were incubated at 37° C in a humidified environment containing 5% CO_2.

To investigate the specificity of NO∗ generation, another series of experiments were performed with or without the addition of the NO∗ synthase inhibitor, N^G-monomethyl-L-arginine (L-NMMA), to the alveolar macrophage cultures (at a concentration of 1 mg/ml). Studies also were performed, to assess whether macrophage-derived O_2^{*-} could affect NO∗ production by AM. To address this issue, another set of parallel experiments were performed in which AM were cultured in the presence or absence of the O_2^{*-} scavenger, superoxide dismutase (150 U/ml).

Nitric oxide formation was quantified in alveolar macrophage supernatant-conditioned medium samples by measuring the absorbance of the NO∗ oxidation product, nitrite (NO_2^-), at 548 nm via the Griess reaction ((Thomas and Ramwell, 1988). The presence of NO∗ in conditioned medium samples was confirmed with the use of a rat aortic smooth muscle bioassay (Thomas and Ramwell, 1986).

Results and Discussion

When rat AM were cultured in the presence of crocidolite asbestos fibers, there was a time-dependent and concentration-dependent increase in NO* formation (measured as NO_2^-) detectable within the conditioned medium samples. Maximal NO* production was noted in 48 h cultures containing 10 μg/ml of crocidolite fibers. Table 1 contrasts the effects of crocidolite and chrysotile asbestos fibers and carbonyl iron particles (10 μg/ml) on NO* formation in the presence or absence of interferon-γ (500 IU/ml). As shown in Table 1, both chrysotile ($p < 0.05$) and crocidolite ($p < 0.01$) significantly increased the production of NO*, whereas carbonyl iron particles had no such effect. Furthermore, the addition of recombinant rat interferon-γ to asbestos-containing cultures also significantly upregulated NO* production ($p < 0.001$). Once again, carbonyl iron particles did not noticeably affect NO* synthesis in the presence of interferon-γ. It is noteworthy that the actions of asbestos and interferon-γ were synergistic with respect to the induction of NO* synthesis by rat AM.

Table 1. Effects of Particulates and Interferon-γ on NO* Formation by Rat AM.

Particulate Added	Without Interferon-γ	With Interferon-γ (500 IU/ml)
	NO_2^- Content (nmol/5 x 10^5 AM)	
None (Control)	1.16 ± 0.58	20.74 ± 2.69
Crocidolite	4.3 ± 0.9**	31.57 ± 2.96***
Chrysotile	2.87 ± 0.66*	28.25 ± 6.38***
Carbonyl iron	1.34 ± 0.44	22.85 ± 3.4

* = $p < 0.05$ compared with corresponding control (in same column).

** = $p < 0.01$ compared with corresponding control (in same column).

*** = $p < 0.001$ compared with corresponding control (in same column).

Collectively, these findings suggest that the ability to induce NO∗ formation in rat AM was a property common to both amphibole and serpentine asbestos fibers. Moreover, this observation appeared to be unique for asbestos fibers since carbonyl iron spheres did not induce a similar effect.

The results depicted in Table 2 demonstrate the effect of L-NMMA on NO_2^- formation by rat AM. It should be noted that L-NMMA is an L-arginine analogue and specific inhibitor of NO∗ synthase (Moncada and Higgs, 1991). As shown in Table 2, L-NMMA significantly abrogated NO∗ synthesis both in particulate-exposed as well as in unexposed (control) cultures containing 500 IU/ml of interferon-γ ($p < 0.001$). Similar results were observed when L-NMMA was added to cultures containing 250 IU/ml of interferon-γ. Since L-NMMA antagonizes NO∗ formation from L-arginine, these observations indicate that asbestos fibers and interferon-γ induce the synthesis of NO∗ not only from the metabolism of a guanidino group of L-arginine (Moncada and Higgs, 1991) but also, possibly, from the guanidino moiety of a related molecule (Thomas and Ramwell, 1988).

Table 2. Effect of L-NMMA on NO∗ Formation by Rat AM Stimulated with Interferon-γ (500 IU/ml).

Particulate Added	Without L-NMMA	With L-NMMA (1 mg/ml)
	NO_2^- Content (nmol/5 x 10^5 AM)	
None (Control)	22.47 ± 0.01	13.8 ± 0.29*
Crocidolite	31.87 ± 1.08	16.55 ± 0.42*
Chrysotile	30.81 ± 0.16	14.55 ± 0.53*
Carbonyl iron	25.06 ± 0.83	11.11 ± 0.05*

* = $p < 0.001$ compared with corresponding cultures without L-NMMA.

Superoxide is known to inactivate NO∗ and to reduce its biologic activity

(Rubanyi and Vanhoutte, 1986). Accordingly, the interaction between O_2^{*-} and NO* should be inhibited by the O_2^{*-} scavenger, superoxide dismutase (SOD). As illustrated in Table 3, the addition of SOD (150 U/ml) to the alveolar macrophage cultures enhanced the generation of NO_2^- by rat AM. However, significantly greater amounts of NO_2^- were generated in SOD-containing cultures after either crocidolite or chrysotile exposure than were produced in unexposed (control) cultures ($P < 0.001$). On the other hand, the effect of carbonyl iron, in the presence of SOD, was not significant. This suggests that both O_2^{*-} and NO* are produced concomitantly by asbestos-activated AM.

Table 3. Effects of Particulates and SOD on NO* Formation by Rat AM.

Particulate Added	Without SOD	With SOD (150 U/ml)
	NO_2^- Content (nmol/5 x 10^5 AM)	
None (Control)	1.5 ± 1	3.82 ± 0.19
Crocidolite	2.43 ± 1.05	10.75 ± 2.07*
Chrysotile	2.8 ± 0.86	7.36 ± 0.3*
Carbonyl iron	2.06 ± 1.56	6 ± 2.93

* = $p < 0.001$ compared with corresponding control (in same column).

It is of interest that NO* can interact with O_2^{*-} to form peroxynitrite ($ONOO^-$) anion (Beckman, *et al.*, 1990). Thus, it is conceivable that asbestos fibers, by upregulating the production of O_2^{*-} as well as NO* in AM, may elicit tissue injury through generation of $ONOO^-$. Furthermore, it has been suggested that decomposition of $ONOO^-$ can generate OH* and NO_2^* radicals (Beckman, *et al.*, 1990) which, in turn can induce membrane oxidation of sulfhydryl groups (Radi, *et al.*, 1991).

When conditioned medium samples from cultured rat AM were tested in a rat aortic smooth muscle bioassay (Thomas and Ramwell, 1986), the samples induced relaxation of de-endothelialized aortic ring segments precontracted with phenylephrine (1×10^{-7} M). This endothelium-derived relaxing factor-like effect was more dramatic after asbestos exposure than was noted in conditioned medium from unexposed (control) rat AM. This observation further confirms that the product released from the AM was an NO*-like molecule (Thomas and Ramwell, 1986; Ignarro, 1990; Moncada and Higgs, 1991).

Previous studies have implicated ROS as the primary cause of asbestos-induced cell injury and death (Kagan, 1988; Mossman and Gee, 1989; Brody, 1993). The findings in the present study provide an alternative or additional hypothesis of asbestos-mediated tissue injury which involves the generation of the NO* radical by inhaled asbestos fibers. Thus, asbestos fibers and interferon-γ, by upregulating NO* production in AM, may mediate cellular injury via several different pathways. These include the nitrosation of important iron-containing enzymes within cells (Kiechle and Malinski, 1993) and the induction of DNA damage and mutations (Nguyen, *et al.*, 1992). Moreover, since NO* can mediate the release of iron from ferritin (Reif and Simmons, 1990), and since ferritin is a component of asbestos bodies (Kagan, 1988), NO*-induced iron mobilization may produce DNA strand breakages (Lund and Aust, 1992). Some NO* metabolites (NO_2^*, N_2O_3 and N_2O_4) also are known to be potent nitrosating agents and can generate potentially carcinogenic nitrosamines (Pryor, 1986; Grisham, 1992). It is also of interest that NO* has been shown to induce mesenchymal cell proliferation (O'Connor, *et al.*, 1991) and to modulate growth factor gene expression (Kourembanas, *et al.*, 1993). The potential importance of interferon-γ in potentiating asbestos-induced injury is underscored by the fact that asbestos workers' bronchoalveolar cells have been shown to spontaneously secrete interferon-γ (Robinson, *et al.*, 1988; Rom and Travis, 1992). Thus, the induction of NO* radical by asbestos and interferon-γ may represent a novel mechanism of asbestos-related pulmonary injury.

References

Beckman JS, Beckman TW, Chen J, Marshall PA, Freeman BA (1990) Apparent hydroxyl radical production by peroxynitrite: implications for endothelial injury from nitric oxide and superoxide. Proc Natl Acad Sci USA 87: 1620-1624

Brody AR (1993) Asbestos-induced lung disease. Environ Hlth Perspect 100: 21-30

Grisham MB (1992) Reactive Metabolites of Oxygen and Nitrogen in Biology and Medicine. R.G. Landes Co., Austin, Texas

Ignarro LJ (1990) Biosynthesis and metabolism of endothelium-derived nitric oxide. Annu Rev Pharmacol Toxicol 30: 535-560

Jorens PG, Van Overveld FJ, Bult H, Vermeire PA, and Herman AG (1991) L-arginine-dependent production of nitrogen oxides by rat pulmonary macrophages. Eur J Pharmacol 200: 205-209

Kagan E (1985) Current perspectives in asbestosis. Ann Allergy 54: 464-473

Kagan E (1988) Current issues regarding the pathobiology of asbestosis: a chronologic perspective. J Thorac Imaging 3 (4): 1-9

Kiechle FL, Malinski T (1993) Nitric oxide. Biochemistry, pathophysiology, and detection. Am J Clin Pathol 100: 567-575

Kourembanas S, McQuillan LP, Leung GK, Faller DV (1993) Nitric oxide regulates the expression of vasoconstrictors and growth factors by vascular endothelium under both normoxia and hypoxia. J Clin Invest 92: 99-104

Lund LG, Aust AE (1992) Iron mobilization from crocidolite asbestos greatly enhances crocidolite-dependent formation of DNA single-strand breaks in phi X174 RFI DNA. Carcinogenesis 13: 637-642

Miller K, Kagan E (1977) Immune adherence reactivity of rat alveolar macrophages following inhalation of crocidolite asbestos. Clin Exp Immunol 29: 152-158

Moncada S, Higgs EA (1991) Endogenous nitric oxide: physiology, pathology and clinical relevance. Eur J Clin Invest 21: 361-374

Mossman BT, Gee JBL (1989) Asbestos-related diseases. N Engl J Med 320: 1721-1730

Mossman BT, Marsh JP, Sesko A, Hill S, Shatos MA, Doherty J, Petruska J, Adler KB, Hemenway D, Mickey R, Vacek P, Kagan E (1990) Inhibition of lung injury, inflammation and interstitial pulmonary fibrosis by polyethylene glycol-conjugated catalase in a rapid inhalational model of asbestosis. Am Rev Respir Dis 141: 1266-1261

Mossman BT, Marsh JP, Shatos MA (1986) Alteration of superoxide dismutase activity in tracheal epithelial cells by asbestos and inhibition of cytotoxicity by antioxidants. Lab Invest 54: 204-212

Mulligan MS, Hevel JM, Marletta MA, Ward PA (1991) Tissue injury caused by deposition of immune complexes is L-arginine dependent. Proc Natl Acad Sci USA 88: 6338-6342

Nguyen T, Brunson D, Crespi CL, Penman BW, Wishnok JS, Tannenbaum SR (1992) DNA damage and mutation in human cells exposed to nitric oxide in vitro. Proc Natl Acad Sci USA 89: 3030-3034

Nussler AK, Billiar TR (1993) Inflammation, immunoregulation, and inducible nitric oxide synthase. J Leukoc Biol 54: 171-178

O'Connor KJ, Knowles RG, Patel KD (1991) Nitrovasodilators have proliferative as well as antiproliferative effects. J Cardiovasc Pharmacol 17 (Suppl): S100-S103

Pryor WA (1986) Oxy-radicals and related species: their formation, lifetimes, and reactions. Annu Rev Physiol 48: 657-667

Radi R, Beckman JS, Bush KM, Freeman BA (1991) Peroxynitrite oxidation of sulfhydryls. J Biol Chem 266: 4244-4250

Reif DW, Simmons RD (1990) Nitric oxide mediates iron release from ferritin. Arch Biochem Biophys 283: 537-541

Robinson, BWS, Rose AH, Hayes A, Musk AW (1988) Increased pulmonary gamma interferon production in asbestosis. Am Rev Respir Dis 138: 278-283

Rom WN, Travis WD (1992) Lymphocyte-macrophage alveolitis in nonsmoking individuals occupationally exposed to asbestos. Chest 101: 779-786

Rom WN, Travis WD, Brody AR (1991) Cellular and molecular basis of the asbestos-related diseases. Am Rev Respir Dis 143: 408-422

Rubanyi GM, Vanhoutte PM (1986) Superoxide anions and hyperoxia inactivate endothelium-derived relaxing factor. Am J Physiol 250: H822-H827

Seekamp A, Mulligan MS, Till GO, Ward PA (1993) Requirements for neutrophil products and L-arginine in ischemia-reperfusion injury. Am J Pathol 142: 1217-1226

Shull S, Manohar M, Marsh JP, Janssen YMW, Mossman BT (1992) Role of iron and reactive oxygen species in asbestos-induced lung injury. In: Moslen MT, Smith CV (eds) Free Radical Mechanisms of Tissue Injury. CRC Press, Boca Raton pp 153-162

Thomas G, Ramwell PW (1986) Induction of vascular relaxation by hydroperoxides. Biochem Biophys Res Commun 139: 102-108

Thomas G, Ramwell PW (1988) Vasodilatory properties of mono-L-arginine-containing compounds. Biochem Biophys Res Commun 154: 332-338

Cytotoxicity and cell stimulating activity of ultrafine titanium dioxide

Mikko Holopainen and Kirsi Luoto
National Public Health Institute
Division of Environmental Health
P.O.Box 95, 70701 Kuopio
Finland and [1]

Introduction

Titanium dioxide is widely used as a white pigment in a variety of applications. The commercially available titanium dioxide polymorphs, rutile and anatase are regarded as inert materials. There is no evidence on pulmonary fibrosis or cancer caused by the exposure to titanium dioxide at the dust levels normally existing at the working places (Ferin et al. 1985, Richards et al 1985). However, exposure to very high concentrations of TiO_2 dust has caused mild fibrosis and malignant tumors in experimental animals (Trochimowicz et al. 1988).

New ultrafine titanium dioxide products with crystal size 20–30 nm and particle size 200–300 nm have shown different biological properties compared with the pigmentary titanium dioxides. Ultrafine anatase particles were translocated into the interstitium of rat lungs from the alveoli to a higher degree and caused more inflammation in rat lungs compared with larger particles (Oberdöster et al. 1992, Ferin et al. 1991 and 1992, Driscoll et al. 1991).

The purpose of this study was to investigate the in vitro cellular effects of ultrafine titanium dioxide. The hemolytic and cytotoxic capacity, the ability of the particles to induce production of reactive oxygen metabolites (ROM) in human polymorphonuclear leukocytes (PMN) and the ability of rat alveolar macrophages to phagocyte ultrafine particles were studied. The results were then compared with larger titanium dioxide particles and, or quartz.

[1] University of Kuopio, Department of General Practice and Primary Health Care, Kuopio, Finland

NATO ASI Series, Vol. H 85
Cellular and Molecular Effects
of Mineral and Synthetic Dusts and Fibres
Edited by J. M. G. Davis and M.-C. Jaurand

Material and methods

The titanium dioxide samples were obtained from Kemira TiO_2, Finland. Silica was a min–u–sil sample < 5µm (Pennsylvania Sand and Glass Corp., Pittsburgh, PA) obtained from NIOSH, Morgantown, WV, USA. The ultrafine TiO_2 samples A, B and C were coated with $Al(OH)_3$ and polydimethylsiloxane (PDMS). Sample A (72% rutile and 28% anatase) had almost spherical particle shape, whereas B and C were needle like rutile. Samples A and B were prepared by Kemira Oy TiO_2, Finland and sample C was a commercially available ultrafine TiO_2 sample. The chemical composition was analyzed with an atomic absorption spectrometer (Perkin Elmer) and with x–ray fluorescence method; the crystal structure was analyzed with X–ray diffraction; the crystal size was measured with transmission electronic microscopy; the particle size distribution was counted with photon correlation spectroscopy (Coulter nano sizer) and the surface areas were analyzed with nitrogen adsorption method (PET). The properties of the titanium dioxide samples are shown in table 1.

Table 1. Crystal and particle size (nm), surface area ($m^{2/g}$), SiO_2 and Al_2O_3 content (%), crystal shape and surface coating of anatase and ultrafine rutile samples A, B and C.

	Anatase	Sample A	Sample B	Sample C
Crystal size nm	170	20	20	20
particle size nm	420	659	247	1745
surface area $m^{2/g}$	8.9	50	50	64
SiO_2 %	0.03	1.0	1.1	0.05
Al_2O_3 %	0.05	1.2	1.3	5.6
crystal shape	pigmentary	spherical	needle like	needle like
coating (PMDS) %	uncoated	coated 0.5	coated 0.6	coated 5.4

Hemolysis was studied by the method described by Harington et al. with minor modifications (Harington et al. 1971, Collan et al 1986). Sheep erythrocytes (1 %) in veronal buffered saline (VBS) solution, pH 7.4 were used. The percentage of hemolysis caused by the dust was calculated proportional to the hemolytic control after reducing the negative control value.

Cytotoxicity was studied by measuring LDH released from dust exposed rat alveolar macrophages (AM) collected by bronchoalveolar lavage from Han: Wistar rats (National Laboratory Animal Center, University of Kuopio, Finland). The method was a modification of methods by Myrvik et al. 1961, Pasanen et al. 1986, Collan et al. 1985. AMs were exposed for 2 hours to 1 mg of dust per 1 ml of AM suspension containing 1 x 10^6 cells per milliliter. LDH activity was measured with spectrophotometer (Phillips, PU 8750, UV/VIS) using the method based on the kinetic oxidation of NADH to NAD^+ according to the Recommendations of the Scandinavian Committee on Enzymes (1974).

The ability of the dusts to induce production of reactive oxygen metabolites in human PMNs was studied by luminometric method. PMNs were isolated from heparinized human venous blood samples taken from apparently healthy adults, as has been described by Hedenborg et al. (Hedenborg et al. 1987), and cells were adjusted to concentrations of 5 x 10^6 cells/ml. Chemiluminescence (CL) was measured for 40 min at 37°C using a microcomputer-controlled luminometer (LKB-Wallac 1251). The reaction mixture consisted of 100 μl of cell su-spension, 700 μl of 1 x 10^{-4} M luminol in PBS, and 200 μl of dust in PBS. Peak Cl values were used for the comparison of the results.

The ability of rat AMs to phagocytize the particles was studied by scanning electron microscopy (SEM). Rat AMs were collected by bronchoalveolar lavage as described above and resuspended in a Regular Medium 1640 containing 100 IU/ml of penicillin, 100 μg/ml of streptomycin, 10% foetal calf serum, and 2 mM of L-glutamine (Gibco Ltd., PO, U.K.), and 10 IU/ml of heparin (Medica Oy, Helsinki, Finland). The cell concentration was adjusted to 1x10^6 cells/ml, and the cells were cultured small cover slips in culture wells in an incubator at 37° C in 5% CO_2 / 95% air mixture. After incubation overnight, the medium was removed, and fresh medium containing 200 μl/ml of dust sample was added to each well. Samples for SEM were taken at 0, 0.5, 1.5, and 24h after the beginning of the exposure. After incubation, the cover slips were fixed in 2% glutaraldehyde (BIO-RAD, Walford, USA) for 10 min at room temperature, for 2 h at 4°C, and then rinsed in phosphate buffer (pH 7.2). The samples were processed through graded ethanol series, and critical point dried using CO_2. The dried cover slips were immediately coated with gold, and analyzed and photographed with SEM (JEOL JSM 35, Japan) at an accelerating voltage of 15 kV at magnifications between 200 to 3600.

Results

Hemolysis

The results of the hemolysis tests are shown in figure 1. The ultrafine TiO_2 samples were non–hemolytic. Sample C and anatase induced minor hemolysis with the highest dose of dust.

Macrophage cytotoxicity

Titanium dioxide samples were relatively non–toxic to alveolar macrophages (fig. 2). The LDH levels released from AMs exposed to ultrafine samples A and B were lower than in the negative control samples. The LDH levels with anatase and sample C were only slightly elevated.

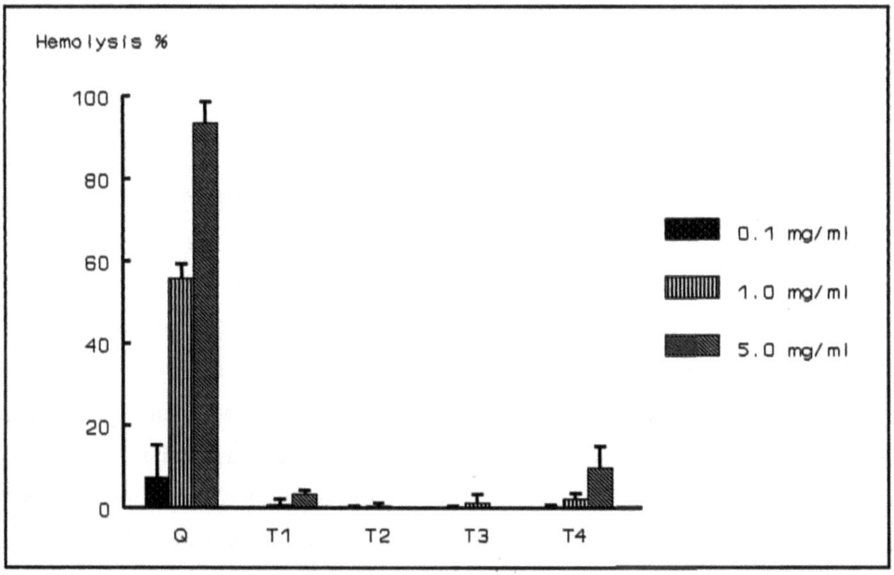

Figure 1. Hemolysis (%) induced by quartz (Q), anatase (T1), and ultrafine TiO_2 samples A (T2), B (T3) and C (T4) at 0.1, 1.0 and 5 mg/ml dose of dust.

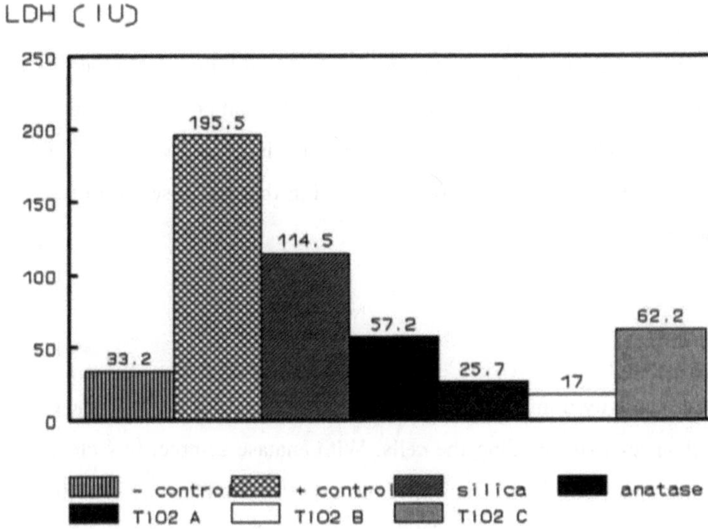

Figure 2. LDH (IU/L) released from rat alveolar macrophages exposed to 1 mg/ml dose of silica, anatase and ultrafine TiO2 samples A, B and C.

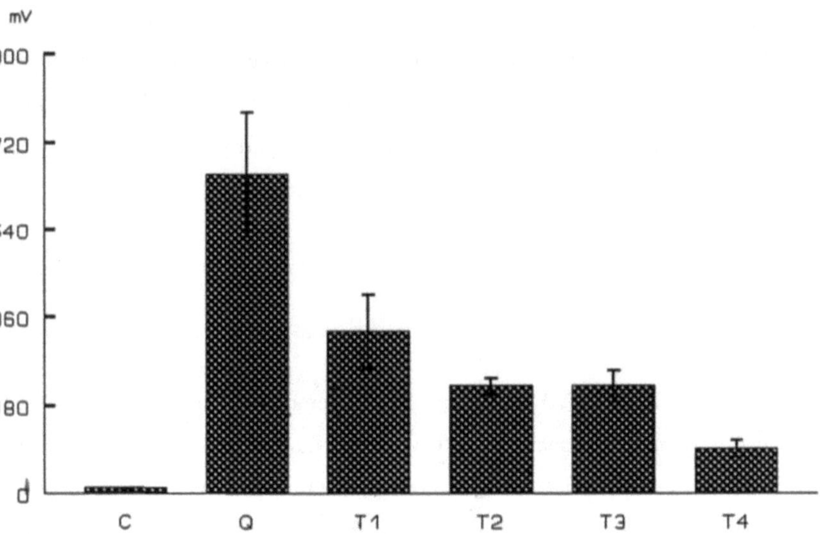

Figure 3. Chemiluminescence induced by 500 μg/ml of dust in human PMNLs. C = control, Q = quartz, Ti = anatase, T2 = sample A, T3 = sample B, T4 = sample C.

Chemiluminescence

The results of the chemiluminescence studies are shown in figure 3. The ability of the dusts to induce production of reactive oxygen metabolites in human polymorphonuclear leukocytes was lower compared with quartz dust. The Cl response increased with 50 – 500 μg/ml dose of dust in a linear pattern (data not shown) indicating that the doses used were stimulating to the cells but not toxic to them.

Phagocytosis

The phagocytotic activity diminished the concentration of particles round the cells which can be seen as a clear area surrounding the cells. With anatase sample, this clearing effect was seen after 30 minutes from the beginning of the experiment. With the ultrafine samples this effect was not so apparent and started later, within 1.5 hours after the beginning of the experiment. At 24 hours, the clearing effect was major with the anatase samples. The same effect but not so powerful was seen with ultrafine sample A. The ultrafine sample B stimulated the cells to make pseudopodies and the clearing effect was also seen. There were differences in the intensity of phagocytosis of different types of titanium dioxide samples. The AMs phagocytosed the particles in the following order of magnitude: Anatase > sample A = sample B > sample C. The samples were not toxic to the cells as there were only few dead cells at the end of the experiment. The sample C was poorly suspended and did not attach to the cover slip. Therefore, the particles of sample C were partly removed during the fixation and only few particles were visible in the SEM photographs.

Discussion

In this study, the ultrafine TiO_2 samples were not hemolytic, they were non–toxic to AMs, and induced only minor production of ROM in PMNLs. The results from our phagocytosis studies show that AMs phagocytized ultrafine TiO_2 less actively than larger particles. Phagocytosis of particles by AMs is a major clearance mechanism of particles from the lung. In case it is deficient, translocation of the particles into the interstitium of the lungs may increase.

Studies of Driscoll and Maurer (1991) comparing pulmonary responses to ultrafine TiO2 (TiO2–D; particle diameter, 0.02 μm) with TiO2–F (particle diameter 0.3 μm) demonstrated that ultrafine particles had a relatively greater toxicity on a mass/lung basis. Exposure to TiO2–D resulted in a persistent increase in AM TNF and fibronectin release which was associated with neutrophil recruitment and fibrosis, respectively. TiO2–D did not stimulate AM IL–1 release and this was consistent with the absence of a granulomatous response to TiO2–D.

It has been proposed that the increased pulmonary toxicity of ultrafine noncoated anatase particles is related to their larger surface area and to their increased interstitial access (Oberdöster et al. 1992). The findings of Ferin et al. (1991 and 1992) and Driscoll and Maurer (1991) that ultrafine particles caused more inflammatory response in rat lungs than larger particles, may be caused by the higher retention of ultrafine particles in the lung. On the other hand, the ultrafine TiO_2 sample used in the studies by Ferin et al. and Driscoll et al. was a noncoated anatase sample and there may be differences in the toxicity of ultrafine anatase and rutile. Furthermore, non–published results from our laboratory indicate that coating of the ultrafine rutile particles with $Al(OH)_3$ and PDMS reduces their ability to induce production of ROM in PMNs.

The ultrafine TiO_2 samples A and B reduced the release of LDH from AMs compared with the negative control samples (AMs not exposed to dust). The mechanism in which ultrafine particles reduced the release of LDH from cells is not known to us and further studies are needed to study the mechanisms of low LDH release induced by the ultrafine TiO_2 dusts.

The results suggest, that the coated ultrafine rutile is non–toxic to AMs and induces only relatively small production of reactive oxygen metabolites in phagocytosing cells. The phagocytosis of ultrafine TiO_2 particles by AMs is relatively inefficient which can lead to higher retention of particles in lung tissue.

Acknowledgements

The authors wish to thank Ms. Jaana Rissanen from National Public Health Institute for her excellent technical assistance.

References

Collan Y, Kosma V–M, Anttonen H, Kulju T (1986) Toxicity of richterite in hemolysis test and macrophage cultures. Arch Toxicol Suppl. 9:292–295.

Driscoll KE, Maurer JK (1991) Cytokine and growth factor release by alveolar macrophages: potential biomarkers of pulmonary toxicity. Toxicol–Pathol 19:398–405.

Ferin J, Oberdöster G (1985) Biological effects and toxicity assessment of titanium dioxides: anatase and rutile. Am Ind Hyg Assoc J 46:69–72

Ferin J, Oberdöster G, Soderholm SC, Gelein R (1991) Pulmonary tissue access of ultrafine particles. J Aerosol Medicine 4:57–68.

Ferin J, Oberdöster G (1992) Pulmonary interstitial access and inflammatory response of ultrafine particles (~20 nm) 8th international conference on occupational lung diseases, Praha 14–17 September 1992.

Harington JS, Miller K, Macnab G. Hemolysis by asbestos. Environ Res 4:95–117 (1971).

Hedenborg M, Klockars M (1987) Production of reactive oxygen metabolites induced by asbestos fibers in human polymorphonuclear leukocytes, J Clin Path 40:1189–1193.

Myrvik QN, Leake ES, Fariss B (1961) Studies on pulmonary alveolar macrophages from a normal rabbit: A technique to produce them in high state of purity. J immunol 86: 128–132.

Oberdöster G, Ferin J, Gelein R, Soderholm SC, Finkelstein J (1992) Role of alveolar macrophages in Lung injury: Studies with ultrafine particles. Environ Health persp 97:193–199.

Pasanen JT, Gustafsson TE, Kalliomäki P–L, Tossavainen A, Järvisalo JO. (1986) Cytotoxic effects of four types of welding fumes on macrophages in vitro: a comparative study. J Toxicol Environ Health 18:143–152.

Recommendations of the Scandinavian Committee on Enzymes (1974) Scand J Lab Invest 33:287.

Richards RJ, White LR, Eik–Nes KB (1985) Biological reactivity of different crystalline forms of titanium dioxide in vitro and in vivo. Scand J Work Environ Health 11:317–320.

Trochimowicz HJ, Lee KP, Reinhardt CF (1988) Chronic inhalation exposure of rats to titanium dioxide dust. J. Appl. Toxicol 8:383–385.

Carcinogenic Effects in Rats of Exposure to Different Minerals from Metallic Mine Ores, Radon and Radon Daughters.

*G. Monchaux, *J.P. Morlier, *M. Morin, **R. Zalma, ***H. Ogata, **H. Pézerat and *R. Masse.

* CEA-DSV-DPTE, BP 6, 92265 Fontenay aux Roses Cedex, France.

Keywords/Abstract: Carcinogenicity/ cocarcinogenicity/ lung cancer/ radon/ radon daughters/ minerals/ metallic mine ores/ activated species of oxygen/ nemalite/ iron pyrite/ biotite/ iron rich chlorite.

Combined exposure to radon and mineral dust is very common in some mine environments. It has been suggested that the carcinogenic activity of some inorganic materials might be related to their surface reducing properties allowing the formation of activated species of oxygen. The potential carcinogenic or cocarcinogenic properties of 4 minerals present in the ores of metallic mines (nemalite, a contaminant of the Quebec chrysotile, biotite, present in many granites and in the french uranium ore, iron pyrite, present in various iron and gold ores, and used here after air-ageing of the powder, and iron rich chlorite, present in iron, tungsten and gold ores), was studied in rats either alone or after radon exposure. The results did not demonstrate a clear carcinogenic or cocarcinogenic effect of minerals from metallic mine ores acting either alone or after radon exposure. However, the slight synergistic effect observed concerns the minerals the most active to generate very electrophilic species in an aerated aqueous medium and might be related to their role in promotion and/or progression steps.

** UA 1106 CNRS, Université Paris VI, 4 Place Jussieu, 75252 Paris Cedex 05, France.
*** National Institute of Public Health, 4-6-1, Shirokanedai, Tokyo 108, Japan.

NATO ASI Series, Vol. H 85
Cellular and Molecular Effects
of Mineral and Synthetic Dusts and Fibres
Edited by J. M. G. Davis and M.-C. Jaurand

Introduction

An increased incidence of lung cancers has been reported in different groups of underground miners: uranium miners (Archer et al., 1973; Sevc et al., 1976; Howe etal., 1986; Samet et al., 1989; Samet et al., 1991), iron miners (Jorgensen, 1984; Radford and Renard St Clair, 1984) and other miners (Fox et al., 1981; Solli et al., 1985). Combined exposure to radon and mineral dust is very common in some mine environments. It has been suggested that the carcinogenic activity of some inorganic materials might be related to their surface reducing properties allowing the formation of activated species of oxygen (Pezerat et al., 1989). The potential carcinogenic or cocarcinogenic properties of 4 minerals present in the ores of metallic mines - nemalite, a contaminant of Quebec chrysotile, biotite, present in many granites and in the French uranium ore, iron pyrite, present in various iron and gold ores, and used here after air-ageing of the powder, and finally iron-rich chlorite, present in iron, tungsten and gold ores -, was studied in rats, either alone or after radon exposure, in relation with potential combined exposure for workers in some of these mines.

Material and Methods

Five groups of 30 rats each were given 4 intratracheal instillations (I.T) of mineral dust, 10 mg each, suspended in phosphate buffered saline (PBS) at weekly interval and five other groups of rats were given 4 I.T instillations of the same mineral dusts, one month after the end of a 1000 Working-Level-Months (WLM) radon exposure. The cumulative radon exposure was obtained after 18 session exposures of 8 hours each (144 hours). The cumulative exposure by session was: 56 ± 11 WLM, and the Potential Alpha Energy Concentration (PAEC): 1213 ± 256 WLM. The unattached fraction of PAEC was 11 ± 3%.

Results

In control rats treated by PBS buffer only, 2 lung carcinomas were observed, a squamous cell carcinoma and an adenocarcinoma. In the groups treated by mineral dust alone, only lung carcinomas were observed, a squamous cell carcinoma in the group treated by air-aged iron pyrite and an adenocarcinoma in the group treated by iron-rich chlorite (Table 1). In the group exposed to radon and PBS buffer, 9 lung carcinomas were observed among 5 rats. In the groups treated by mineral dusts after previous radon inhalation exposure, lung carcinomas but also one pleural mesothelioma were observed. A slight non significant increase in the incidence of lung carcinomas was observed in rats exposed to both radon and minerals, especially, nemalite, air-aged iron pyrite and iron-rich chlorite compared to rats exposed to radon and PBS buffer. The occurrence of a pleural mesothelioma in the group exposed to biotite might be related to a specific carcinogenic effect of mineral dust at the level of the pleura. In the groups injected with mineral dusts after radon exposure, lung carcinomas were mostly large and more invasive and classified as T3 and T4 according to a classification derived from the tumor-node-metastasis (TNM) classification (Renaud and Merlier, 1975) compared with those observed in the group treated by radon and PBS buffer. There were also more tumors which had spread to the pleura, and were classified as P 1, and more intrapulmonary metastases or multiple lung tumors classified as M 2 in the group exposed to both radon and mineral dusts than in the group exposed to radon and PBS buffer.

In control rats exposed to PBS buffer alone, survival times of rats with lung carcinomas were not shortened compared with those without thoracic tumors (Table 2). In the groups exposed to mineral dusts alone, survival times of rats with squamous cell carcinomas were markedly shortened compared with controls. In the groups of rats exposed to radon and mineral dusts, rats with no thoracic tumors had shorter survival times than those with lung carcinomas, as previously reported for rats exposed to various doses of radon and its daughters. However, survival times of rats with very large and invasive squamous cell carcinoma were markedly shortened compared with controls.

Conclusion:

These results, as those previously observed with crocidolite asbestos administered by inhalation (Wagner et al., 1974; Wagner et al., 1984), did not demonstrate a clear carcinogenic effect of the minerals from metallic mine ores injected intratracheally. Moreover, they did not demonstrate a strong cocarcinogenic effect of these minerals compared with other airborne pollutants, such as tobacco smoke, which have been shown to induce a markedly increased incidence of lung carcinomas after radon exposure (Monchaux et al., 1994). However, the slight synergistic effect observed concerns the minerals the most active to generate very electrophilic species in an aerated aqueous medium (Costa et al., 1990) and might be related to their role in promotion and or progression steps.

References

Archer, V.E., Wagoner, J.K. and Lundin, F.E. (1973) Lung cancer among uranium miners in the United States. Health Physics, 25: 351-371.

Costa, D., Guignard, J. and Pezerat, H. (1990) Oxidizing surface properties of divalent iron-rich phyllosilicates in relation to their toxicity by oxidization stress mechanism. In: *Related Health Effects of Phyllosilicates*, J. Bignon, Editor, NATO ASI Series Vol. G 21, 1990, Springer Verlag, pp. 47-58.

Fox, A.J., Goldblatt, P. and Kinlen, L.J. (1981) A study of the mortality of Cornish tin miners. Br. J. ind. Med., 38: 378-380.

Howe, G.R., Nair, R.C., Newcombe, H.B., Miller, A.B. and Abbat, J.D. (1986) Lung cancer mortality (1950-80) in relation to radon daughter exposure in a cohort of workers at the Eldorado Beaverlodge uranium mine. J. Natl Cancer Inst., 77: 357-362.

Jorgensen, H.S. (1984) Lung cancer among underground workers in the iron ore mines of Kiruna based on thirty years of observation. Ann. Acad. Med., 13 (suppl.): 371-377.

Monchaux, G., Morlier, J.P., Morin, M., Chameaud, J., Lafuma, J. and Masse, R. (1994) Carcinogenic and cocarcinogenic effects of radon and radon daughters in rats: an overview of 20 years of study. Environ. Health Perspect., 102: 000-000 (in press).

Pezerat, H., Zalma, R., Guignard, J. and Jaurand, M.C. (1989) Production of oxygen radicals by the reduction of oxygen arising from the surface activity of mineral fibres. In: *Non-occupational Exposure to Mineral Fibres*. J. Bignon, J. Peto and R. Saracci, Editors, IARC Scientific Publications N° 90, 1989, pp. 100-111.

Radford, E.P. and Renard St Clair, K.G. (1984) Lung cancer in Swedish iron ore miners exposed to low doses of radon daughters. N. Engl. J. Med., 310: 1485-1494.

Renault, P. and Merlier, M. (1975) La codification TNM appliquée aux cancers bronchopulmonaires opérés. Rev. Fr. Mal. Resp., 3: 59-64.

Samet, J.M., Pathak, R., Morgan, M.V., Marbury, M.C., Key, C.R. and Valdivia, A.A. (1989) Radon progeny exposure and lung cancer risk in New Mexico U miners: a case-control study. Health Phys. 56: 415-421.

Samet, J.M., Pathak, R., Morgan, M.V., Key, C.R., Valdivia, A.A. and Lubin, J.H. (1991) Lung cancer mortality and exposure to radon progeny in a cohort of New Mexico underground uranium miners. Health Phys., 61: 745-752.

Sevc, J., Kunz, E. and Placek, V. (1976) Lung cancer in uranium miners and long term exposure to radon daughter products. Health Phys., 30: 433-437.

Solli, L.H., Andersen, A., Stranden, E., and Langard, S. (1985) Cancer incidence among workers exposed to radon and thoron daughters at a niobium mine. Scand. J. Work Environ. Health, 11: 7-13.

Wagner, J.C., Berry, G., Skidmore, J.W. and Timbrell, V. (1974) The effects of the inhalation of asbestos in rats. Br. J. Cancer, 28: 173-185.

Wagner, J.C., Griffiths, D.M. and Hill, R.J. (1984) The effect of fibre size on the in vivo activity of UICC crocidolite. Br. J. Cancer, 49: 453-458.

Table 1. Incidence of the different histological types of malignant thoracic tumor in rats after exposure to radon and minerals from metallic mine ores.

	Bronchopulmonary carcinoma			Pleural tumors	Total number of rats with tumors
	Squamous cell carcinoma	Bronchiolo alveolar carcinoma	Adeno carcinoma	Pleural mesothelioma	
Controls (PBS, I.T)	1	–	1	–	2/32
Nemalite (I.T)	–	–	–	–	0/20
Iron pyrite (I.T)	1	–	–	–	1/22
Biotite (I.T)	–	–	–	–	0/30
Chlorite (I.T)	–	–	1	–	1/30
* Radon (1000 WLM) and PBS (I.T)	1	6	2	--	5/30
* Radon (1000 WLM) and nemalite (I.T)	4	2	3	–	9/30
* Radon (1000 WLM) and iron pyrite (I.T)	1	2	6	–	8/30
* Radon (1000 WLM) and biotite (I.T)	2	3	3	1	7/30
* Radon (1000 WLM) and chlorite (I.T)	1	2	6	–	8/30

* Radon exposed groups
WLM: working-level-month
PBS: phosphate buffered saline
I.T: intratracheal

Table 2. Mean survival times (days) of rats (± standard deviation) according to histological tumor type after exposure to radon and minerals from metallic mine ores.

	Bronchopulmonary carcinoma			Pleural tumors	No thoracic tumor
	Squamous cell carcinoma	Bronchiolo alveolar carcinoma	Adeno carcinoma	Pleural mesothelioma	
Controls (PBS, I.T)	758	–	782	–	757 ± 123
Nemalite (I.T)	–	–	–	–	768 ± 93
Iron pyrite (I.T)	618	–	–	–	712 ± 135
Biotite (I.T)	–	–	–	–	785 ± 79
Chlorite (I.T)	369	–	–	–	799 ± 83
*Radon (1000 WLM) and PBS (I.T)	807	863 ± 43	888	–	721 ± 145
*Radon (1000 WLM) and nemalite (I.T)	729 ± 65	865 ± 15	746 ± 38	–	753 ± 93
*Radon (1000 WLM) iron pyrite (I.T)	535	896 ± 11	768 ± 174	–	716 ± 140
*Radon (1000WLM) and biotite (I.T)	869 ± 16	789 ± 93	864 ± 16	846	692 ± 121
*Radon (100WLM) and chlorite (I.T)	456	775 ±1 27	751 ± 140	--	735 ± 116

* Radon exposed groups
WLM: working-level-month
PBS: phosphate buffered saline
I.T: intratracheal

Biological Effects of a New Type of Pure Carbon (Fullerenes) *in vitro*

Thomas Baierl, Elisabeth Drosselmeyer and Arnulf Seidel

Kernforschungszentrum Karlsruhe (KfK) - Institut für Toxikologie
Postfach 3640 - D 76021 Karlsruhe - Germany

Introduction

Fullerenes are a new modification of pure carbon in addition to graphite and diamond. The first types to become known were called "Buckyballs". They consist of cage-like molecules with a well defined number of carbon atoms (e.g. 60 atoms: C_{60}). The Fullerenes are readily soluble in many organic solvents and vaporizable. A wide range of technological and also biomedical applications is expected for these molecules. First studies on *in vitro* effects of C_{60} Fullerenes showed that the particles could induce the release of pro-inflammatory mediators by alveolar-macrophages (AM). However, this early material still might have contained about 10% of C_{70} and solvents due to the preparation procedure used at this time. Meanwhile, pure and solvent free C_{60} particles became available. One of the aims of the present study was to reevaluate our previous data with really pure material. In addition, the hemolytic properties were tested and the interaction of C_{60} with superoxide anion production was investigated.

Material and Methods

C_{60} was obtained from the Institute for Nuclear Solid State Physics of the KfK. The material was prepared by elution from raw soot from a light arc procedure (according to the method established by [Krätschmer et al. (1990)]), separated and purified by HLPC and dried by evaporation. The **graphite** used for incubation was also obtained from the same institute. It was chemicaly pure carbon dust with a particle size < 10 µm. The **quartz** was DQ12 material with 50 % of the particles < 2 µm. **Latex** with a particle size of 1 µm was used as a standard negative control. The standard positive control for stimulation was lipopolysaccharide (*Abortus equi*, **LPS** (10 µg/ml)).

Material	Particle size	Function	Tested for
Fullerene C_{60}	< 5µm	test material	hemolysis, TNF, O_2^-
Graphite	< 10 µm	control	hemolysis, TNF
Quartz DQ12	50% < 2µm	positive control	hemolysis, TNF
Latex	1 µm	negative control	TNF

Table 1: Materials for Incubation

Bovine alveolar macrophages (BAM) were obtained by lavage of bovine lungs (0.9 % NaCl) within less than 1 hour *post mortem* [Fox (1973), Fisher et al. (1983)]. $3x10^6$ cells/ml (2 ml/well, 6 well Macroplate, Greiner) were cultured overnight (RPMI 1640, P/S (100 I.U.

NATO ASI Series, Vol. H 85
Cellular and Molecular Effects
of Mineral and Synthetic Dusts and Fibres
Edited by J. M. G. Davis and M.-C. Jaurand

Penicilin/ml, 100 µg Streptomycin/ml), 2% FCS) to reduce pre-stimulation from the lavage.
HL-60 cells were cultured in RPMI 1640 (with P/S, 15% FCS, 1.5% L-Glutamine, 1% MEM 100x and 10 mM Sodiumpyruvate). For the differentiation to macrophage like cells (HL-60-M) $0.5x10^6$ HL-60 cells/ml were incubated with Vitamin D_3 (Calcitriol, 10^{-6} M) [Rossi et al. (1987)]. The HL-60-M were used at the 4th day of differentiation. For incubation with the materials about $1.5x10^6$ HL-60-M/ml were used.
For the **hemolysis assay** (adapted from [Nolan et al. (1981)]) fresh bovine blood with 10 000 I.U. Sodium-Heparin/l was centrifugated three times (1000 g, 4 °C, 10 min) and resuspensated with 0.9% NaCl. The erythrocyte pellet was diluted to a 4.2% suspension. After incubation with the dusts (2 hours in waterbath, 36 °C) and centrifugation the concentration of the hemoglobin in the supernatant was measured (530 nm).
Tumor necrosis factor (TNF) as an indicator for the immune response of the cells was tested by the use of TNF-sensitive WEHI 164 cells [Espevik and Nissen-Meyer (1986), Eskandari et al. (1990)]. For the test $2.5x10^5$ cells/ml in RPMI 1640 (with P/S, 10% FCS, 1% L-Glutamine, 2 µg Actinomycin D/ml) were used. After 18 h of incubation with the supernatants from the incubation of BAM, survival of WEHI cells was assayed with MTT (5 mg/ml, PBS without Ca^+/Mg^+) and plotted against a standard curve with human recombinant TNF (yielding concentration values in U/ml). TNF concentration from HL-60-M (in pg/ml) was tested with ELISA (Quantikine). **Viability** of the cells was tested by the use of Trypanblue [Hanks and Wallace (1958)]. The **superoxide anion production** was measured by the reduction of Cytochrome c [Johnston (1981)].

Results and Discussion

In Fig. 1 an electron microscopic photograph of the C_{60} material used in our experiments is presented. It forms crystals with various shapes, most of them with a geometric diameter of < 5 µm. These crystals can agglomerate to form needle like structures.
When HL-60-M are incubated with Fullerenes or other powders they release TNF into cell culture medium (Fig. 2). The reaction towards LPS is shown as positive control. After 4 hours the effect reaches or even surpasses (in case of graphite) that of LPS.

Fig. 1: Native C_{60} crystals

There is a distinctive reduction of the TNF concentration in the culture medium after 24 hours of incubation. This is not due to a loss of viability of the cells which remained about 100% of the control over the whole period. According to data obtained in our laboratory [Geiger and Polzer, unpublished] the HL-60-M either reinternalize the TNF molecules or they release a factor which eventually leads to a neutralization of the cytokine.

The results presented in Fig. 3 show that also BAM can be triggered by C_{60} Fullerenes to produce TNF. In this respect Fullerenes have the same effects as graphite or quartz. When Latex particles (used as negative controls for the effect of mere phagocytosis of anything) were applied, no TNF was released by BAM (data not shown). The TNF concentration in the culture medium decreases with the period of incubation also in this experiment, most probably also due to the reasons mentioned above. It is not surprising that quartz particles induce the release of TNF by AM. Analogous results have been reported already in the literature [Dubois (1989)] as well as by our own group [Brückner-Nieder et al. (1992)]. The release of TNF after incubation with the carbon particles is not only the consequence of mere phagocytosis but obviously a real effect. It can almost certainly be excluded that it is due to impurities of the C_{60} material.

HL-60-M respond to Zymosan with production and release of O_2^- (Fig. 4). Fullerene particles alone do not induce the release of reactive oxygen species by the cells. However, when they were applied together with Zymosan, they caused a distinctly enhanced reaction. Fullerenes alone without cells also did not produce any reactive oxygen species in the darkness of the CO_2 incubator.

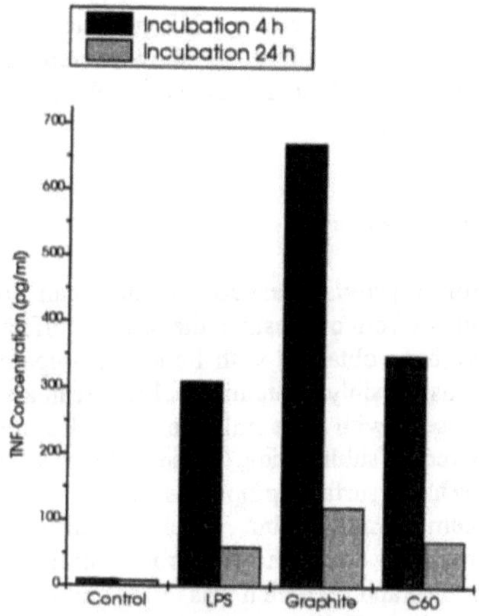

Fig. 2: Release of TNF by HL-60-M after 4 and 24 h incubation with particles (representative for three experiments).

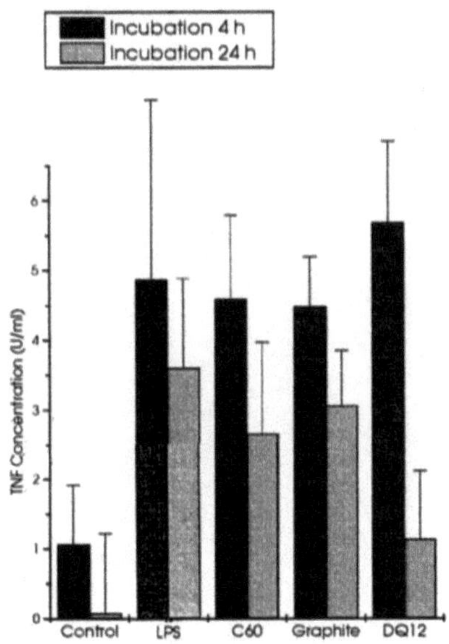

Fig. 3: Release of TNF by BAM after 4 and 24 h incubation with C_{60}, Graphite and DQ12. Arithmetic means ± S.E., n=3

In Fig. 5 the results of the hemolysis experiments are presented which allow a qualitative statement at least. In contrast to the effects concerning TNF graphite is not active but Fullerenes show hemolytic properties which are already well known for quartz.

Summary

Some provisional conclusions can be drawn from our results: the *in vitro* effects we have obtained with Fullerene preparations possibly containing solvents can also be seen with pure Fullerene particles prepared by sublimation. These particles seem to have surface properties which render them hemolytic but which are different from the other known carbon configuration (graphite) which was tested and which was inert. Fullerenes are no scavengers of reactive oxygen species under the conditions used. In summary Fullerenes are not inert *in vitro*.

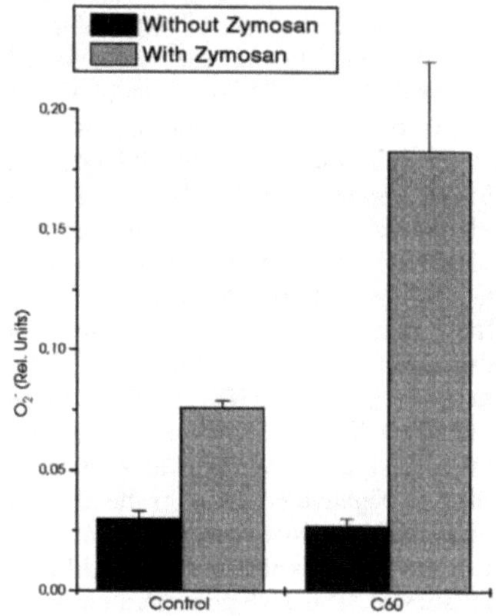

Fig. 4: O_2^--production by HL-60-M during incubation with C_{60}. Arithmetic means ±S.E., n=3

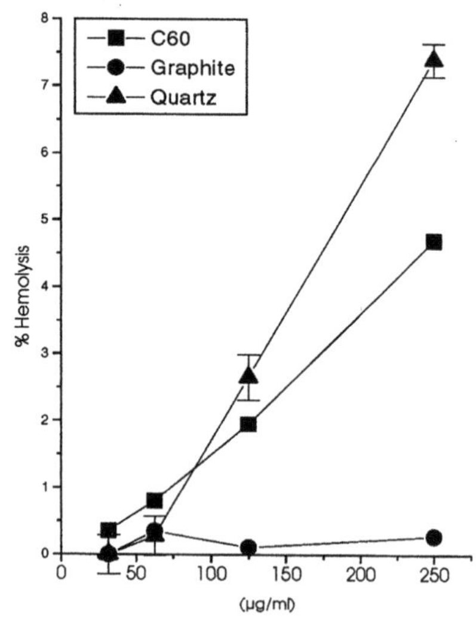

Fig. 5: Hemolysis of bovine erythrocytes by particles. Arithmetic means ± S.E., n =4

Literature (in alphabetical order):

Brückner-Nieder U, Schmidt A, Schimmelpfeng J and Seidel A (1992)
HL-60-derived Macrophages Are Resistant to Cytotoxic Effects of Quartz
in: Environmental Hygiene III (Ed.: Seemayer, N.H. and Hadngy, W.), Springer, Berlin

Dubois CM Bissonette B and Rola-Pleszczynski M (1989)
Asbestos Fibres and Silica Particles Stiumlate Rat Alveolar Macrophages to Release Tumor Necrosis Factor
Am. Rev. Resp. Dis., 139:1257-1264

Eskandari MK, Nguyen DT, Kunkel SL and Remick DG (1990)
WEHI 164 Subclone 13 Assay for TNF: Sensitivity, Specificity and Reliability
Immuno. Invest., 19:69-79

Espevik T and Nissen-Meyer M (1986)
A Highly Sensitve Cell Line, WEHI 164 Clone 13 for Measuring Cytotoxic Factor/Tumor Necrosis Factor from Human Monocytes
J. Immunol. Meth., 95:99-105

Fisher GL, McNeill KL and Democko CJ (1983)
Application of Bovine Alveolar Macrophage Bioassays in the Analysis of Toxic Agents in Complex Environmental Mixtures
in: Short Term Bioassays in the Analysis of Complex Environmental Mixtures III (Ed.: Waters MD, Sandhu SS, Lewtas J, Claxton L, Chernoff N and Nesnow S), Plenum Press, N.Y., 257-268

Fox ML (1973)
The Bovine Alveolar Macrophage.. I. Isolation, *in vitro* Cultivation, Ultrastructure and Phagocytosis
Can. J. Microbiol., 19:1207-1210

Hanks JH and Wallace JH (1958)
Determination of Cell Viability
Proc. Soc. Exptl. Biol. Med., 98:156-160

Johnston RBJr. (1981)
Secretion of Superoxide Anion
in: Methods for Studying Mononuclear Phagocytes (Ed.: Adams PO, Edelson PJ and Karen H), Academic Press, N.Y., 489-497

Krätschmer W, Lamb LD, Fostiropoulos K and Huffman DR (1990)
Solid C_{60}: A New Form of Carbon
Nature, 318:354-358

Nolan RP, Langer AM, Harington JS, Oster G and Selikoff IJ (1981)
Quartz Hemolysis as Related to its Surface Functionalities
Environmental Research, 26:503-520

Rossi P, Chini L, Fattorossi A, Gidlund M, Galli E, Laan K, Jondal Mand Wigzell H (1987)
1,25-Dihydroxyvitamin D_3 and Phorbolesters (TPA) May Induce Select *in vivo* Differentiation Pathways in the HL-60 Promyelocytic Cell Line
Clin. Immunol. Immunopathol., 44:308-316

Augmentation in the Differential Oxidative DNA-Damage by Asbestos in Presence of H_2O_2 and Organic Peroxide/Hydroperoxide

Q. Rahman+, N. Mahmood, S.G. Khan and M. Athar*
Industrial Toxicology Research Centre
Post Box 80, Lucknow 226001 INDIA

*Department of Medical Elementology & Toxicology
Faculty of Science, Jamia Hamdard (Hamdard University)
Hamdard Nagar, New Delhi 110062 INDIA

Abstract

DNA-damaging effects of asbestos in the presence of hydrogen-peroxide, organic peroxides and hydroperoxides were investigated. The DNA damage was assessed in terms of the destabilization of secondary structure of DNA, damage to deoxyribose sugar and DNA fidelity which were measured respectively by S-1 nuclease hydrolysis, formation of thiobarbituric acid (TBA) - reacting species and melting temperature (Tm) profile using calf thymus DNA. S-1 nuclease hydrolysis and Tm determinations have shown that the presence of hydrogen peroxide (H_2O_2), benzoyl peroxide (BOOB), Cumene hydroperoxide (COOH) or tertiarybutyl hydroperoxide (t-BOOH) augmented asbestos-mediated DNA damage many folds compared to either asbestos alone or H_2O_2/peroxide/hydroperoxide alone. H_2O_2 also damaged deoxyribose sugar in DNA whereas no formation of TBA-reacting species could be observed in case of any other peroxide/hydroperoxide. The quenchers of reactive oxygen species (ROS) afforded protection against these DNA damage. These results suggest that asbestos in the presence of H_2O_2/organic peroxides damages DNA which is mediated by the generation of oxygen free radicals. The implications have been made for the significance of these results in relation to the observed development of cancer of respiratory tract among the asbestos-exposed population.

+ To whom the correspondence should be addressed

NATO ASI Series, Vol. H 85
Cellular and Molecular Effects
of Mineral and Synthetic Dusts and Fibres
Edited by J. M. G. Davis and M.-C. Jaurand

Introduction

The potential role of oxygen derived or xenobiotic derived free radicals in the pathogenesis of several human diseases has stimulated extensive research linking the toxicity of various xenobiotics and diseased processes to a free radical mechanism (Halliwell and Gutteridge, 1985). Asbestos is one of the important industrially used mineral fibre which has been proposed to generate free radicals in exposed animal tissues. Further, the occupational exposure to asbestos fibre results in an increased risk of cancer of pleura and peritoneum or cancinoma of lung, oesophagus or stomach (Campbell et al, 1980). These manifestations of asbestos have been duplicated in experimental animals. However, the mechanism by which asbestos, causes cancer induction remains elusive to a great extent (Rahman et al, 1990).

Substantial experimental evidence exists which implicates the role of both oxygen and organic-free radical intermediates in the multistep process of carcinogenesis (Trush and Kenster, 1991). The generation of reactive oxygen species (ROS) by asbestos has been very well demonstrated and documented in both *in vivo* and *in vitro* test systems (Maples and Johnson, 1992). Asbestos mediated generation of ROS ultimately leads to the initiation of lipid peroxidation (LPO) in biomembranes with the concomitant accumulation of peroxides and hydroperoxides in the target organs. It seems plausible that asbestos-induced malignencies specially mesothelioma may be a consequence of oxidative damage to DNA molecules by organic and inorganic free-radicals generated from the accumulated peroxides and hydroperoxides probably catalyzed by transition metals, such as iron present in the composition of asbestos involving Haber-weiss and Fenton reactions (Weitzman and Witberg, 1985). To ascertain whether or not various inorganic or organic peroxides and hydroperoxides accumulated in the target organs as a consequence of asbestos mediated LPO or as a result of aerobic metabolism (Fridovich, 1986, Halliwell & Gutteridge, 1985) have distinct DNA damaging effects in the presence of asbestos, the effect of inorganic peroxide (H_2O_2), and organic peroxides/hydroperoxides was studied. Benzoyl peroxide

(BOOB), cumene hydroproxide (COOH) and t-butyl hydroperoxide (t-BOOH) were chosen as mimics of lipid peroxides to study their effects on the *in vitro* damage to calf thymes DNA in the presence of different varieties of asbestos. The various parameters used to detect the DNA damage were S-1 nuclease hydrolysis; damage to deoxyribose sugar moiety as determined by formation of thiobarbituric acid (TBA) reacting species and DNA fidelity as measured by melting temperature profile (Tm). The quenchers of reactive oxygen species (ROS) and specific iron chelator (desferal) were also tried to identify the specific role of ROS and iron in DNA damage.

Material and Methods

Calf thymus deoxyribonucleic acid (Sodium salt, average molecular weight one million) was obtained from Sigma Chemical Company, U.S.A. and S-1 nuclease was procured from Bethesda Research Laboratory, U.S.A. All the other chemicals and reagents used were either procured from Sigma Chemical Company, U.S.A. or Sisco Research Laboratory, Bombay, India in the highest purity commercially available. Chrysotile, crocidolite asbestos and UICC standard reference samples (Particle size below 30 micron) were given as gift by Dr J.B. Leinweber, John-Manville Mills, U.S.A. Indian chrysotile asbestos was obtained from a local asbestos factory (particle size below 30 micron) (Zaidi, S.H., 1969). Reaction mixtures in the final volume of 2ml contained 2mg DNA in TNE (0.01 M Tris-HCl, pH 7.4, 0.01 M NaCl and $2x10^{-4}$ M EDTA), 5 mg asbestos (suspension), 40mM final concentration of peroxides or hydroperoxides (Ochi and Miyaura, 1989 references therein) and 0.1M phosphate buffer (pH 7.4). These reaction mixtures with well-matched controls were incubated at 37°C for 20 hour in 25ml conical flasks with constant shaking in metabolic shaker. The doses of various constituents of reaction mixture and time of reaction were decided on the basis of dose and time-response studies (data not shown). The asbestos fibers were removed by centrifugation. S1 nuclease hydrolysis of control and treated DNA (0.5 mg) was carried out in a total volume of 1 ml containing final concentration of 0.1 M acetate buffer (pH 4.5) and 1mM $ZnSO_4$. The reaction mixture contained 80 units of S-1 nuclease. After incubation for 2 hours at 40°C, solution of bovine serum (10 mg/ml, 0.2 ml) and ice cold perchloric acid (14%, v/v, 1.0 ml) were added. Deoxyribonucleotides in the supernatant were determined calorimetrically (Schneider, 1957). Thermal melting profiles (Tm) of control and treated DNA were measured by the degree of S-1 nuclease digestion (case and Baker, 1975). Damage to deoxyribose sugar in DNA was measured according to the method as described by Gutteridge and Wilkins (1983).

Results

The effect of various varieties of asbestos in the presence of organic peroxide and hydroperoxide on DNA susceptibilities to S-1 nuclease hydrolysis is shown in table-1. Crocidolite rendered DNA more susceptible to S-1 nuclease hydrolysis in the presence of hydrogen peroxides, organic peroxides and hydroperoxides compared to other varieties of asbestos. However, H_2O_2/peroxide/hydroperoxide alone or asbestos alone had a little effect compared to untreated control (Table-1). The mid range melting temperature (Tm) of control DNA was 85°C , while the treatment of DNA with COOH, BOOB and t-BOOH only slightly altered Tm to 84, 82 and 86°C respectively. All the peroxides and hydroperoxides when incubated with DNA in the presence of asbestos showed a marked decrease in Tm. The effect of H_2O_2, COOH, BOOH, t-BOOH and ABIN on asbestos-mediated damage to deoxyribose sugar moiety of DNA is shown in table-2. Only H_2O_2 and to some extent ABIN damage the sugar moiety. However, organic peroxides/hydroperoxides showed no interaction with it. Various quenches of oxygen radicals such as dimethylsulfoxide (DMSO), histidine, sodium azide, cytochrome 'C', mannitol and sodium benzoate afforded different degrees of protection against the augmentation of S-1 nuclease hydrolysis or decrease in Tm of DNA. Similarly, desferal, a chelator of iron also provided protection against DNA damage-mediated by asbestos and peroxide or hydroperoxide (data not shown).

Discussion

Even though oxygen is essential for the aerobic life, the generation of reactive oxygen species (ROS) by asbestos mineral fibre has gained considerable attention with respect to its carcinogen responses in the target organs (Mossman et al, 1990). The particular mineral fibre so called "solid state" carcinogenesis unlike other carcinogens do not either alkylate or make adducts with DNA molecule, suggesting that asbestos mediated formation of ROS may cause DNA damage, which could be the initiation step in the multistep process of carcinogenesis (Mengehini, 1988).

There are numerous mechanism(s) by which asbestos can generate ROS as bye-products of oxygen metabolism in biological systems. In addition to phagocytosis of asbestos by alveolar macrophages (AMs) that ultimately results in the generation of oxygen free radicals, the agonist stimulation of superoxide anion radical ($O_2^{\cdot-}$) production by this mineral fibre in AMs seems to be via activation of phospholipase C pathways (Roney and Holian, 1989). While the presence of reducing sites on the surface of asbestos may contribute to the initiation or progression of oxidant generation (Rahman and Casciano, 1985), the generation of $O_2^{\cdot-}$ directly by asbestos is also possibel. Inaddition, its interaction with cell membrane resulting in the stimulation of NADPH or NADH oxidase has also been reported (Cox et al, 1985). Another important possibility for the production of ROS may be through the interaction of asbestos with various cellular oxidants generated as side products of normal metabolism. H_2O_2 and $O_2^{\cdot-}$ are known to be produced in mammalian cells during aerobic metabolism (Fridovich, 1986; Halliwell and Gutteridge, 1985). Once any one of the oxidant is generated, a series of other more nasty agents may be generated from the interconversion of these ROS. For example, $O_2^{\cdot-}$ is dismutated both enzymatically and non-enzymatically into hydrogen peroxide. The lipid peroxidative process is also reported to be in a self-sustaining chain reaction (propagation phase) which could convert poly-unsaturated falty acids (PUFA) to lipid hydroperoxides (Lutz 1990). The LPO of PUFA in cellular membrane not only causes increased rates of non-enzymic free radical generation reactions but also simultaneously activates enzymes such as cyclooxygenase and lipooxygenase that favour the formation of arachidonic acid peroxides. Asbestos exposure also leads to the enhancement of LPO, which is also a physiological process, for instance in prostagladin synthesis (Lutz, 1990). These processes in turn favour the excessive accumulation of lipid peroxides in the target organs which could results *in vivo* damage to DNA (Philips et al, 1984). The various peroxides and hydroperoxides are not only produced via redox reaction in the living organism but may also be introduced artificially into the living cells (Freeman, 1984). Thus, it seems logical that the presence of catalytic sites on asbestos can decompose the accumulated peroxides and hydroperoxides to generate free radicals in the target organs via involving Haber-weiss and Fenton reaction that ultimately may cause injury to DNA (Haber and Weiss, 1934; Ehrenfeld et al, 1987). The results of the present study strengthen the above said hypothesis and clearly demonstrate that asbestos via generating free radicals from peroxides/ hydroperoxides cause structural changes in DNA.

The individual effect of either asbestos or peroxides and hydroperoxides in damaging DNA was meager when compared the cumulative effects of asbestos and peroxides and hydroperoxides. The *in vitro* incubation of calf thymus DNA with asbestos and either inorganic peroxide viz. hydrogen peroxide (H_2O_2) or organic peroxides and hydroperoxides such as tert-butyl hydroperoxide, cumene hydroperoxide and benzoyl peroxide, the mimics of various lipid peroxides predominantly increased the S-1 nuclease hydrolysis as compared to asbestos alone or peroxides-or hydroperoxides-alone, suggesting the creation of single stranded region. The presence of H_2O_2 induced damage to the deoxyribose sugar moiety of DNA, while organic peroxide and hydropedroxides were completely ineffective in this regards suggesting the more pronounced effect of asbestos and H_2O_2 on both primary and secondary structure of DNA. The sugar damage in the DNA by the attack of hydroxyl radical ($\cdot OH$) probably on carbon 3′ and 4′ is known to produce stand breaks in this macromolecule. This suggests that asbestos via generating $\cdot OH$ from H_2O_2 caused the sugar damage and subsequently the strand breaks which seems to have important implications in terms of the development of asbestos-induced pathological changes in the target organs (Floyd, 1990 and references therein). These observations led us to believe the behaviour of inorganic and organic peroxides and hydroperoxide-mediated DNA damage is differential in the presence of asbestos. The similarity between H_2O_2 and 2,2'-azobis (isobutyronitrile) (ABIN) a lipid soluble agents that generate superoxide anion radicals while causing/producing damage to deoxyribose sugar moiety of DNA in the presence of asbestos. Similarly all these peroxides and hydroperoxides were effective in increasing DNA fidelity as measured by the decrease in thermal melting profile of DNA but again the effects produced by H_2O_2 and ABIN were more gnonounced in this respect.

To ascertain the involvement of various factors in causing differential DNA damage, the effect of various quenchers of ROS was studied on the set parameters taken as tools to defect the damage in DNA molecule. Quenchers of singlet oxygen and ·OH were only partially effective in diminishing DNA damage induced by asbestos and organic peroxides/hydroperoxides, but in the case of asbestos and H_2O_2/AIBN mediated DNA damage, ·OH quenchers showed significant protection, suggest that asbestos-mediated generation of OH radicals in the presence of H_2O_2/AIBN cause DNA damage while the effects produced by organic peroxides/hydroperoxides involve different other free radicals. However, desferal a specific chelator of iron was effective in suppressing DNA damage caused by these agents, suggesting the role of iron in these reactions. Thus, the presence of transition metal, iron to the levels as high as 36% by weight in the asbestos composition (Campbell et al., 1980) revealed/indicate its significance with respect to DNA damage in the presence of peroxides and hydroperoxides. The similarity of inorganic and organic peroxides and hydroperoxides in the presence of asbestos to cause single stranded regions in DNA and to alter its fidelity while the marked differences in the behaviour of the two groups of peroxides in inducing damage to the sugar moiety of DNA may be attributed to the production of hydroxyl radicals from H_2O_2 and $O_2^{\cdot-}$ and alkoxy or aryloxy radicals from organic peroxides respectively. The stearic factors or the relative stability of the free radicals generated could be the determinant of these anomalies, which also suggest that alkoxy or aryloxy radicals produced by organic peroxides/hydroperoxides due to these reasons seems to be incompetent to cause DNA sugar damage. The present studies also substantiate the previous reported data which suggests the formation of single strand breaks in rat embryo cells and enhanced hydroxylation of DNA bases particularly the formation of 8-hydroxyguanosine residues (197, 198 reference in the chapter).

In conclusion, these results suggest that asbestos by generating free radicals from peroxides and hydroperoxides catalyzed by iron present at its surface via Haber-weiss and Fenton reactions may cause clastogenic changes in DNA. Further, these changes in DNA seems to be important to contribute to the ability of asbestos to induce mesotheliomas in the exposed population as reported by Maples and Johnson (1992). However, detailed investigations defining the exact role of asbestos deregulation of DNA (genes) and thereby causing malignanicies remains to be elucidated.

References

Campbell, W J, Huggins C W and Wylie A G (1980) Chemical and physical characterization of amosite, chrysotile, crocidolite, and non-fibrous tremolite for oral ingestion studies by the National Institute of Environmental Health Sciences, Bureau of Mines Report of Investigations, 8452.

Case S T and Baker R F (1975) Investigation into the use of Aspergillus oryzae S-1 nuclease in the presence of solvents which destabilize or prevent DNA secondary structure: Formaldehyde, formamide and glyoxal. Anal. Biochem. **64**: 477-484.

Cox J A, Jeng, A Y, Sharkey N A, Blumberg P M and Tauber A I (1985). Activation of the human neutrophil nicotinamide adenine dinucleotide phosphate (NADPH) -oxidase by protein kinase C. J. Clin. Invest. **76**: 1932-1938.

Ehrenfeld G M, Shipley J B, Hembrook D C, Sugiyama H, Long E C, VanBoom J H, Vander Marel G A, Oppenheiner N J and Hecht S M (1987) Copper-dependent cleavage of DNA by bleomycin. Biochemistry **26**: 931-942.

Floyd R A (1990) Role of oxygen free radicals in carcinogenesis and brain ischemia. FASEB J. **4**: 2587-2597.

Freeman B A (1984) In: *Free Radicals in Molecular Biology, Aging and Disease*, Ed by D. Armstrong et al., Ranuem Press, New York, pp 43-52.

Fridovich I (1986) Biological effects of the superoxide radical. Arch. Biochem. Biophys. **247**: 1-11.

Gutteridge J M C and Wilkins S (1983) Copper salt-dependent hydroxyl radical formation damage to proteins acting as antioxidants. Biochem. Biophys. Acta **759**: 38-41.

Haber R and Weiss J (1934) The catalytic decomposition of hydrogen peroxide by iron salts. Proc. R. Soc. London Ser A **147**: 332-351.

Halliwell B and Gutteridge J M C (1985) The importance of free radicals and catalytic metal ions in human diseases. Mol. Aspects Med. **8**: 89-193.

Lutz W K (1990) Endogenous genotoxic agents and processess as basis of spontaneous carcinogenesis. Mutant Res. **238**: 287-295.

Maples K R and Johnson N F (1992) Fibre-induced hydroxyl radical formation: Correlation with mesothelioma induction in rats and humans. Carcinogenesis **13**: 2035-2039.

Mengehini R (1988) Genotoxicity of active oxygen species in mammalian cells. Mutat. Res. **195**: 215-230.

Mossman B T, Bignon J, Corn M, Seaton A and Gee J B L (1990) Asbestos: Scientific Developments and implications for public policy. Science **247**: 294-301.

Ochi T and Miyaura S (1989) Cytotoxicity of organic hydroperoxides and cellular antioxidant defense system against hydroperoxides in cultured mammalian cells. Toxicology **55**: 69-62.

Philips B J, James T E B and Andersen D (1984) Genetic damage in CHO cells exposed in enzymically generated active oxygen species. Mutat. Res. **126**: 265-271.

Rahman Q and Casciano D A (1985) Involvement of superoxide radical in the toxicity of mineral fibers. In: *In vitro effects of mineral dusts, vol 3*, (eds) Beck E G, and Bignon J, Springer-Verlag, New York, pp 483-486.

Rahman Q, Khan S G and Ali S (1990) Effect of chrysotile asbestos on cytochrome P-450-dependent monooxygenase and glutathione-S-transferase activities in rat lung. Chem. Biol. Interact. **75**: 305-314.

Roney P L and Holian A (1989) Possible mechanism of chrysotile asbestos-stimulated superoxide anion production in guinea pig alveolar macrophages. Toxicol. Appl. Pharmacol. **100**: 132-144.

Schneider W C (1957) Determination of nucleic acids in tissues by pentose analysis. In: *Methods in Enzymology, vol. 3*, (eds) Colowick S P and Kaplan O N. Academic Press, New York, pp 680-684.

Trush M A and Kensler T W (1991) An overview of the relationship between oxidative stress and chemical carcinogenesis. Free Rad. Biol. Med. **10**: 201-209.

Weitzman S A and Witberg A B (1985) Asbestos-Catalyzed lipid peroxidation and its inhibition by deferoxamine. Biochem. J. **225**: 259-262.

Zaidi S H (1969) Experimental pneumoconiosis, John Hopkins, Baltimore, pp 94-109.

Table 1:

S_1 nuclease hydrolysis of asbestos-treated DNA in the presence of hydrogen peroxide (H_2O_2), cumene hydroperoxide (COOH), benzoyl peroxide (BOOB), and tertiary-butyl hydroperoxide (t-BOOH) and 2,2′-azobis (isobutyronirile) (ABIN)

Incubation system	%DNA hydrolysed				
	Control	+ TiO_2	+Indian Chrysotile	+UICC Chrysotile	+UICC Crocidolite
DNA	2.0 ± 0.12	2.0 ± 0.10	3.0 ± 0.17	2.5 ± 0.13	4.0 ± 0.21
DNA + H_2O_2 (40mM)	6.0 ± 0.34	7.6 ± 0.51	22.5 ± 1.50	34.9 ± 2.16	50.2 ± 3.51
DNA + COOH (40mM)	5.3 ± 0.28	6.1 ± 0.36	18.6 ± 1.12	28.3 ± 1.69	42.5 ± 2.42
DNA + BOOB (40mM)	4.5 ± 0.27	5.4 ± 0.21	16.2 ± 1.13	20.4 ± 1.40	37.8 ± 2.11
DNA + t-BOOH (40mM)	4.9 ± 0.24	5.0 ± 0.21	12.9 ± 1.03	18.1 ± 1.09	25.7 ± 1.49
DNA + ABIN (40mM)	2.0 ± 0.15	2.7 ± 0.33	10.4 ± 1.18	18.0 ± 3.12	25.9 ± 4.13

The data represent mean ± S.E. (n = 6).
Treatment of DNA samples with dusts in presence of peroxides and hydroperoxides followed by S_1 nuclease hydrolysis and assay of acid soluble nucleotides is described in the text.

Table 2:

Effect of hydrogen peroxide (H_2O_2), cumene hydroperoxide (COOH), benzoyl peroxide (BOOB), tertiary-butyl hydroperoxide (t-BOOH) 2,2′-azobis (isobutyronirile) on asbestos-mediated damage to deoxy ribose sugar moiety of DNA

Reaction mixture components	n moles of TBA reactive products formed after 20 hours incubation at 37°C				
	Control	+ TiO_2	+ Indian Chrysotile	+ UICC Chrysotile	+ UICC Crocidolite
DNA	0.0	0.0	0.0	0.0	0.0
DNA + H_2O_2 (40mM)	0.0	0.0	2.52 ± 0.10	4.93 ± 0.18	8.29 ± 0.23
DNA + COOH (40mM)	0.0	0.0	0.0	0.0	0.0
DNA + BOOB (40mM)	0.0	0.0	0.0	0.0	0.0
DNA + t-BOOH (40mM)	0.0	0.0	0.0	0.0	0.0
DNA + ABIN (40mM)	0.0	0.0	1.71 ± 0.06	2.59 ± 1.0	3.42 ± 0.12

The data represent mean ± S.E. (n = 6).
Treatment of DNA samples with dusts in presence of peroxides and hydroperoxides followed by determination of TBA reacting species is described in the text.

IN VITRO DNA AND CHROMOSOME DAMAGE PRODUCED BY SOME MINERALS AND MAN MADE PARTICLES ON RAT PLEURAL MESOTHELIAL CELLS (RPMC) - MECHANISMS AND RELATIONSHIP WITH *IN VIVO* EXPERIMENTAL FINDINGS

M.C. Jaurand°, M. Yegles°, H.Y. Dong, A. Renier°, L. Saint Etienne°, L. Kheuang°, X. Janson* and J. Bignon°

° INSERM Unité 139
CHU Henri Mondor
94010. Créteil Cedex
France
et
* LEPI, DASES
11 rue George Eastman
75013 Paris
France

INTRODUCTION

It is now well established that neoplastic transformation results from progressive cell changes involving genes (Boyd and Barrett, 1990) ; it is then relevant to investigate the effects of particles on the cell structures carrying the genes.
Several *in vitro* assays can be used allowing the detection of effects of potential carcinogens at the DNA and chromosome level. To use these assays as predictive or mechanistic models, one important point is to determine whether differences exist between cell responses to carcinogenic/non carcinogenic particles and why ?

METHODS AND RESULTS

The model developed in the laboratory consisted in the determination of the *in vitro* response of RPMC (Jaurand et al., 1981) treated with different sorts of particles. The assays were selected because of their ability to give informations on the genotoxic and nongenotoxic effects of the tested agents (Jackson et al., 1993). We always studied at least 2 different cell strains, each obtained from 2 different rats. To relate the effects to the sample characteristics (number of fibres per unit weight, fibre dimensions), transmission electron microscopy (TEM) studies were carried out with each sample. For this purpose the fibres were suspended in the culture medium used for cell growth and maintenance, i.e. Ham' F10 supplemented with 10% fetal calf serum and dispersed by ultrasonic treatment

NATO ASI Series, Vol. H 85
Cellular and Molecular Effects
of Mineral and Synthetic Dusts and Fibres
Edited by J. M. G. Davis and M.-C. Jaurand

(Renier et al., 1991). The measurement of man made fibres (MMF) length was determined by scanning electron microscopy because their size exceeded those possibly determined by TEM.

In vitro tests consisted in the determination of DNA repair of damage using the unscheduled DNA synthesis assay (Renier et al., 1990), formation of numerical and structural chromosome abnormalities (Jaurand et al., 1986) and of abnormal anaphase/telophase (Yegles et al., 1993). *In vivo* tests unless specified consisted in the inoculation of 20 mg of fibres in the pleural cavity of Sprague Dawley rats maintained for all their life time and scored for histologically confirmed pleural mesothelioma. Data have been reported earlier (Monchaux et al., 1981, Jaurand et al., 1987, Van der Meeren et al., 1992).

Asbestos fibres (several samples of chrysotile, crocidolite and amosite) produced both UDS *in vitro* and mesothelioma *in vivo* but another fibre type, attapulgite, did not produce UDS nor mesothelioma. It must be emphasized that the same batches of samples were used in these *in vitro* and *in vivo* assays. DNA repair in asbestos-treated cells likely results from DNA damage as demonstrated by 3 results obtained with asbestos-treated cells : an activation of the poly ADP ribosyl polymerase, a strong decrease in the incorporation of labelled thymidine during exponential growth and a blockage of the cells in S phase (data not shown).
Recently, we had the opportunity to determine the effects of some MMF either organic (*p*-aramid) or mineral fibres (MMMF) : RCF 1, RCF 4 and MMVF 11. A non fibrous non processed p-aramid was also tested. RCC chrysotile and TiO_2 were used as controls. All the fibrous particles tested produced UDS while the non fibrous did not. The MMF samples have not been tested by intrapleural inoculation but by inhalation (Hesterberg et al., 1993 ; Lee et al., 1988). RCF 1 and a chrysotile used as control were carcinogenic. The groups treated with RCF 4 and MMVF 11 showed more tumors than the untreated control but the enhancement was not significant (Hesterberg et al., 1993). The carcinogenic potential of *p*-aramid is debated because some lung tumours, only detected in female rats, were cystic keratinizing squamous cell carcinoma (Lee et al., 1988). From these results, a good correlation appeared between positivity in the UDS assay and tumourigenicity by intrapleural inoculation but it was not evident when the inhalation results were taken into consideration.

We also tested these particles in the anaphase/telophase assay. Table 1 represents results obtained with different types of MMF and the RCC chrysotile control sample with mention of the lowest concentration producing a significant increase in the percentage of abnormal cells or the highest concentration tested (if there is no significant enhancement). No significant enhancement of anaphase/telophase abnormalities was observed with MMF in contrast with chrysotile (Table 2).

At first sight, no correlation between *in vitro/in vivo* findings appears since RCF 1 which is carcinogenic does not produce abnormal anaphases/telophases. However, a remark can be made : both the total number of fibres and the number of "Stanton" fibres (length > 8µm ; Ø ≤ 0.25 µm) are much higher with chrysotile than with RCF (1.7 millions versus around or less than 50 thousands).

When data obtained with asbestos and other fibres are taken into consideration (Table 2) it appears that all fibres but amosite and attapulgite produced a significant enhancement of anaphase/telophase abnormalities. The differences here cannot stand in the total number of particles which are reasonably similar ; however both amosite and attapulgite lacked "Stanton" fibres. With the other samples, the number of Stanton fibres exhibiting a significant effect was around five hundred thousands to 1 million of fibres/cm^2.

Thus the conclusions concerning this second type of test may suggest that the cell response in the anaphase/telophase assay is dependent on the number of "Stanton" fibres present in the sample. It can be assumed that *in vitro* responses can be observed only when the non observable effect level (NOEL) is exceeded ; this one should be determined for every test.

Table 3 reports data on aneuploidy obtained in experiments using asbestos, glass microbeads and TiO_2. It is observed that all asbestos samples produced aneuploid cells, in contrast with non fibrous particles. However, when RCF 1, RCF 4 and MMVF 11 were tested, no significant enhancement of aneuploid cells were detected at the concentrations of 2, 4 and 8 µg/cm^2. Aneuploid cells likely result from chromosome missegregation during earlier mitoses. Thus, the absence of aneuploid cells agrees with the absence of anaphases/telophases abnormalities. This result may be also related to the poor internalisation of the MMF samples by RPMC (data not shown).

To assess the relevance of *in vitro* tests in terms of the amount of particles applied to the cells, comparison with *in vivo* conditions can be made. The doses received by inhalation are difficult to compare to the concentrations applied *in vitro* because of the epuration processes. Nevertheless if we look at the intrapleural inoculation, we are in a very similar situation. In the intrapleural model, we inject the fibres in the pleural cavity, i.e. between the 2 mesothelial cells layers. In the alternative *in vitro* model, we deposit fibres at the surface of the mesothelial cells. One difference stands in the fact that *in vivo*, in contrast to *in vitro*, macrophages can take up the particles and reduce the local concentration. As mentioned above, *in vivo*, we use 20 mg of fibres ; if the pleural surface is about 200 cm^2 (probably much less), this corresponds to 100 $\mu g/cm^2$. *In vitro* we use about 1-10 $\mu g/cm^2$; this represent about 100/10 times less Stanton fibres in our *in vitro* model than in the intrapleural model.

DISCUSSION

The absence of effect exhibited by some carcinogenic MMF in *in vitro* cell tests might be due to the too low number of "efficient" fibres applied to the cells. To solve this problem it should be theoritically possible to use more fibres. However, this is inconsistent with our culture conditions because of the dimensions of MMF. Very long and thick fibres are observed ; to increase the amount would have given a carpet of fibres. This is not observed with asbestos.

We have looked at the fibre dimensions of MMF samples as well as chrysotile in the culture medium. The preparation method did not change the size distribution of the samples used in the aerosols in inhalation experiments (Hesterberg et al., 1993) and both geometric mean length (GML) and geometric mean diameter (GMD) are very similar (Table 4). However, the size of the fibres retained in the lung is different (Hesterberg et al., 1993) since both shorter and thinner fibres are found. This is due to the fibre selection by the airways that occurs with MMF but not with chrysotile because of the differences in the dimensions of the aerosolized fibres. It would have been better to use samples of these dimensions to study the cellular effects of the MMF.

Differences exist between the cell responses to fibres non carcinogenic and carcinogenic in rodent bioassays. The non carcinogenic fibres, as well as non fibrous particles, generally do not produce UDS enhancement or chromosome abnormalities. Carcinogenic fibres, especially asbestos, are generally active but the responses given by MMF are less clear-cut. Two types of investigations should be made to improve the relevance of the in vitro tests : to determine the fibre parameters directing the responses and establish the NOEL for each assay.

ACKNOWLEDGEMENTS

This work has been suppoerted by INSERM funds, MRE grant 89 D0889, EURIMA and AKZO/DUPONT grants.

REFERENCES

Boyd J A, Barrett J C (1990) Genetic and molecular basis of multistep carcinogenesis. Pharmacol Ther 46:469-486

Hesterberg T W, Miller W C, Mc Connell E E, Chevalier J, Hadley J G, Bernstein D M, Thevenaz P, Anderson R (1993) Chronic Inhalation Toxicity of Size-Separated Glass Fibers in Fischer-344 Rats. Fund Appl Toxicol 20:464-476

Jackson M A, Stack H F, Waters M D (1993) The Genetic Toxicology of Putative Nongenotoxic Carcinogens. Mutat Res 296:241-277

Jaurand M C, Bernaudin J F, Renier A, Kaplan H, Bignon J (1981) Rat pleural mesothelial cells in culture. In vitro 17:98-106

Jaurand M C, Fleury J, Monchaux G, Nebut M, Bignon J (1987) Pleural carcinogenic potency of mineral fibers (Asbestos, attapulgite) and their cytotoxicity on cultured cells. J Natl Cancer Inst 79:797-804

Jaurand M C, Kheuang L, Magne L, Bignon J (1986) Chromosomal changes induced by chrysotile fibres or benzo(3-4)pyrene in rat pleural mesothelial cells. Mutat Res 169:141-148

Lee K P, Kelly D P, O'Neal F O, Stadler J C, Kennedy G L Jr (1988) Lung response to ultra fine Kevlar aramid synthetic fibrils following 2-year inhalation exposure in rats. Fundam Appl Toxicol 11:1-20

Monchaux G, Bignon J, Jaurand M C, Lafuma J, Sebastien P, Masse R, Hirsch A, Goni J (1981) Mesotheliomas in rats following inoculation with acid-leached chrysotile asbestos and other mineral fibres. Carcinogenesis 2:229-236

Renier A, Levy F, Pilliere F, Jaurand M C (1990) Uncheduled DNA synthesis in rat pleural mesothelial cells treated with mineral fibres or benzo[a]pyrene. Mutat Res 241:361-367

Renier A, Yegles M, Buard A, Dong H Y, Kheuang L, Saint-Etienne L, Laurent Ph, Jaurand M C (1992) Use of mesothelial cell cultures to assess the carcinogenic potency of mineral or man made fibers. Cell Biol Toxicol 8:133-139

Van der Meeren A, Fleury J, Nebut M, Monchaux G, Janson X, Jaurand M C (1992) Mesothelioma in rats following intrapleural injection of chrysotile and phosphorylated chrysotile (chrysophosphate). Int J Cancer 50:937-942

Yegles M, Saint-Etienne L, Renier A, Janson X, Jaurand M C (1993) Induction of Metaphase and Anaphase/Telophase Abnormalities by Asbestos Fibers in Rat Pleural Mesothelial Cells Invitro. Amer J Respir Cell Mol Biol 9:186-191

Table 1 : ANAPHASE/TELOPHASE ABNORMALITIES IN RPMC TREATED WITH DIFFERENT SORTS OF MMF

Particle	Concentration*	Abnormal	Nb of particles/cm ^{2} x 10 ^{5}	
	µg/cm ^{2}	cells (%)**	Total	"Stanton"
RCF 1	10	5	0.78	< 0.17°
RCF 3	10	1	0.59	-
RCF 4	10	5	0.96	< 0.52°
MMVF 10	10	2	-	-
MMVF 11	8	6	0.80	< 0.21°
Chrysotile	<u>0.2</u>	9.8	17.00	17.00

* Lowest concentration producing a significant increase (underlined) in % of abnormal cells or highest concentration tested if no significant enhancement
** % in treated cells - % in control cells [Control cells : 9% - 1%]
° Nunber of fibres < 8 µm of length

Table 2 : ANAPHASE/TELOPHASE ABNORMALITIES IN RPMC TREATED WITH DIFFERENT SORTS OF FIBRES

Particle	Concentration* $\mu g/cm^2$	Abnormal cells (%)**	Nb of particles/cm^2 x10^6 Total	Stanton
UICC crocidolite	<u>7.0</u>	16.0	37.0	0.5
UICC amosite	8.0	3.0	7.2	0.0
UICC chrysotile	<u>0.5</u>	21.3	11.0	0.3
Canadian chrysotile	<u>1.0</u>	8.6	34.5	1.5
RCC chrysotile	<u>0.2</u>	9.8	1.7	1.7
p-aramid	<u>4.0</u>	8.3	9.6	0.2
Attapulgite	8.0	0.0	25.6	0.0

* Lowest concentration producing a significant increase (underlined) in % of abnormal cells or highest concentration tested if no significant enhancement

** % in treated cells - % in control cells [Control cells : 10% - 2% ; μ =3±2, n =16]

Table 3 : ANEUPLOIDY IN RPMC TREATED WITH DIFFERENT SORTS OF PARTICLES

Particle	Concentrations tested µg/cm 2	Aneuploid cells (%treat - %ctrl)*	Nb of total particles**x10 6
Crocidolite	1 2 <u>4</u> 6	36	21
Canadian chrysotile	<u>.4</u> 1 2 3	14	7.6
Chrysophosphate	.4 <u>1</u> 2 3	28	17
Glass µbeads [SER, 3%]	2 5 10	0.3	0.7
Glass µbeads [GH, 12.7%]	2 5 10	0.1	0 .6
Glass µbeads [C3, 0.22%]	2 5 10	0.0	0 .8
Titanium dioxide	2 5 10	2	22,000

* Control cells : 80% - 60% metaphases n = 42 ; µ = 14±15, n = 15
** Nb of particles corresponding to the lowest of the concentration (underlined) poducing a significant increase of aneuploid cells (n≠42) when relevant or corresponding to the highest concentration tested

Table 4 : FIBRE DIMENSIONS [μm] OF MMF SAMPLES IN DIFFERENT CONDITIONS
[GML (D) : geometic mean length (diameter)]

FIBRE	AEROSOL*	IN THE RAT LUNG**	IN THE CULTURE MEDIUM*°
	GML/GMD	GML/GMD	GML/GMD
RCF 1	15.9/0.82	7.28/0.44	14.6/1.0
MMVF 11	13.7/0.69	6.89/0.45	13.1/0.9
Chrysotile	1.2/0.08	1.43/0.45	2.6/<0.2

* From Hesterberg et al., Fund. App. Toxicol, 1993, 20, 464-476
** 3 months exposure (no recovery). RCF : 30 mg per cubic meter ; chrysotile : 10 mg per cubic meter
*° Prior to application to RPMC in culture

EVALUATION OF PROINFLAMMATORY CYTOKINE EXPRESSION IN ALVEOLAR MACROPHAGES AND IN THE LUNG OF PNEUMOCONIOTIC PATIENTS (1).

Vanhée D, Gosset P, Wallaert B, Voisin C, Tonnel AB.

INSERM-CJF 90-06 and Laboratoire de Pathologie Respiratoire Expérimentale et de Pollution Atmosphèrique, Institut Pasteur,

BP 245,

F-59019 Lille, France.

SUMMARY

In order to confirm their *in vivo* participation in the pathogenesis of Coal Workers' Pneumoconiosis (CWP), TNF and IL-6 secretion and expression were evaluated in bronchoalveolar lavage (BAL) and lung tissue from pneumoconiotic patients. BAL was collected from 19 simple pneumoconiosis (SP) and 8 progressive massive fibrosis (PMF), while lung parenchyma specimen were obtained from 1 SP and 3 PMF. TNF and IL-6 mRNA expression was assessed using *in situ* hybridization. To identify the cellular source of these monokines, immunocytochemistry was performed on lung tissue. TNF and IL-6 concentrations were quantified using bioassay. *In situ* hybridization was performed using ^{35}S-labelled RNA probes encoding for TNF and IL-6.

The number of positive BAL cells was significantly higher for patients with PMF (TNF : 52% ± 2.8; IL-6 : 45% ± 4.8) than in those with SP (TNF : 30% ± 4.4; IL-6 : 27% ± 3) or control subjects (TNF : 12.4% ± 1; IL-6 : 13.1% ± 1.6). The percentage of positive BAL cells was correlated with the level of biologically active cytokines present in AM-supernatants after 3hr of culture. In lung biopsies, hybridization signals were only observed in patients with PMF. TNF and IL-6 mRNA expression was closely associated with the presence of coal dust, and was limited to lung macrophages; IL-6 mRNA was detected in

[1] This work was supported in part by a grant n° 7280-03-018 from the Commission of European Communities.

NATO ASI Series, Vol. H 85
Cellular and Molecular Effects
of Mineral and Synthetic Dusts and Fibres
Edited by J. M. G. Davis and M.-C. Jaurand

mononuclear phagocytes, but also in other cell types like endothelial cells. Synthesis of the related protein was confirmed by immunohistochemistry for both TNF and IL-6. These data confirm that TNF and IL-6 production is increased in the lung from patients with CWP. Moreover TNF and IL-6 mRNA expression was associated with the presence of coal mine dust particles suggesting a direct role of mineral particles in the cytokine production and thus in the development of pneumoconiotic lesions.

INTRODUCTION

Coal worker's pneumoconiosis (CWP) is one of the most widespread occupational lung diseases in the North of France : this disease is usually divided into two groups : simple pneumoconiosis (SP) and progressive massive fibrosis (PMF)(Cochrane 1973, Naeye 1972).

Although pathophysiology of pneumoconiosis remains uncompletely understood, several lines of evidence suggest that alveolar macrophages (AM) participate at least in the initiation of the alveolitis (Lugano 1982). Phagocytosis of mineral dust by human or murine AM induces the secretion of cytokines which might be relevant to the development of interstitial lung disorders (Gritter 1986, Driscoll 1990). Bronchoalveolar lavage (BAL) studies show an influx of mononuclear phagocytes in the lower respiratory tract of humans exposed to chronic inhalation of inorganic dusts. In addition AM from pneumoconiotic patients are known to spontaneously produce large amounts of oxidants, neutrophil chemotactic factor, fibronectin and monokines (Lassalle 1990, Wallaert 1990).

At the present time, the relationship between the severity of CWP and the composition of inhaled mine dust remains a subject of controversy. Some pathological studies showed the absence or only the presence of small quantities of silica in the lung of coal miners with PMF (Naeye 1972). We recently demonstrated that coal dust was able to induce the release of alveolar macrophage-derived mediators whereas silica alone at the same concentration had no effect suggesting that AM-secretion could be related to the combined action of different components of coal mine dust (Gosset 1991).

Although *in vitro* exposure of AM to coal mine dust has been shown to induce the production of TNF and IL-6 (Gosset 1991) which may potentially be involved in the development of the fibrotic process, relationships between the presence of coal mine dust and monokine synthesis have not yet been definitively established in coal miners. In a mouse model, Piguet *et al* (1990) demonstrated the direct role of TNF in the development of silica-induced fibrosis. So, the aim of this study was to determine a possible relationship between the presence of coal dust and monokine synthesis in the lung from pneumoconiotic patients and to extend data concerning cytokine production, which were only performed on AM, to the lung tissue itself. For this, TNF and IL-6 production and mRNA expression was assessed either on AM collected by BAL or on tissue lung fragments, concomitantly to the optical detection of coal dust particles.

MATERIALS AND METHODS

Patients

Twenty seven coal miners were studied : 19 patients had SP and 8 had PMF. Ten control subjects not exposed to coal dust were also selected. For histological analysis, lung tissue fragments were obtained from lobectomy performed either on patients presenting a lung cancer but without CWP (n=2), or on pneumoconiotic patients with neoplasia (n=4, 1 with SP and 3 with PMF). Biopsies were taken as far as possible of the neoplasic area. Oral informed consent was obtained for BAL from both pneumoconiotic patients and normal subjects. None of the patients and controls was under corticosteroids at the time of BAL or surgery.

Cell isolation, alveolar macrophage culture and lung tissue section preparation.

The lavage fluid was filtered through sterile surgical gauze and centrifuged at 400g for 10min at 4°C. After three washings, the pellet was resuspended at a cell concentration of 1.5 10^6/ml in RPMI 1640 containing 5% of heat inactivated fetal calf serum and 2mM of L-Glutamine (Gibco, Coubevoie, France). Cytospins were prepared on TESPA-coated slides (3-Aminopropyl-triethoxysilane; Aldrich), fixed in 4% paraformaldehyde in PBS, dehydrated through graded alcohols and stored at -20°C. For culture, cells were allowed to adhere to plastic petri dishes (2ml in 35mm diameter well) for 2hr at 37°C. The non-adherent cells were removed by three washings with RPMI. Adherent cells contained more than 95% AM and less than 1% lymphocytes. The viability of the AM was evaluated by trypan blue exclusion test and was never less than 90%. AM supernatants were collected after 3 hours of culture.

Lung tissue obtained after surgical resection was cooled in liquid nitrogen. Cryostat sections (6µm) were cut from the biopsies on TESPA-coated slides and then treated as the cytospins.

Tumor Necrosis Factor α assay

TNF concentration was evaluated by L929 cytotoxic test. Biological activity was estimated from a standard curve as the amount necessary to kill 50% of actinomycin-treated L929 mouse fibroblasts after 18hr culture. TNF concentration was expressed in U/ml.

Interleukin-6 assay

IL-6 concentration was assayed as previously described (Gosset 1991). Briefly the hybridoma cells 7TD1 (a generous gift of Dr Van Snick, Institut Ludwig, Brussels, Belgium) was incubated with serial dilution (1/3 to 1/2187) of cell supernatants in microtiter plates. After four days of incubation, the dilution corresponding to 50% of maximun proliferation was evaluated by MTT colorimetric method. IL-6 concentration was expressed in U/ml.

Probes

A cDNA for TNF (0.3Kb fragment length) was inserted into a pBluescript vector, linearized with Pst 1 or Eco R1 to produce antisense or sense probes respectively before transcription. The IL-6 probe (a generous gift of Dr Emilie D, INSERM U131, Clamart, France) corresponding to the Xba 1-Eco R1 517 bp long fragment inserted into pBluescript vector was linearized with Xba 1 or Eco R1. Transcription was performed in the presence of ^{35}S-CTP and the appropriate T3 or T7 RNA polymerases for antisense and sense riboprobes.

In situ hybridization

In situ hybridization was performed as previously described (Queva 1992). Briefly, cells and tissue sections were permeabilized with proteinase K, and non-specific binding of ^{35}S was inhibited by triethanolamine and acetic anhydride. Antisense or sense probes (10^6cpm/slide) in hybridization buffer containing dithiothreitol and formamide were heated to 80°C for 2min before hybridization (18hr at 55°C). During high stringency (2-0.1X SSC at 60°C) washings, unhybridized single-stranded RNA was removed by RNase-A. Slides after dipping in NTB2 nuclear track emulsion (Kodak) were exposed for 10 days. Autoradiography was followed by Hoechst staining. Hoechst molecule intercalated in the DNA molecule and under UV light led to a blue coloration of nucleus. Microscopic observations were made in dark field : silver grains and dust particles appeared birefringent. The variation of focus level allowed the distinction between dust particles and hybridization signals. Background signal is determined by the number of silver grains detected into emulsion in absence of cells, non-specific signals are defined as the number of silver grains detected over the preparation which was hybridized with the sense probes. This observation lead to the definition of positivity threshold for hybridization with the antisense probes.

Immunohistochemistry

Immunoreactive TNF and IL-6 were identified in cryostat sections with rabbit polyclonal sera to recombinant human cytokines. Anti-TNF immune serum was obtained from Genzyme and anti-IL-6 immune serum was a gift from Dr Wietzerbin J (INSERM U196, Institut

Curie, Paris, France). To demonstrate antibody specificity, immunostaining for both TNF and IL-6 were completely inhibited by the addition of the exogenous recombinant molecule.

RESULTS

Expression of TNF and IL-6 mRNA by cells isolated from BAL

In situ hybridization technique was used to determine the percentage of cells expressing TNF and IL-6 mRNA on cytospins of the total BAL cell population. The separation of the two groups of patients with CWP showed an overexpression of mRNA in cells from patients with PMF (TNF : 52% ± 2.8; IL-6 : 45% ± 4.8; m±SEM) when compared with those with SP (TNF : 30% ± 4.4; IL-6 : 27% ± 3) and control subjects (TNF : 12.4% ± 1; IL-6 : 13.1% ± 1.6)(Fig 1). In addition the percentage of positive cells for both TNF and IL-6 mRNA was significantly correlated ($r=0.88$, $p<10^{-4}$)(data not shown).

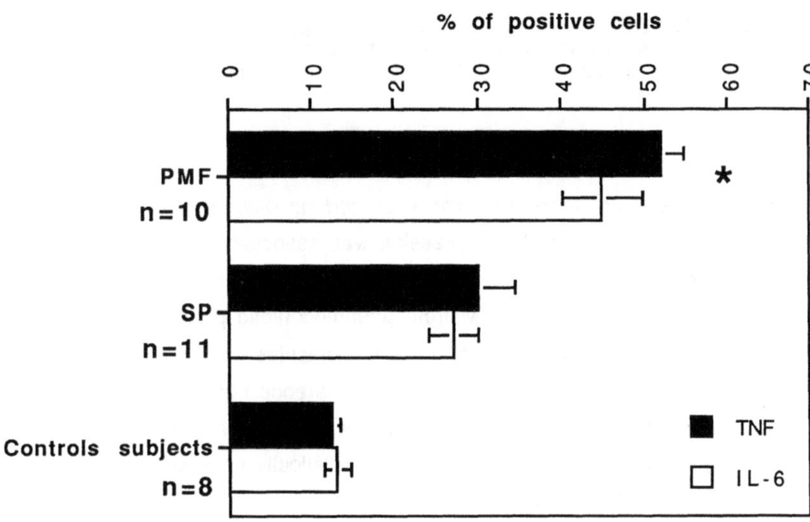

Figure 1 : Spontaneous expression of TNF and IL-6 mRNA in BAL macrophages from control subjects and pneumoconiotic patients with either SP or PMF (m±SEM). Preparations were evaluated in parallel and the number of messenger RNA positive cells was expressed as the percentage of total BAL cells (as described in Materials and Methods). * $p<0.01$ compared to both control and SP groups. Data from the SP group were not statistically different from the data from controls.

Correlation between the presence of coal mine dust particles and the expression of TNF mRNA.

Using this technique of *in situ* hybridization it was possible to determine the correlation between the mRNA expression and the presence of coal dust particles. The number of positive alveolar macrophages increased with the number of particles per cell. This observation was true for the two groups of patients, although an higher percentage was detectable in PMF compared to SP (Table 1). Similar results were obtained for the expression of IL-6 mRNA (data not shown).

Table 1 : Correlation between the presence of intracellular coal mine dust particles and the expression of TNF mRNA.

Particles per cells	0	1 - 5	6 - 20	>20
Patient with SP				
percentage of positive cells	13	17	23	34
Patient with PMF				
percentage of positive cells	25	25	49	82

Counts were performed by studying 2 cytospin preparations (200 cells per slide) in 3 cases from each group of patients (SP, PMF). We present one representative case from each group.

Correlation between messenger expression and monokine secretion

To elucidate if TNF and IL-6 mRNA expression was associated with an enhanced protein secretion, AM from pneumoconiotic patients were cultured and monokine release was evaluated after 3hr culture (Fig 2). AM from pneumoconiotic patients with PMF secreted significantly higher amounts of biologically active molecules compared to AM from control subjects or from patients with SP. There was no difference between patients with SP and control subjects. Furthermore, levels of TNF and IL-6 in supernatants and the number of cells expressing mRNA for both cytokines were statistically correlated ($r=0.71$, $p<0.05$ for TNF; $r=0.62$, $p<0.018$ for IL-6).

In situ hybridization in normal and pneumoconiotic lung

In situ hybridization using TNF and IL-6 RNA probes was performed in pneumoconiotic patients (n=4) and in 2 patients with lung cancer without any previous exposure to coal dust. Sections of open lung biopsies from control subjects or pneumoconiotic patients with SP, hybridized with the antisense probes did not show any positive signal when compared to those hybridized with the sense probes. When lung tissue sections from pneumoconiotic patients with PMF were examined, the TNF transcript was always detected in relation with an intracellular deposition of coal dust. The TNF mRNA expressing cells were localized in the alveoli (particularly inside AM), in the interstitial connective tissue, and in perivascular areas where an abundant inflammatory infiltrate was detected in relation with coal dust particle deposition. For IL-6 mRNA expression, hybridization signals were observed with the same localization. Messengers for IL-6 could be also detected in the vascular walls, and as testified by coloration of an adjacent section, in endothelial cells in the fibroconiotic areas, a positive signal for IL-6 was exclusively observed at the edge of the fibrotic span in relation with mineral dust deposition, but no transcript for TNF was detected.

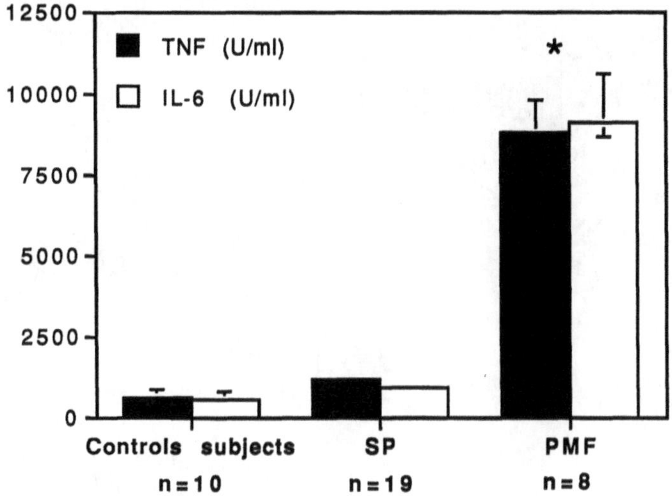

Figure 2 : Spontaneous production of TNF and IL-6 by human AM (m±SEM). Supernatants were collected after 3hr of culture. * p<0.01 compared to both control and SP groups. Data from SP were not statistically different from data from controls.

To confirm that mRNA expression was followed by the synthesis of the related protein, immunohistochemistry was used to localize the presence of TNF and IL-6 antigen. In lung tissue, we detected the presence of antigenic TNF and IL-6 that was predominantly observed into lung macrophages in relation with the presence of intracellular coal mine dust. The synthesis of IL-6 was also confirmed in the same sites as observed by *in situ* hybridization (*ie.* in alveolar macrophages, endothelial cells, fibroblasts).

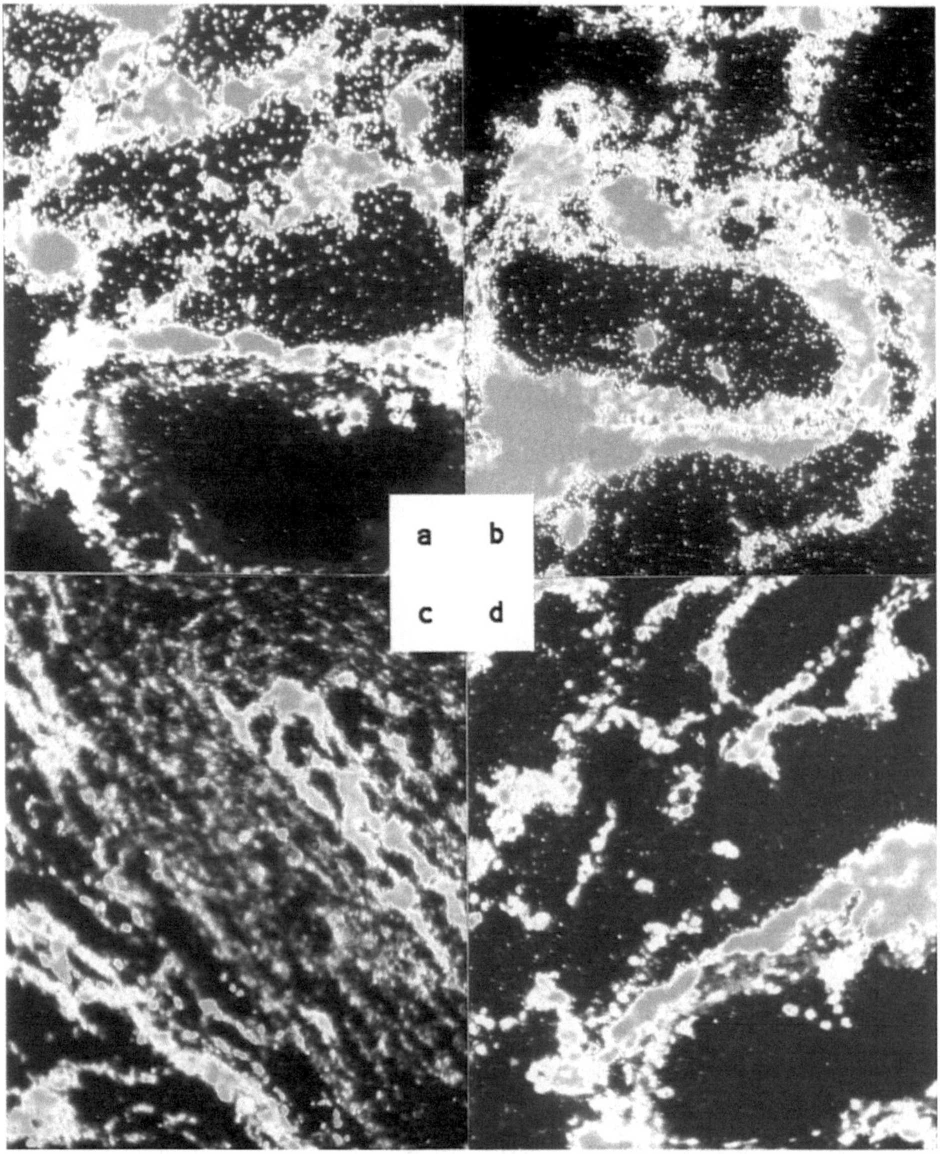

Figure 3 : Cryostat sections showing TNF and IL-6 mRNA localization in tissue from pneumoconiotic patients with PMF. a-b : perivascular zone of dust deposition (a : section hybridized with TNF anti-sense probe. b : section hybridized with the IL-6 anti-sense probe). c : fibroconiotic area (sections hybridized with the IL-6 antisense probes. d : section hybridized with TNF sense probe.

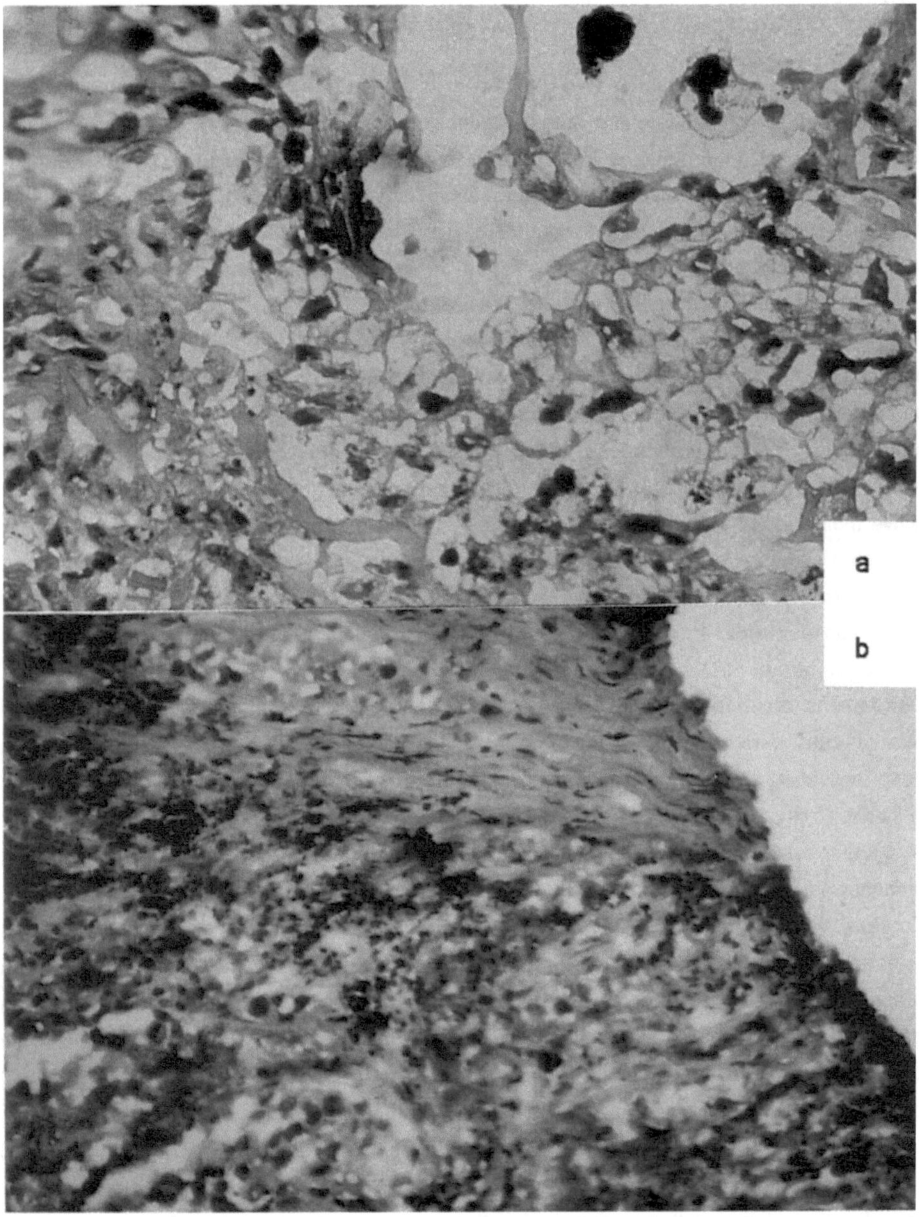

Figure 4 : Localization of TNF and IL-6 immunoreactive molecules by anti-TNF (a; x200) or IL-6 (b; x100) antibodies in lung biopsies from pneumoconiotic patients.

DISCUSSION

In this study, the reality of an *in vivo* production of cytokines, TNFα and IL-6 was evidenced by both *in situ* hybridization and immunostaining methods. In previous works we already showed that BAL macrophages from patients with CWP were activated to release various mediators such as superoxide anion, fibronectin, and a chemotactic factor for neutrophils (Wallaert 1987). In this study we confirm our previous work where we reported that AM from coal miners and patients with CWP had an increased spontaneous secretion of TNF compared to unexposed control subjects (Lassalle 1990).

The present study evidenced that AM recovered by BAL from control subjects as well as from pneumoconiotic patients spontaneously expressed mRNA for TNF and IL-6, and that the percentage of positive macrophages was significantly higher in BAL from pneumoconiotic patientswith PMF in comparison with SP and control subjects. This increased mRNA expression was associated with an increased secretion of the final products : biologically active molecules of TNF and IL-6. Nevertheless there was a certain discrepancy between AM monokine secretion and mRNA expression in the group of patients with SP. The number of mRNA positive cells was only increased 2-fold between patients with SP and PMF, while there was a 10-fold increase in TNF and IL-6 secretion suggesting a difference in the catabolism rate of mRNA between both groups of patients.

Correlations observed between the cellular content in coal dust and the percentage of positive cells were consistent with the observation by Gosset *et al* (1991) demonstrating a direct activation of AM from control subjects after coal mine dust exposure. However all the cells expressing mRNA for cytokines did not contain coal mine dust particles, suggesting that this might result from a paracrine stimulation elicited by TNF produced by neighbouring mononuclear phagocytes.

To study the monokine production at the level of lung parenchyma, *in situ* hybridization was performed on lung sections derived from pneumoconiotic patients and control subjects. The presence of positive hybridization signals were only observed in lung parenchyma from patients with PMF, mRNA expression for monokines being more closely associated with the presence of coal dust than in macrophages collected by BAL. The mRNA expressing cells could be considered as lung macrophages as determined morphologically. Cells expressing mRNA for TNF were also positive for IL-6, but other cell types such as endothelial cells and fibroblasts appeared positive for IL-6. The endothelial cell activation, here shown by IL-6 mRNA expression and the presence of antigenic IL-6, could result from TNF secretion by surrounding macrophages. This was suggested by the effective synthesis of monokines evidenced by the recognition of the related protein by anti-TNF or anti-IL-6 antibodies. However, processes leading to the increased and sustained TNF and IL-6 production, particularly in PMF, remain unclear.

The results, presented here, confirm the macrophage activation and the resulting TNF and IL-6 secretion in pneumoconiosis. The implication of TNF at least in the inflammatory and fibrotic process is suggested by several evidences. TNF induces the recruitment of different inflammatory cells : lymphocytes, eosinophils, and neutrophils putatively involved in interstitial lung diseases. The mechanisms leading to this cell infiltration can be sustained by TNF that is able to induce a chemotactic factor production (like IL-8) (Rolfe 1991) or an enhanced adhesion molecule expression on the vascular endothelium (Springer 1990). The involvement of TNF in lung fibrotic process has been demonstrated in an animal model (Piguet 1990). Monokines, such as TNF are known to activate or stimulate fibroblast growth *in vitro*, directly or by induction of growth factors potentially active in the fibroblast proliferation (PDGF, IGF-1, TGFβ)(Kovacs 1991).

A role of IL-6 in pneumoconiotic disorder has been suspected after the observation of its secretion by AM exposed to coal mine dust (Gosset 1991). IL-6 can be implicated in inflammatory processes by its ability to induce cellular adhesion molecule on monocyte which facilitates its infiltration into the lung (Duits 1991). The demonstration of IL-6 secretion in the pneumoconiotic lung might explain the frequent association with autoimmune diseases as well as the hypergammaglobulinemia observed in CWP (Pearson 1981).

Finally, the positive *in situ* hybridization of lung tissue with TNF and IL-6 probes and the presence of immunoreactive molecules of TNF and IL-6 confirm the direct role of these cytokines in the pathogenesis of CWP. But exact mechanisms by which the expression of monokines is exaggerated in PMF *versus* SP remain debatable.

REFERENCES

Cochrane AL. 1973. Relation between radiographic categories of pneumoconiosis and expectation of life. *Br Med J.* 2 : 532-534.

Driscol KE, Linderschmidt RC, Maurer JK, Higgins JM, Ridder G. 1990. Pulmonary response to silica or titanium dioxide : inflammatory cells, alveolar macrophage-derived cytokines, and histopathology. *Am J Respir Cell Mol Biol.* 2 : 381-390.

Duits AJ, Jainandunsing SM, Van de Winkel GJ, Capel PJA. 1991. Selective enhancement of Leu-CAM expression by interleukin-6 during differentiation of human promocytic U937 cells. *Scan J Immunol.* 33 : 151-159.

Gosset P, Lassalle P, Vanhée D, Wallaert B, Aerts C, Voisin C, Tonnel AB. 1991. Production of Tumor Necrosis Factor alpha and Interleukin-6 by human alveolar macrophages exposed to *in vitro* coal mine dust. *Am J Respir Cell Mol Biol.* 5 : 431-436.

Gritter HL, Adamson IYR, King GM. 1986. Modulation of fibroblast activity by normal and silica exposed alveolar macrophages. *J Pathol.* 148 : 263-271.

Kovacs EJ. 1991. Fibrogenic cytokines : the role of immune mediators in the development of scar tissue. *Immunol Today.* 12 : 17-23.

Lassalle P, Gosset P, Aerts C, Fournier E, Lafitte JJ, Degreeef JM, Wallaert B, Tonnel AB, Voisin C. 1990. Abnormal secretion of Interleukin-1 and Tumor Necrosis Factor alpha by alveolar macrophages in coal worker pneumoconiosis : comparison between simple pneumoconiosis and progressive massive fibrosis. *Exp Lung Res.* 16 : 73-80.

Lugano EM, Dauber JH, Daniele RP. 1982. Acute experimental silicosis. Lung morphology, histology and macrophage chemotaxis secretion. *Am J Pathol.* 109 : 27-36.

Naeye RL. 1972. Types of fibrosis in coal worker pneumoconiosis. *Ann NY Acad Sci.* 200 : 381-400.

Pearson DJ, Mentnech MS, Elliott JA, Taylor G, Major PC. 1981. Serologic changes in pneumoconiosis and progressive massive fibrosis. *Am Rev Respir Dis.* 124 : 696-699.

Piguet PF, Collart MA, Grau CE, Sappinot AP, Vassali P. 1990. Requirement of Tumor Necrosis Factor for development of silica-induced pulmonary fibrosis. *Nature.* 344 : 245-247.

Queva C, Ness SA, Graf T, Vandenbunder B, Stehelin D. 1992. Expression patterns of c-myb and of v-myb induced myeloid-1 (mim 1) gene during the development of the chick embrio. *Development.* 114 : 125-133.

Rolfe MW, Kunkel SL, Standiford TJ, Chensue SW, Allen RM, Evanoff HL, Phan SH, Strieter RM. 1991. Pulmonary fibroblast expression of Interleukin-8 : a model for alveolar macrophage-derived cytokine networking. *Am J Respir Cell Mol Biol.* 5 : 493-501.

Springer TA. Adhesion receptors of the immune system. *Nature.* 1990. 346 : 425-434.

Wallaert B, Bart F, Aerts C, Ouaissi MA, Meyrink JL, Voisin C. 1987. Sécrétion spontanée d'anion superoxyde, de facteurs chimiotactiques des polynucléaires neutrophiles et de fibronectine par les macrophages alvéolaires de patients pneumoconiotiques. In "Silicose et pneumoconiose à poussières mixtes". Monographie INSERM. 155 : 115-121.

Wallaert B, Lassalle P, Fortin F, Aerts C, Bart F, Fournier E, Voisin C. 1990. Superoxide anion generation by inflammatory cells in simple pneumoconiosis and in progressive massive fibrosis of nonsmoking coal workers. *Am Rev Respir Dis.* 141 : 129-133.

ACTIVATION OF EARLY CELLULAR RESPONSES BY ASBESTOS: INDUCTION OF c-*FOS* and c-*JUN* PROTOONCOGENE EXPRESSION IN RAT PLEURAL MESOTHELIAL CELLS.

Yvonne Janssen, Joanne Marsh, Timothy Quinlan, Cindy Timblin, Kelly BéruBé Lois Jimenez, Hua Fung, Ping Taishi, Christine Zanella, Nicholas Heintz, and Brooke Mossman.

Department of Pathology
University of Vermont, College of Medicine
Medical Alumni Building
Burlington, VT 05405 U.S.A.

Introduction

Asbestos fibers belong to a family of ubiquitous mineral silicates. Two types of malignancies have been documented in respiratory tract and pleura following inhalation of asbestos. Bronchogenic carcinoma, arising in the tracheobronchial epithelium, is a tumor found in individuals exposed occupationally to asbestos. Since this tumor is rarely found in non-smoking asbestos workers, asbestos is thought to act mainly as a co-carcinogen or tumor promoter in the development of this disease (Mossman et al. 1990). However, the development of malignant mesothelioma in the pleura is not associated with cigarette smoking. The generally accepted view is that amphibole asbestos acts as a complete carcinogen in the development of mesothelioma (Mossman et al.1990).

Bronchogenic carcinoma and mesothelioma arise after long latency periods. For instance, bronchogenic carcinoma typically develops on an average of 20-30 years after initial exposure to asbestos, whereas mesothelioma can arise 30-40 years after initial exposure to asbestos (reviewed in Mossman and Gee, 1989).

NATO ASI Series, Vol. H 85
Cellular and Molecular Effects
of Mineral and Synthetic Dusts and Fibres
Edited by J. M. G. Davis and M.-C. Jaurand

The mechanisms involved in asbestos-induced cancers are poorly understood. In order to study these mechanisms our laboratory has focused on the early events induced by asbestos in progenitor cells of disease, i.e tracheobronchial epithelial cells or mesothelial cells, which can give rise to bronchogenic carcinoma or mesothelioma respectively. Knowledge of the early processes triggered in these cells by asbestos may provide insight in those events which are important in the causation of these diseases. Thus far we have focused on early proliferative alterations in hamster tracheal epithelial cells and rat pleural mesothelial cells as chronic cell division is associated causally with many human tumors (Preston-Martin et al., 1990) triggered by asbestos exposure. We have focused on c-*fos* and c-*jun* protooncogenes, since these genes are involved in cell proliferation and transformation (Angel and Karin, 1991). The fos/jun gene family encodes proteins that dimerize to form the transcription factor, AP-1. Fos/*jun* heterodimers or *jun*/*jun* homodimers then bind to DNA at the AP-1 sequence and trigger transcription of a number of AP-1 dependent genes. Determination of gene expression of c-*fos* and c-*jun* by Northern Blot analysis and DNA binding activity of cellular extracts (containing c-*fos* and c-*jun* proteins) to the AP-1 consensus sequence has provided important insights in early events which may be important in asbestos-induced disease formation in a model involving chronic cell proliferation (Heintz et al, 1993). Here we present data regarding c-*fos* and c-*jun* mRNA levels and AP-1 DNA binding activity in rat pleural mesothelial cells (RPM) after exposure to asbestos or related mineral dusts. The importance of active oxygen species (AOS) and cell signalling cascades in asbestos-induced protooncogene expression are discussed.

Materials and Methods

Mineral dusts

Reference samples of National Institute of Environmental Health Sciences (NIEHS) crocidolite or chrysotile (Jeffrey Mines, Quebec), obtained from the Thermal Insulation Manufacturers Association (TIMA) Fiber Repository

(Littleton, CO) were used in studies described here. Polystyrene Beads (average diameter 1.05 µm, Polyscience, Warrington, PA) were used as negative control particles.

Cell culture

Rat pleural mesothelial cells (RPM) were isolated from the parietal pleura of Fischer 344 rats and propagated in cultures for 4-6 passages as described elsewhere (Jaurand et al, 1981). Cells were grown to confluence in 10% fetal bovine serum containing medium. Twenty-four hours prior to addition of asbestos or polystyrene beads, cells were switched to 2% serum-containing medium. Asbestos fibers were triturated 8 times through a 18 gauge needle and added to RPM cultures at increasing concentrations. In separate experiments, the protein synthesis inhibitor, cycloheximide, was added to cultures in the presence or absence of asbestos to determine the role of protein synthesis in protooncogene induction by asbestos.

Northern blot analysis

After selected time periods of exposure to asbestos fibers, RPM were harvested for Northern blot analysis as described previously (Shull et al., 1991). Total RNA was isolated, separated on 3-[N-morpholino]-propanesulfonic acid-formaldelyde gels, transferred onto nitrocellulose filters. Blots were probed with [^{32}P]-labeled cDNAs encoding *c-fos* or *c-jun* (generously provided by Richard Gaynor, University of California, Los Angeles). Hybridization signals were visualized by exposure to Kodak-X-Omat film and quantitated using a Betascope Blot analyzer (Betagen Corp., Waltham MA).

Gel mobility shift assays

Whole cell extracts were prepared from RPM after 4, 8, or 24 hours of exposure to asbestos as described previously (Heintz et al, 1993). A double stranded oligonucleotide containing an AP-1 consensus sequence was end-labeled with [^{32}P], incubated with cell extracts, and electrophoresed in a polyacrylamide gel. Gels were dried and exposed to Kodak-X-Omat film at -80° C, and examined for a shift in the electrophoretic mobility, representing binding of the AP-1 protein complex to the consensus sequence. Results were quantitated using a Betascope blot analyzer (Betagen Corp., Waltham, MA) and expressed in counts per minute (cpm).

Results

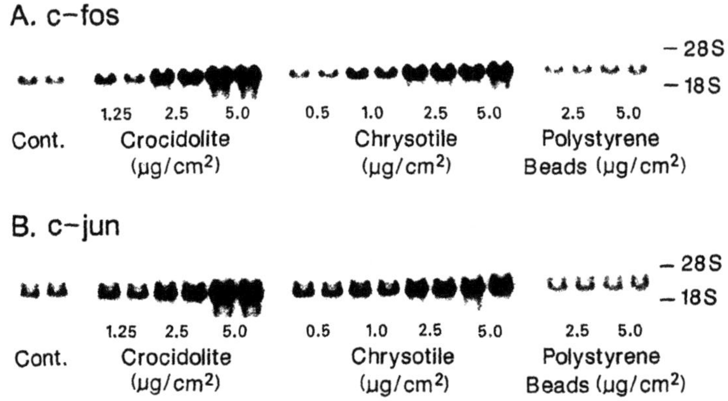

Figure 1. Northern blot analysis of c-*fos* and c-*jun* mRNAs in rat pleural mesothelial cells (RPM). RPM cells were exposed to increasing amounts of NIEHS crocidolite or chrysotile asbestos or polystyrene beads for 8 hours. RNA was extracted and prepared for Nothern blot analysis of c-*fos* or c-*jun* mRNA levels. Hybridization signals were visualized by exposure to Kodak-X-Omat film.

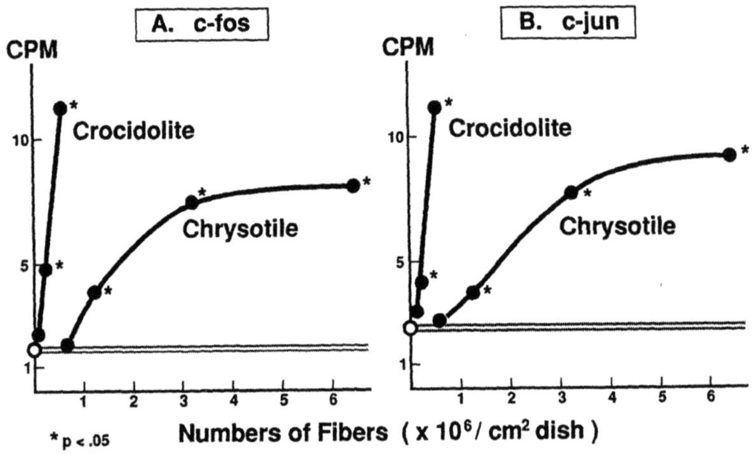

Figure 2. Quantitation of Northern blots examining c-*fos* and c-*jun* mRNA in RPM cells exposed to crocidolite or chrysotile asbestos. Hybridization signals were quantitated using a Betascope blot analyzer and expressed in counts per minute (cpm). Fiber numbers of crocidolite and chrysotile asbestos in each exposure group are indicated.

Figure 1 shows the results from a Northern blot study examining gene expression of c-*fos* and c-*jun* protooncogenes in RPM cells after exposure to asbestos or polystyrene particles. As can be seen, addition of increasing amounts of crocidolite or chrysotile asbestos, but not polystyrene beads, leads to dose-dependent increases in steady state mRNA levels of these protooncogenes. Although results here were obtained at 8 hours after exposure of cells to asbestos, similar findings are observed after 24 hours of exposure (Heintz et al., 1993). When evaluated on a fiber mass basis, crocidolite and chrysotile appear equivalent in upregulating c-*fos* or c-*jun* gene expression (figure 1). However, on an equal fiber number basis after characterzation of fibers (Heintz et al, 1993), crocidolite is far more potent in increasing mRNA levels of c-*fos* and c-*jun* protooncogenes as compared to chrysotile. These striking differences are visualized in figure 2. As shown in figure 3, asbestos increased c-*fos* and c-*jun* mRNA levels in RPM cells in the presence of the protein synthesis inhibitor, cycloheximide. This indicates that the effect of asbestos on c-*fos* and c-*jun* mRNA is direct at the level of gene tran-

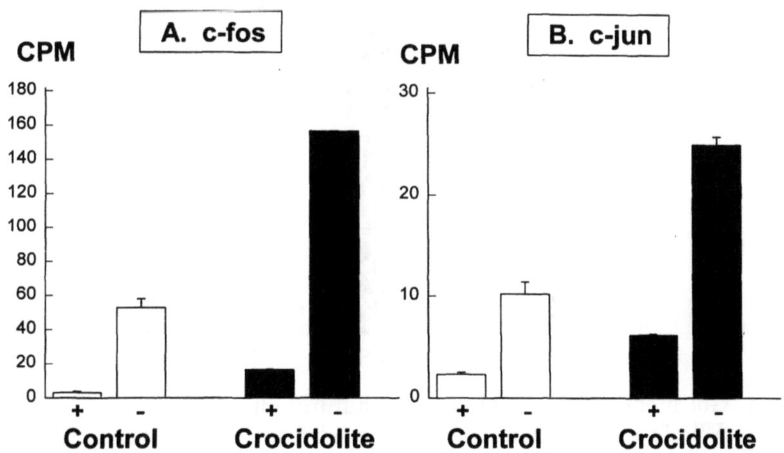

Figure 3. Northern blot analysis of c-*fos* and c-*jun* mRNA levels in RPM cells. RPM cells were exposed to 2.5 $\mu g/cm^2$ of crocidolite asbestos for 8 hours in the presence or absence of the protein synthesis inhibitor, cycloheximide (5µM). RNA was extracted, electrophoresed, blotted and hybridized with cDNAs encoding c-*fos* or c-*jun*. Blots were quantitated using a betascope blot analyzer and results expressed in cpm.

scription and does not require synthesis of intermediate proteins prior to induction of these protooncogenes. In order to determine whether increases in c-*fos* and c-*jun* mRNA levels were accompanied by increases in AP-1 DNA binding activity, gel shift analyses employing an oligonucleotide containing the AP-1 consensus sequence were performed. As shown in figure 4, asbestos causes increases in DNA binding to the AP-1 sequence in RPM cells, suggesting that upregulation of c-*fos* and c-*jun* gene expression leads to formation of the transcription factor, AP-1.

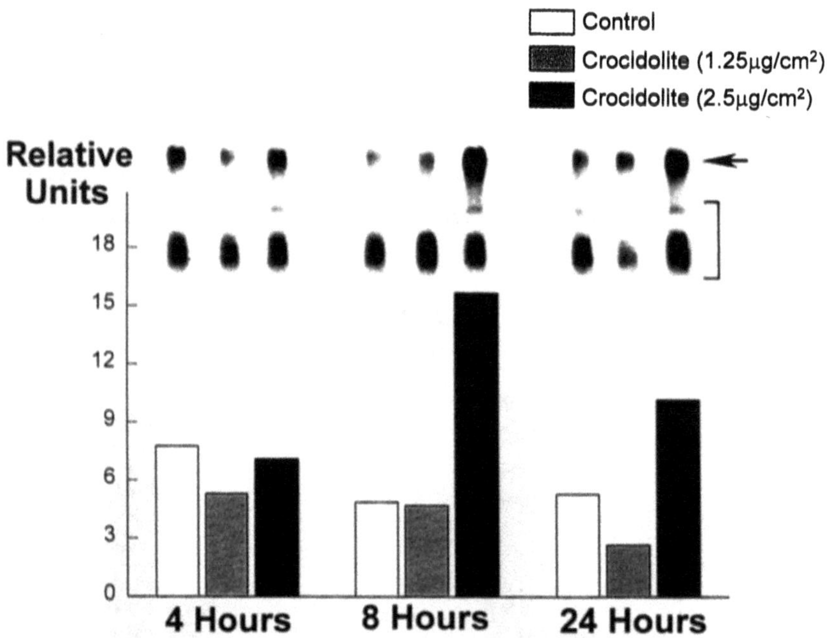

Figure 4. Gel mobility shift analysis examining binding to the AP-1 consensus sequence. Whole cell extracts were prepared from RPM exposed to crocidolite asbestos for 4, 8 or 24 hours. 5 μg of protein was electrophoresed in the presence of a [^{32}P] end-labeled oligonucleotide containing a consensus AP-1 sequence. After electrophoresis, the gel was dried, exposed to Kodak-X-Omat film, and quantitated on a Betascope blot analyzer. The site of the specific AP-1 DNA complex is indicated by the arrow whereas non-specific protein DNA complexes are indicated by a bracket. Quantitated results, expressed in relative units, represent hybridization signals of the specific AP-1 DNA complex normalized to the hybridization signals of the non-specific complex.

Discussion

In the present paper we show that asbestos induces gene expression of c-*fos* and c-*jun* protooncogenes in rat pleural mesothelial cells, the progenitor cell type of asbestos induced mesothelioma. Increases in c-*fos* and c-*jun* mRNA levels are accompanied by increases in AP-1 DNA binding activity, suggesting increased production of the transcription factor AP-1, which might lead to enhanced activation of AP-1 dependent genes. The negative particle control, polystyrene beads, does not alter c-*fos* or c-*jun* mRNA levels. Similar results are obtained with riebeckite, a particle chemically similar to crocidolite but lacking a fibrous structure (Heintz et al., 1993). Thus our studies suggest that the fibrous structure of asbestos is important in induction of c-*fos* and c-*jun* protooncogenes. Examination of other fibrous dusts, including man made vitreous fibers (MMVF) or refractory ceramic fibers (RCF) reveals minor alterations in c-*fos* and c-*jun* mRNA expression in comparison to crocidolite (Janssen et al, submitted). However, erionite, a zeolite fiber associated with environmental mesothelioma in man, (Artvinli and Baris, 1979) is extremely potent in inducing c-*fos* and c-*jun* gene expression. Persistent induction of c-*fos* and c-*jun* protooncogenes by asbestos suggests that asbestos may act through a model involving chronic proliferation in induction of mesothelioma. The tumor promotor, 12-O-tetradecanoylphorbol-13-acetate (TPA), and growth factors also induce upregulation of these c-*fos* and c-*jun* protooncogenes. However, in comparison to asbestos, these agents cause more rapid, transient increases in mRNA expression of these protooncogenes (Heintz et al., 1993). This suggests that the signalling pathways triggered by asbestos differ in time from those triggered by TPA or receptor-mediated growth factors.

DNA damage, (Hollander and Fornace Jr, 1989) /AOS/ (Amstad et al., 1992) and activation of protein kinase C (PKC) (Angel and Karin, 1991) are distinct events/agents capable of inducing gene expression of members of the *fos/jun* gene family in a variety of models and cell systems. Since asbestos is known to cause DNA damage, generation of AOS and activation of PKC in rodent mesothelial cells (reviewed in Mossman, 1993), these processes might also be involved in c-*fos*/c-*jun* protooncogene induction by asbestos.

Since AOS appear to be second messengers of many cellular effects induced by asbestos (Mossman and Marsh, 1991), work in our laboratory has focused on the role of AOS in asbestos-induced upregulation of c-*fos* and c-*jun*. In RPM cells, chemical generating systems of AOS, H_2O_2 or xanthine plus xanthine oxidase, failed to alter c-*fos* or c-*jun* gene expression (Janssen et al., manuscript in preparation). However, increased expression of manganese containing superoxide dismutase or heme oxygenase was observed after addition of these AOS, indicating an oxidant stress response. These results suggest that AOS and asbestos trigger distinct signalling events in RPM cells that may or may not result in increases in c-*fos* or c-*jun*. Related studies show that cellular glutathione (GSH) levels appear to govern c-*fos* and c-*jun* mRNA levels in RPM cells. For example, asbestos fibers cause depletion of cellular GSH pools at time periods showing protooncogene induction. Modulation of GSH pools by N-acetylcysteine (NAC), a GSH precursor, or alternatively, buthionine sulfoximine (BSO), a GSH depletor, reversed and enhanced the GSH depletion by asbestos in RPM cells, respectively. Interestingly, preexposure of RPM cells to NAC abolished c-*fos* and c-*jun* protooncogene induction by asbestos, whereas BSO augmented asbestos-induced c-*fos* and c-*jun* mRNA levels, but did not cause induction of protooncogenes in and of itself. In summary, these results suggest a redox-sensitive component in the signalling cascade triggered by asbestos in RPM cells which leads to increases in c-*fos* and c-*jun* mRNA levels. We are currently delineating the signalling cascades triggered by asbestos at the cell membrane which may lead to induction of protooncogenes. Elucidation of growth factor-induced receptor activation and subsequent phosphorylation cascades activating ras and mitogen-activated protein (MAP) kinases, which may be coupled to upregulation of c-*fos*, c-*jun* or c-*myc*, will provide important avenues for investigation.

Acknowledgements

This work was funded by grants from the NHLBI, NIEHS and a grant from the Environmental Protection Agency.

References

Angel, P, Karin, M (1991) The role Jun, Fos and the AP-1 complex in cell-proliferation and transformation. Biochim. Biophys. Acta., 1072:129-157

Amstad PA, Krupitza G, Cerutti PA. (1992) Mechanism of *c-fos* village in the Anatolian region in Turkey: an epidemiological study. J Natl Cancer Inst. 63:17-22

Artvinli M, Baris YI. (1979) Malignant mesothelioma in a small induction by active oxygen. Cancer Research 52:3952-3960

Heintz, NH, Janssen, YMW, Mossman, BT (1993) Persistent induction of *c-fos* and *c-jun* expression by asbestos. Proc. Natl. Acad. Sci. USA 90:3299-3303

Hollander MC, Fornace Jr AJ. (1989) Induction of *fos* RNA by DNA-damaging agents. Cancer Res 49:1687-1692

Jaurand MC, Renier A, Meeren van der A, Magne J, Pinchon L, Bignon J. (1981) Rat pleural mesothelial cells in culture. In Vitro 17:98-108

Mossman BT. (1993) Mechanisms of asbestos carcinogenesis and toxicity. The amphibole hypothesis revisited. Br. J. Ind. Med., 50:673-676.

Mossman BT, Marsh JP. (1991) Role of active oxygen species in asbestos-induced cytotoxicity, cell proliferation and carcinogenesis. In: Harris CC, Lechner JF, Brinkley BR (Eds.) Cellular and Molecular Aspects of FiberCarcinogenesis. Cold Spring Harbor, 1991,pp159-168.

Mossman BT, Bignon J, Corn M, Seaton, A, Gee JBL (1990) Science 247:294-301

Mossman BT, Gee JBL. (1989) Medical progress. Asbestos-related diseases. N Engl J Med 320:1721-30

Preston-Martin, S, Pike, MC, Ross, RK Jones, PA & Henderson, BE (1990) Increased cell division as a cause of human cancer. Cancer Res. 50:7415-7421

Shull S, Heintz NH, Periasamy M, Manohar M, Janseen YMW, Marsh J, and Mossman BT. (1991) Differential regulation of antioxidant enzymes in response to oxidants. J. Biol. Chem. 266:24398

MOLECULAR ALTERATIONS IN MURINE MESOTHELIOMAS INDUCED BY CROCIDOLITE ASBESTOS

V.D. Moyer, C.A. Vaslet, and A.B. Kane

Department of Pathology and Laboratory Medicine
Brown University
Providence, RI 02912
U.S.A.

INTRODUCTION

Repeated weekly intraperitoneal injections of crocidolite asbestos lead to the development of malignant mesothelioma in mice. Acute asbestos-induced mesothelial injury is accompanied by an inflammatory reaction characterized by nonspecifically activated macrophages (Branchaud *et al.*, 1993). Macrophages surround asbestos fibers and generate reactive oxygen species during phagocytosis (Goodglick and Kane, 1990). A single episode of asbestos-induced mesothelial cell injury is repaired by regeneration after 14-21 days. Weekly exposure to asbestos results in repeated episodes of injury, inflammation, and regeneration (Moalli *et al.*, 1987). We hypothesize that constitutive activation and/or deregulation of growth factor signaling pathways may contribute to the development of malignant mesothelioma.

Growth factor signaling pathways enable a cell to respond to its environment by coupling cellular proliferation and differentiation. Peptide growth factors are important in regulating normal tissue development and homeostasis. They are also crucial in mediating wound healing by stimulating cellular proliferation and tissue regeneration or fibrosis (Sporn and Roberts, 1992). Aberrations in growth factor signaling pathways can alter differentiation and lead to persistent, unregulated cell proliferation (Antoniades, 1992).

Transforming growth factor alpha (TGF-α) and epidermal growth factor (EGF) are peptide growth factors expressed by normal and neoplastic cells. These growth factors regulate development and proliferation of mesenchymal and epithelial cells (Massague, 1990). Both peptides bind to the epidermal growth factor receptor (EGF-R; a

NATO ASI Series, Vol. H 85
Cellular and Molecular Effects
of Mineral and Synthetic Dusts and Fibres
Edited by J. M. G. Davis and M.-C. Jaurand

transmembrane tyrosine kinase) and stimulate similar biological activities (Derynck, 1988). TGF-α is synthesized as a 159 amino-acid precursor, pro-TGF-α. Mature TGF-α is released from the extracellular domain of pro-TGF-α by proteolytic cleavage. Both pro-TGF-α and the mature secreted peptide can bind to the EGF-R and stimulate proliferation (Lyons and Moses, 1990).

Protooncogenes are involved in the cascade of events linking growth factor receptor activation to DNA synthesis. Protooncogenes encode growth factors, growth factor receptors, signal transduction proteins and transcription factors that mediate growth factor-induced gene expression (Cantley *et al.*, 1991). Growth factors stimulate cellular proliferation by increasing protooncogene expression and transcriptional activity. Mutations in protooncogenes can alter their expression and/or function resulting in deregulated or aberrant growth factor signaling.

Expression of the c-myc early response gene is associated with EGF and TGF-α stimulation (Shibuya *et al.*, 1992). The c-myc protooncogene facilitates cellular progression from the G_0 to G_1 phase of the cell cycle and is expressed only during G_1 (Marcu, 1992). However, deregulated or constitutive expression of c-myc is observed in a variety of malignant cell lines (Xu *et al.*, 1993).

Expression of TGF-α and EGF-R following asbestos exposure was examined in regenerating mesothelial cells *in vivo*. Altered expression of the c-myc protooncogene expression was also investigated in murine malignant mesothelial cell lines *in vitro*.

METHODS AND RESULTS

Model System

We developed a model system of malignant mesotheliomas using weekly intraperitoneal injections of 200ug (5.8×10^8 fibers) of UICC crocidolite asbestos fibers in male C57Bl/6 mice. Malignant mesothelial cells were isolated after 40 to 60 weekly injections. Proliferating mesothelial cells were obtained after 3 weekly injections. Normal mesothelial cells require EGF and ITS (insulin-transferrin-selenium) in addition to 10% serum for proliferation in culture. Malignant mesothelial cell lines have reduced EGF requirements when cultured in 10% serum.

EGF may have species-specific effects on mesothelial cells. Normal human mesothelial cells proliferate in response to EGF and murine mesothelial cells require EGF for growth *in vitro*. However, EGF is not mitogenic for normal rat mesothelial cells (Walker *et al.*, 1992). Therefore, this murine model system of malignant mesotheliomas more closely resembles the human disease.

Activation of the TGF-α Pathway

Expression of TGF-α and EGF-R in asbestos-induced mesothelial proliferation was investigated using immunohistochemistry. The following antisera were used: rat pro-TGF-α (Dr. L. Gentry, Medical College of Ohio) monoclonal antibody used at a dilution of 1:1500 and anti-human EGF-R polyclonal antibody (Upstate Biotechnology, Inc, New York) used at 8ug/ml. Uninjured, quiescent mesothelium does not express TGF-α or EGF-R protein. However, a coordinate rise in TGF-α and EGF-R expression in reactive mesothelial cells near fiber clusters was observed within 4 days following a single intraperitoneal injection of 200ug of crocidolite asbestos. Expression subsided by 14-21 days. Stromal cells within the submesothelium also expressed the EGF-R protein. Mesothelial proliferation as assessed by BrdU incorporation was more widely distributed throughout the mesothelium. However, the kinetics of TGF-α and EGF-R induction coincided with the proliferative response of mesothelial cells. Pro-TGF–α and EGF-R proteins were also expressed in malignant mesotheliomas.

Overexpression of c-myc

Northern blot analysis of murine malignant mesothelial cell lines indicated a 3-10 fold increased expression of c-myc. No amplification of the c-myc gene was detected by Southern blot analysis. We investigated whether c-myc expression was deregulated in malignant mesothelial cell lines. Normal and malignant mesothelial cell lines were synchronized by serum starvation for 36 and 48 hours respectively. Upon serum addition, total RNA was isolated at intervals throughout the cell cycle and proliferation was assessed by bromodeoxyuridine (BrdU) incorporation. Semiquantitative RT-PCR using actin amplification as an internal control was used to analyze c-myc mRNA levels. Normal mesothelial cells demonstrated regulated expression of c-myc with a transient increase at 6 and 12 hours before entry into S phase at 18 hours. Malignant mesothelial cell lines expressed high levels of c-myc throughout the cell cycle.

DISCUSSION

We have demonstrated induction of TGF-α and EGF-R expression in mesothelial cells *in vivo* in response to asbestos-induced injury. The pattern of expression correlates with mesothelial proliferation indicating a potential regulatory role for TGF-α in the regenerative process. TGF-α expression subsides or is down-regulated following a single injection of asbestos. We hypothesize that chronic stimulation of the TGF-α pathway may contribute to the development of malignant mesotheliomas. Chronic TGF-α stimulation may result from persistence of fibers in the mesothelium or from repeated exposures. Alternatively, this pathway may become deregulated or constitutively activated in malignant mesotheliomas (Antoniades, 1992).

TGF-α overexpression is frequently associated with c-myc overexpression in malignant cells (Lee *et al.*, 1991). Our results indicate that malignant mesothelial cells constitutively overexpress c-myc mRNA. Additional experiments will be necessary to determine the mechanism of c-myc deregulation and its effects on the mesothelial cell response to TGF-α. Constitutive expression of c-myc has been shown to increase the response to TGF-α in 3T3 cells (Stern *et al.*, 1986).

Asbestos-induced malignant mesotheliomas exhibit a variety of histological phenotypes ranging from epithelioid to fibroblastic or mixed morphology (Davis, 1984). The EGF pathway has been demonstrated to be important in regulating the human mesothelial phenotype *in vitro*. Mesothelial cells express keratins while maintaining the ability to express vimentin (Kim *et al.*, 1987). EGF stimulation induces a fibroblastic morphology while increasing vimentin expression (Kim *et al.*, 1987). Vitamin A or retinoic acid (via cAMP dependent protein kinase activation) induces an epithelial morphology and increased keratin expression. Keratin proteins are characteristic of epithelial differentiation; however, vimentin is typically expressed in fibroblasts (Kim *et al.*, 1987). The plasticity of the mesothelial phenotype may stem from the common mesenchymal origin of mesothelial cells and fibroblasts. Differentiation of mesothelial cells may involve the expression of an intermediate fibroblastic phenotype. Unregulated TGF-α stimulation following asbestos injury may further inhibit mesothelial differentiation in favor of the fibroblastic phenotype. Further studies will be needed to identify whether asbestos-induced injury increases vimentin expression in surface mesothelial cells.

WORKING HYPOTHESIS

Fibers and macrophages generate oxidants following asbestos exposure *in vivo*. These oxidants cause mesothelial injury that is repaired by mesothelial regeneration. Activation of the TGF-α growth pathway and c-myc expression accompany the regenerative response. Repeated exposures to asbestos cause repeated episodes of mesothelial injury and regeneration. Chronic mesothelial cell proliferation may result in the accumulation of genetic lesions that confer a growth advantage. In the early stages of tumor development, additional growth factors may be necessary *in vivo* to sustain mesothelial proliferation following asbestos-induced injury. At later stages of tumor progression, constitutive TGF-α expression and deregulated c-myc expression may be sufficient to induce autonomous proliferation. Decreased expression of the p53 tumor suppressor gene has been demonstrated in murine malignant mesothelial cell lines derived from invasive mesotheliomas. Alterations in this tumor suppressor gene appear to occur at late stages in tumor progression in this murine model system (Cora and Kane, 1992).

ACKNOWLEDGEMENTS

This research was supported by grants from the National Institute of Environmental Health Sciences: ROI ES 03721 and T32 ES 07272.

REFERENCES

Antoniades, H.N. (1992) Linking Cellular Injury to Gene Expression and Human Proliferative Disorders: Examples With the PDGF Genes. Mol. Carcinogenesis, 6: 175-181.

Branchaud, R.M., L.J. Garant, and A.B. Kane. (1993) Pathogenesis of Mesothelial Reactions to Asbestos Fibers. Pathobiology, 61: 154-163.

Cantley, L.C. *et al.* (1991) Oncogenes and Signal Transduction. Cell, 64: 281-302.

Cora, E.M. and A.B. Kane. (1992) Alterations in a tumour suppressor gene, p53, in mouse mesotheliomas induced by crocidolite asbestos. Eur. Respir. Rev., 2: 1-3.

Davis, JMG. (1984) The pathology of asbestos related disease. Thorax, 39: 801-808.

Derynck, R. (1988) Transforming growth factor α. Cell., 54: 593-595.

Goodglick, L.A. and A.B. Kane. (1990) Cytotoxicity of Long and Short Crocidolite Asbestos Fibers *in Vitro* and *in Vivo*. Cancer Res., 50: 5153-5163.

Kim, K.H., V. Stellmach, J. Javors and E. Fuchs. (1987) Regulation of Human Mesothelial Cell Differentiation: Opposing Roles of Retinoids and Epidermal Growth Factor in the Expression of Intermediate Filament Proteins. J. Cell Biol., 105: 3039-3051.

Lee, L.W. *et al.* (1991) Clonal Cosegregation of Tumorigenicity with Overexpression of c-myc and Transforming Growth Factor α Genes in Chemically Transformed Rat Liver Epithelial Cells. Cancer Res., 51: 5238-5244.

Lyons, R.M. and H.L. Moses. (1990) Transforming growth factors and the regulation of proliferation. Eur. J. Biochem., 187: 467-473.

Marcu, K.B. (1992) Myc function and regulation. Annu. Rev. Biochem., 61: 809-860.

Massague, J. (1990) Transforming growth factor-α. J. Biol. Chem., 265: 21393-21396.

Moalli, P.A., J.L. Macdonald, L.A. Goodglick and A.B. Kane. (1987) Acute injury and regeneration of the mesothelium in response to asbestos fibers. Am. J. Pathol., 128: 425-445.

Shibuya, H. *et al.*, (1992) IL-2 and EGF Receptors Stimulate the Hematopoietic Cell Cycle via Different Signaling Pathways: Demonstration of a Novel Role for c-myc. Cell, 70: 57-67.

Sporn, M.B. and A.B. Roberts. (1988) Autocrine secretion-10 years later. Ann. Int. Med., 1117: 408-413.

Stern, D.F. *et al.* (1986) Differential Responsiveness of myc- and ras-Transfected cells to Growth Factors: Selective Stimulation of myc-Transfected Cells by Epidermal Growth Factor. Mol. and Cell Biol., 6: 870-877.

Walker, C., J. Everitt, and J.C. Barrett. (1992) Possible Cellular and Molecular Mechanisms for Asbestos Carcinogenicity. Am. J. Industrial Med., 21: 253-273.

Xu, L., R. Wallen, V. Patel and R.A. DePinho. (1993) Role of first exon/intron sequences in the regulation of *myc* family oncogenic potency. Oncogene, 8: 2547-2553.

Long and short amosite asbestos samples: comparison of chromosome-damaging effects to cells in culture with *in vivo* pathogenicity.

Kenneth Donaldson [1,2], Nadya Golyasnya[3] and John MG Davis[2],
[1]Biomedicine and Biotechnology Research Group, Napier University, Edinburgh, Scotland, [2]Institute of Occupational Medicine, Edinburgh,Scotland and [3]Laboratory of Chemical Mutagenesis, Institute of Ecology and Genetics of Micro-organisms, Russian Academy of Sciences, Perm, Russia.

INTRODUCTION

The pathogenic effects of asbestos in humans and laboratory animals are well documented , with lung tumours and mesothelioma amongst the major pathologies being produced (Davis and Donaldson 1993). The ability of asbestos and other fibres to damage chromosomes of cells in culture has also been demonstrated (eg Hesterberg and Barrett 1990) and this has been considered to be a likely factor in its toxicity, particularly its carcinogenic action. However, few laboratories have had the opportunity to study the pathogenic effects of defined fibre samples and, using the same samples, examine their clastogenic effects.

In the 1980s the Institute of Occupational Medicine in Edinburgh carried out a large-scale rat inhalation and instillation study on two amosite asbestos samples that differed only in their length. These studies demonstrated that the short and long samples had dramatically differing pathogenic potentials, with the long sample producing many lung tumours and mesotheliomas whilst the short sample was virtually without activity (Davis et al 1986). Studies on the ability of the same two fibre samples to cause inflammation (Donaldson et al 1988) again demonstrated the long fibre sample to be many times more active than the short sample. We report here on the ability of these two amosite asbestos samples to cause chromosome damage to cells in culture.

We have previously been interested in the anti-oxidant activities of the free radical scavenger reduced glutathione (GSH), in defending the lung against the oxidant damage caused by cigarette smoke (MacNee et al 1991) . It has been suggested that asbestos may cause chromosome damage by oxidant radical-mediated pathways (Kamp et al 1992) . We therefore utilised buthionine sulphoximine (BSO), an inhibitor of the GSH synthetic pathway, to deplete GSH in the target cells and assessed whether this had any effect on the chromosome damage caused by amosite asbestos.

MATERIALS AND METHODS

Amosite asbestos samples The long and short amosite asbestos samples utilised in these experiments have been used extensively and are described in detail elsewhere (Davis et al 1986). In brief the short sample was derived from the long sample by milling and was found to be elementally and crystallographically identical to the parent sample. The diameters of the two samples were almost identical but the long sample was substantially longer as shown in Figure 1.

NATO ASI Series, Vol. H 85
Cellular and Molecular Effects
of Mineral and Synthetic Dusts and Fibres
Edited by J. M. G. Davis and M.-C. Jaurand

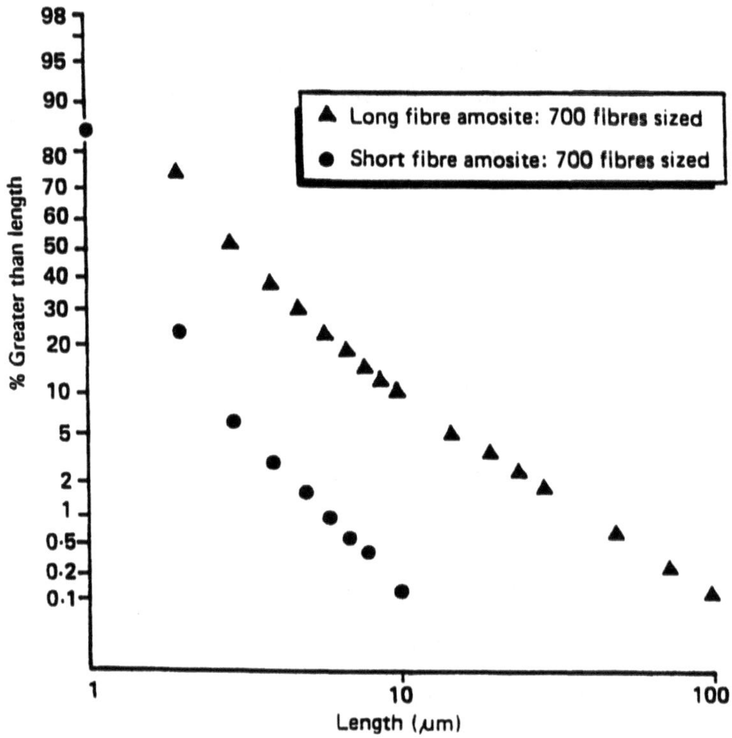

Figure 1 . Length distribution of the long and short fibre amosite samples; 700 fibres counted by SEM.

Experimental Pathology The experimental pathology data summarised below is derived from the published account of experiments where rats were exposed to airborne fibres and the long-term pathological responses were monitored; full details of the exposure systems and analysis can be obtained from Davis et al (1986). Details of the mouse peritoneal inflammation assay can be obtained from Donaldson et al (1989); in brief, mice were injected intra-peritoneally with varying doses of long or short fibre amosite asbestos and the 4 day inflammatory response was assessed by peritoneal lavage and counting of the numbers of inflammatory cells.

Chromosome-damaging effects of fibres Chinese hamster ovary cells (CHO) at 1 x 10^6 cells/ml of F10 medium + 10% foetal calf serum were seeded onto 24-well plates and incubated overnight. The medium was replaced with 1 ml of 10μg/ml of long or short amosite asbestos in medium and incubated for a further 48 hours. Cultures were then treated with 0.1μg/ml colchicine for the final 2 hours of culture and trypsinised off the culture plates. The cells were then spun down and treated with 0.75M KCl for 10 minutes and stained with aceto-orcein. Chromosomal aberrations and hyperdiploidy was assessed in 100 metaphases in 3 separate experiments.

RESULTS

Experimental pathology The results of the experimental pathology experiments are shown in Table 1

Table 1. Results obtained from pathology studies* with the amosite samples

	SHORT AMOSITE	LONG AMOSITE
LUNG TUMOURS (% BY INHALATION)	0	32.5
FIBROSIS (BY INHALATION, ARBRITARY UNITS)	0.5	11.0
MESOTHELIOMA (% BY INSTILLATION)	4	94
INFLAMMATION SCORE (INSTILLATION)	+	++++

*for further details see Davis et al (1986) and Donaldson et al (1989)

Chromosomal and mitotic effects The effects of the amosite samples on chromosomes of cells in culture are shown in Table 2

Table 2. Effects of amosite samples on chromosomes of cells in culture

	CONTROL	SHORT AMOSITE	LONG AMOSITE
CELLS WITH ABERRATIONS (%)	4.5 (0,5)	6.3(0.3)	12.1(0.2)
ABERRATIONS PER CELL	0.06(0.003)	0.07(0.05)	0.19(0.02)
HYPERDIPLOID CELLS (%)	3.3(0.3)	5.7(0.3)	9.5(0.5)
MITOTIC INDEX	7.5(0.2)	6.3(0.2)	4.5(0.05)

All values mean(sd) of 100 cells in 3 separate experiments

EFFECTS OF TREATMENT WITH BSO The effects of treatment of cells with BSO on chromosome aberrations in cells treated with amosite asbestos are shown in Table 3.

Table 3. Effect of BSO on chromosome aberrations

	NO BSO	BSO
CONTROL	4.5(0.5)	9.0(0.5)
SHORT AMOSITE	6.3(0.3)	16.2(1.5)
LONG AMOSITE	12.1(0.2)	18.4(2.3)

All values mean(sd) of 100 cells in 3 separate experiments

DISCUSSION

The increased number of chromosomal aberrations in cells treated with long fibres is consistent with the pathogenic potential of the two samples, with the long sample being substantially more potent than the short in causing lung fibrosis and tumours. Previous workers have shown a relationship between fibre length and pathogenicity of fibres (Hesterberg and Barrett 1990) and Stanton's work with implanted fibres revealed increasing pathogenic potential as the length approached 8μm and over (Stanton et al 1977). Many fibres in the long amosite sample were in this range and longer, whilst there were effectively no 'Stanton' fibres in the short fibre sample. it has been considered that the longer fibres can physically interfere with the migration of chromosomes at anaphase (Hesterberg and Barrett 1990) although this was not observed in the present study. In general, therefore, this study has confirmed the induction of fibre length-related chromosomal abnormalities to cells *in vitro* with two samples of amosite asbestos that have been shown to exhibit the same length -related ability to cause pathology and inflammation in vivo.

GSH is a ubiquitous anti-oxidant found in the lining fluid and cells of the lung and recent studies within our group have suggested that extra-cellular GSH can defend cells in culture against the oxidant injury caused to cells by cigarette smoke (Lannan et al in press). Reactive oxygen species and free radicals have been shown, in many studies, to have a role in asbestos pathogenicity (Kamp et al 1990). The exact role of intra-cellular GSH in maintaining chromosome integrity is unknown but treatment of CHO cells with BSO, an inhibitor of GSH synthesis, increased the amount of chromosomal damage caused to the cells by both short and long amosite. This suggests that free radicals may have role to play in the production of chromosomal abnormalities.

We have found previously that acute exposure of cells in culture to cigarette smoke causes a sharp fall in intracellular GSH (Lannan et al in press). The synergism between asbestos exposure and cigarette smoke in leading to lung cancer is well documented. Effects of cigarette smoke on the GSH levels of the epithelial lining of the lung could be important in reducing the defences against chromosomal damage and these could be important in enhancing fibre toxicity in smokers.

This research was funded by The Pollution Research Unit, Napier University and the Colt Fibre Research Programme

References

Davis, J.M.G. Bolton, R.E. Donaldson, K. Jones, A.D. Smith, T. (1985). The pathogenicity of long versus short fibres of amosite asbestos administered to rats by inhalation and intraperitoneal injection. Br. J. Exp. Pathol. 67, 415-430.

Donaldson, K. Brown, G.M. Brown, D.M. Bolton, R.E. Davis, J.M.G. (1989). The inflammation generating potential of long and short fibre amosite asbestos samples. Br. J. Ind. Med. 46, 271-276.

Donaldson, K Brown, R.C. Brown. G.M. Respirable industrial fibres: mechanisms of pathogenicity. Thorax 1993 48 390-395

Hesterberg, T.W. Barrett, J.C. (1990). Dependence of asbestos and mineral dust induced transformation of mammalian cells on fibres dimension. Cancer Res. 44, 115-124.

Kamp, D.W. Gracefa, P. Pryor, W.A. Weitzman,S.A. (1992). The role of free radicals in asbestos-induced diseases. Free Radical Biol. 12, 293-315.

Lannan S Donaldson K Brown D and MacNee W The role of oxidant in the epithelial injury caused by cigarette smoke products *in vitro.* Amer J Physiology (in press)

MacNee W, Bridgeman MME, Marsden M, Drost E, Lannan S, Selby C, Donaldson K. (1991) The effects of N-Acetylcysteine and glutathione on smoke-induced changes in lung phagocytes and epithelial cells. Am J Med 91(3C) 60S-66S

Stanton, M.F. Layard, M. Tegeris, A. (1977). Carcinogenesis of fibrous glass: Pleural response in the rat in relation to fibre dimension. J. Natl. Cancer Inst. 58, 587-604.

Induction of DNA - Damage after Rats Exposure to Crocidolite Asbestos Fibers

B. Marczynski, T. Kerenyi*,
W. Marek and X. Baur
Research Institute of Occupational
Medicine, University of Bochum
Gilsingstr.14, 44789 Bochum
Germany

ABSTRACT

Several experiments suggest that reactive oxygen metabolites generated after asbestos phagocytosis cause DNA damage. According to recent findings, crocidolite is able to induce DNA strand breaks in vitro. The aim of this study was to find out whether crocidolite asbestos fibers induce in vivo double-strand breaks in the DNA of white blood cells and in the nuclei of lung and liver. Male Wistar rats received a single tracheostoma instillation of 2.5 mg UICC crocidolite asbestos into the right lung. The same dose was injected intraperitoneally. The animals were sacrificed 4 and 13 months after instillation.

Compared to controls, we did not find double-strand breaks in chromosomal DNA in fresh white blood cells of rats 4 and 13 months after crocidolite fibers instillation. Liver and lung nuclei DNA damage could be demonstrated in crocidolite asbestos exposed rats. It remains to be clarified whether the apparent DNA damage is directly due to chronic exposure of asbestos fibers in liver or lung or due to an accelerated or impaired apoptosis. The pertinent increased incidence of double-strand DNA fragments may indicate an important mechanism of asbestos-associated carcinogenesis. Iron ions present in crocidolite fibers might at least partly be responsible for the crocidolite-dependent generation of oxygen radicals which obviously cause DNA double-strand breaks.

INTRODUCTION

Asbestos is an ubiquitous environmental substance clearly carcinogenic to humans and animals. Experiments suggest reactive oxygen metabolites generated after asbestos phagocytosis to induce DNA damage (Kamp et al., 1992). Recent findings showed that

* Medical University, Robert-Karoly krt. 44, Budapest, Hungary

NATO ASI Series, Vol. H 85
Cellular and Molecular Effects
of Mineral and Synthetic Dusts and Fibres
Edited by J. M. G. Davis and M.-C. Jaurand

crocidolite can induce DNA strand breaks in vitro (Lund and Aust, 1992).

The aim of the present work was to find out whether crocidolite asbestos fibers induce in vivo double-strand breaks in the DNA of white blood cells and in the nuclei of lung and liver. The effect of crocidolite on chromosomal DNA of rats exposed to crocidolite fibers was analysed by pulsed-field gel electrophoresis (PFGE) using the "chromosomal DNA-protect agarose-plugs" method and laser densitometric evaluation of stained DNA bands.

METHODS

Rat exposure to crocidolite. Male Wistar rats received a single tracheostoma instillation of 2.5 mg (0.2 ml crocidolite suspension of 10 mg crocidolite/1 ml of physiological saline solution) UICC crocidolite asbestos fibers (BIO-RAD) into the lower lobe of the right lung under barbital anesthesia. The same dose was injected intraperitoneally. The animals were sacrificed 4 and 13 months after instillation, and blood, lungs and livers were taken for DNA double-strand breaks investigation. The blood was directly investigated, but lungs and liver were quickly frozen at -70 °C until the preparation of DNA. Each experiment was repeated three times and the percentages reported (Tab. I and II) represent results from a single experiment.

Isolation of fresh white blood cells. The white blood cells (WBC) from fresh blood samples were immediately separated using technique of Fichtinger-Schepman et al. (1987).

Isolation of liver and lung nuclei. Liver and three parts of lungs (asbestos lung, para-asbestos lung and contra-lateral lung) from rats exposed to crocidolite asbestos fibers and from non-exposed control rats of the same age were quickly frozen at -70 °C until preparation of DNA. For liver and lungs nuclei isolation the method of Rickwood and Birnie (1976) was used.

Preparation of chromosomal DNA in agarose plugs for PFGE. Procedures were essentially like those of Gardiner et al.(1986).

Standard gel conditions. The gel was placed vertically into the gel box (Beckman Gene Line II) for pulsed-field gel electrophoresis with 16 cm between electrodes and a 120° angle between fields. Saccharomyces cerevisiae chromosomes in agarose plugs used as size markers were from Beckman (Munich, FRG). Electrophoresis was carried out under high-voltage conditions (350 V, 60 sec pulse time for 8 hrs 120 sec pulse time for 8 hrs and 180 sec pulse time for 4 hrs at 14 °C) in TAFE II buffer, and the gels were stained with ethidium bromide (0.4 μg/ml). The stained gels were then photographed (Kodak, Polaroid) and the negatives evaluated using "The Personal Densitometer" (Molecular Dynamics, USA).

RESULTS

We investigated DNA double-strand breaks in white blood cells, liver and lung nuclei of rats exposed to crocidolite asbestos fibers. White blood cells und lung and liver nuclei were embedded in an agarose block, lysed and deproteinized as described by Gardiner et al. (1986). Electrophoresis (PFGE) of these isolated chromosomal DNA was carried out, and laser densitometric evaluation of the stained DNA bands was performed.

Using the chromosomal DNA protective method of agarose-plugs, no DNA fragments in the range of 200 to about 10000 kb were found in the fresh white blood cells of non-exposed rats as well as of rats exposed to crocidolite asbestos fibers 4 and 13 months after crocidolite instillation.

Liver nuclei DNA fragments in the region between 250 and 700 kb showed four months after instillation multiple double-strand breaks at random sites within chromosomes of rats exposed to crocidolite asbestos fibers. Comparison with non-exposed rats showed DNA fragments in the range from 250 to 850 kb indicating longer double-strand DNA fragments as compared to chromosomal DNA of Saccharomyces cerevisiae standard marker. Similar results were found in liver nuclei DNA 13 months after exposure (figure).

We also examined three different parts of rat lungs 4 and 13 months after crocidolite fibers instillation. Lung nuclei DNA of control rats showed three peaks, one at the start (> 10000 kb), the second at about 2.200 kb and the third one between 250 and 1.000 kb. 4 and 13 months after crocidolite fiber instillation, we did not find significant changes in the contra-lateral lung. However in the asbestos lung and para-asbestos lung fewer large DNA fragments and many smaller DNA fragments were detected (Tab. I and II).

DISCUSSION

We did not find any double-strand breaks in chromosomal DNA in white blood cells of rats 4 and 13 months after instillation with crocidolite fibers. DNA from the liver of non-exposed rats showed one peak in the range of 250 to 850 kb. Compared to this DNA of rats showed 4 and 13 months after crocidolite instillation shorter DNA fragments in the range of 250 to 700 kb (figure). These findings indicate a more intact DNA in controls.

4 and 13 months after crocidolite instillation, we also found DNA fragments in exposed lung regions between 250 and 2200 kb produced by multiple double-strand breaks at random sites of the same rats. In this case, DNA double-strand breaks were found mainly in asbestos lungs and only partly in para-asbestos lungs. No DNA double-strand breaks could be observed in the contra-lateral lung (Tab. I and II).

Crocidolite-induced DNA damage could be mediated by oxygen-free radicals which

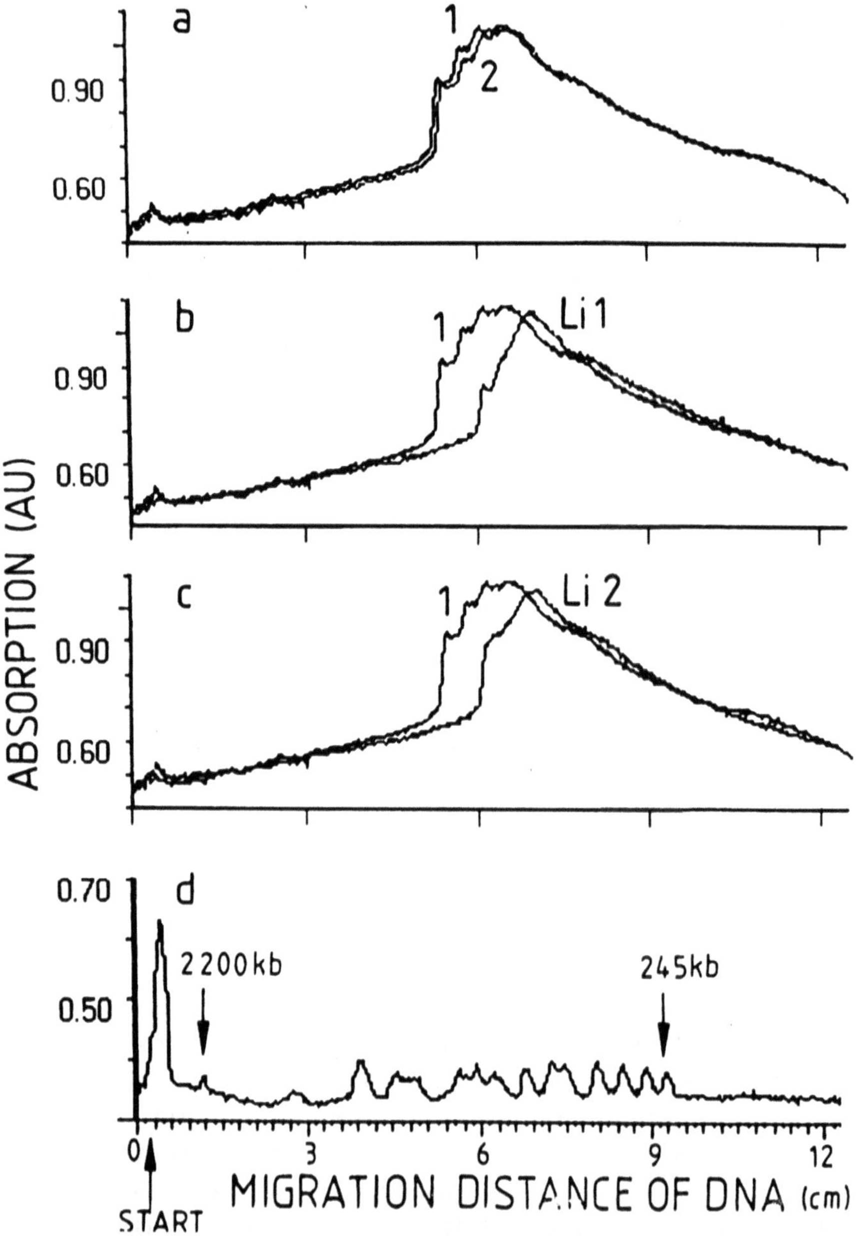

Typical laser densitometric scanning of photograph negatives showing PFGE of liver nuclei DNA from two control rats (a/1 and 2) and two rats exposed to crocidolite asbestos fibers after 13 months of instillation (b/Li1 and c/Li2). Chromosomal DNA of Saccharomyces cerevisiae was used as a standard marker (d).

Table I

Quantitative DNA double-strand breaks analysis of 3 different parts of a rat lung 4 months after crocidolite asbestos instillation in comparison with the lungs of 2 control rats.

Lung	Percentage of DNA fragments in the range of:		
	> 10000 kb	~2200 kb	250 to 1000 kb
controls:			
right lung	18	19	63
left lung	19	18	63
after exposure:			
contra lateral lung	20	24	56
para-asbestos lung	15	17	68
asbestos lung	10	14	76

Table II

Quantitative DNA double-strand breaks analysis of different parts of the right lungs of three rats 13 months after crocidolite asbestos instillation in comparison with the lung of a control rat.

Right lung	Percentage of DNA fragments in the range of:		
	> 10000 kb	~2200 kb	250 to 1000 kb
control	57	9	34
after exposure:			
para-asbestos lung	38	27	35
para-asbestos lung	34	29	37
asbestos lung	19	19	62

are generated following exposure of rats to these fibers. This would account for the induction of DNA double-strand breaks as documented in this report and might explain the clastogenic and tumorigenic effects of crocidolite asbestos fibers.

The results clearly demonstrate that liver and lung nuclei DNA damage are induced by crocidolite asbestos fibers. It remains to be clarified whether the apparent DNA damage is directly due to chronic exposure of asbestos fibers in liver or lung or due to an accelerated or impaired apoptosis (programmed cell death). However, the increased incidence of double-strand DNA fragments can be an important indicator of the chronic genotoxic effect finally resulting in asbestos-associated carcinogenesis.

Apparently, the chronic effects of crocidolite asbestos observed here appear to be similar to that of previously reported in vitro effects (Lund and Aust, 1992) indicating that crocidolite asbestos fibers induced DNA strand breaks. It was found that the generation of hydroxyl radicals by crocidolite asbestos is proportional to the surface concentration of iron (Ghio et al., 1992). The results of this in vivo study suggest that iron present in crocidolite fibers could at least partly be responsible for the crocidolite-dependent generation of oxygen radicals resulting in the formation of DNA double-strand breaks.

REFERENCES

Fichtinger-Schepman AMJ, Van Oosterom AT, Lohman PHM, Berends F (1987) cis-Diamminedichloroplatinum(II)-induced DNA adducts in peripheral leukocytes from seven cancer patients: quantitative immunochemical detection of the adduct induction and removal after a single dose of cis-diamminedichloroplatinum(II). Cancer Res. 47: 3000-3004

Gardiner K, Laas W, Patterson D (1986) Fractionation of large mammalian DNA restriction fragments using vertical pulsed-field gradient gel electrophoresis. Somatic Cell Mol. Genet. 12: 185-195

Ghio AJ, Zhang J, Piantadosi CA (1992) Generation of hydroxyl radical by crocidolite asbestos is proportional to surface [Fe^{3+}]. Arch. Biochem. Biophys. 298: 646-650

Kamp DW, Graceffa P, Pryor WA, Weitzman SA (1992) The role of free radicals in asbestos-induced diseases. Free Rad. Biol. Med., 12: 293-315

Lund LG, Aust AE (1992) Iron mobilization from crocidolite asbestos greatly enhances crocidolite-dependent formation of DNA single-strand breaks in ϕX174 RFI DNA. Carcinogenesis, 13: 637-642

Rickwood D, Birnie GD (1976) Preparation, characterization and fractionation of chromatin. In: G.D. Birnie, editor, Subnuclear Components, Preparation and Fractionation, pages 129-180. Butterworth, London

Unscheduled DNA synthesis (UDS) induced *in vitro* by cadmium in tracheal epithelial cells of rat and hamster.

J.L. Poncy , M. Dhilly and R. Masse
CEA/DSV/DPTE/Laboratoire de Radiobiologie Appliquée
91190 Gif sur Yvette Cedex.
France

Cadmium (Cd) is an ubiquitous pollutant of industry and environment. It is known to damage various organs such as bones, lungs and kidneys in man and experimental animals (Paterson 1947; Friberg et al., 1986; Foulke 1986). The carcinogenic effects of cadmium exhibit species specificity. Pulmonary tumors induction has been demonstrated after inhalation of several cadmium compounds in rats (Hadley et al., 1979; Takenaka et al., 1983; Oldiges et al., 1989). In contrast, no carcinogenic response has been found in mice and hamsters treated by similar compounds (Glaser, 1989). This discrepancy raises the problem of extrapolation of experimental results to humans and obviously, there is a need to determine why animal species differ in their response to the metal. However, this response is modulated substantially with the cadmium compound used that indicates the predominant influence of the toxicokinetics of the different forms of inhaled substances. Several toxicokinetic parameters can indeed interact on the induction of carcinogenicity. One of them is the uptake of pollutant particles by target cells which induces directly the cytotoxicity and the genotoxic effects in these cells. These observations may be especially important for carcinogenesis induced by Cd and ask to be explore.d
Genotoxic effects of cadmium were previously demonstrated in *in vitro* systems (Swierenga et al., 1987; Coogan et al, 1992). Cadmium was shown to induce DNA damage in cultured cells by the action of reactive free radicals (Ochi et al., 1983; Snyder et al., 1988).
The present *in vitro* study was undertaken to investigate the cellular cytotoxic and genotoxic effects of cadmium chloride on tracheal epithelial cells of various rodent species, rat and hamster, exposed *in vitro*. Moreover, the genotoxic effects induced in epithelial cells from rat and hamster tracheal explants exposed to cadmium chloride, were studied using

NATO ASI Series, Vol. H 85
Cellular and Molecular Effects
of Mineral and Synthetic Dusts and Fibres
Edited by J. M. G. Davis and M.-C. Jaurand

unscheduled DNA synthesis (UDS) assay, including autoradiography. The aim of this work was also to identify the potential effect of the cadmium on cell repair process.

Methods

Cell-culture techniques

Epithelial cells were isolated from tracheal explant outgrowths by dissociating the primary culture with trypsin-EDTA solution. Cells were plated on irradiated human fibroblast layer.

Trachea were isolated from 3 month-old male Sprague Dawley rats (Iffa Credo, France) or Syrian golden hamsters (Laboratoire de Physiologie Sensorielle, CNRZ Jouy, France), sacrified by an intraperitoneal injection of Phenobarbital. Trachea were cleaned to remove any adherent connective tissue and cut longitudinally. Explants were obtained by thin transverse sections of each hemi-trachea between cartilagenous rings. At least, three explants were planted on the bottom of plastic dishes. Less than 1 ml of culture medium was added to just immerge the mucosal surface of the explants but not permit them to float. HD medium (Ham F12-DMEM, 2vol/1vol) was complemented by antibiotics, insulin ($5\mu g.ml^{-1}$), retinol ($0.2\mu g.ml^{-1}$), hydrocortison ($0.1\mu g.ml^{-1}$), EGF ($10ng.ml^{-1}$), cholera toxin ($10ng.ml^{-1}$) and 1% of foetal calf serum. Explants were maintained at 37° C in an humidified atmosphere containing 5% carbon dioxide.

Cytotoxic assay

Cadmium chloride ($CdCl_2$, 2.5 H2O; Merck AG) was dissolved in sterile distillated water and diluted with HD medium immediately before use to the concentration required.

Epithelial cells in culture were treated with $CdCl_2$ (0, 5, 11, 22, 33, 55 μM) for 6 hours, washed with medium and harvested by trypsin. Replicated dishes were seeded at 10^3 and 2.10^3 cells per dish. After 10 days, the number of colonies were determined after staining with giemsa and the relative cell survival curve after the different dose of Cd was established. This result was important for the relation with the genotoxic effect induced by the Cd.

Genotoxic assay : Unscheduled DNA synthesis (UDS)

Tracheal epithelial cell cultures were carried out by plating 5 to 10.10^4 dispersed cells from tracheal explant outgrowths, to 35 mm culture dishes. Cells were incubated for 6 hours to allow cell attachment and then the culture medium was removed and replaced by a medium containing appropriate concentrations of $CdCl_2$ (0, 11, 28, 56 and 84 μM) and (^3H)thymidine (370 $KBq.ml^{-1}$ final concentration). After an incubation time of 20 hr., cells were abundantly washed and incubated for 2 hr. in a fresh medium. Cells were removed by trypsin and spread on glass slides by cytospin. The autoradiography was performed using Ilford K5 emulsion and developped and fixed by classical methods after one week of exposure. Grain counting was done on more than 50 nuclei of epithelial cells using objective (x100) under immersion oil. Results were expressed as net grains per nucleus (NG) which correspond to the number of grains over the nucleus minus the number of grains observed in an equal appropriate area of cell cytoplasm. Heavily labelled cells (S-phase cells) were not confounded with cells in UDS phase.

Irradiation of epithelial cells and UDS

15 to 20.10^4 cells were seeded on glass dishes (labteks, Nunc) in normal HD medium and incubated 6 hours before radiation exposure. Adherent cells were irradiated by gamma rays (^{60}Co) at a dose of 0.5 Gy. Immediatly after exposure, cells were placed in a medium containing different concentrations of $CdCl_2$ (0, 2.2, 5.5 and 6.3 μM) and (3H)thymidine (370 $KBq.ml^{-1}$). Cells were incubated during 4 hours and finally washed 3 times with a fresh medium and fixed with a solution of methanol/acetic acid (3v/1v). The autoradiographs were performed as described above. The number of grains over the nuclei were counted for each dose of $CdCl_2$.

Results and Discussion

Cytotoxic assay

This assay was important to define the relative survival of tracheal epithelial cells exposed *in vitro* to $CdCl_2$. The toxic effect of Cd appearing at $CdCl_2$ concentrations as low as 5 μM and 10 μM reduced the relative survival of rat epithelial cells to 50%. These results confirm previous data (Maximilien et al, 1992) which have shown that the proliferation of the tracheal cells of rats was more inhibited than hamster epithelial cells for $CdCl_2$ doses of 5 and 10 μM. A quasi threshold has been observed in hamster up to a concentration of 10 μM and a rapid decrease for a higher concentration.

Genotoxic assay

A direct genotoxic effect was demonstrated in rat tracheal epithelial cells by the induction of UDS after $CdCl_2$ treatment in culture.

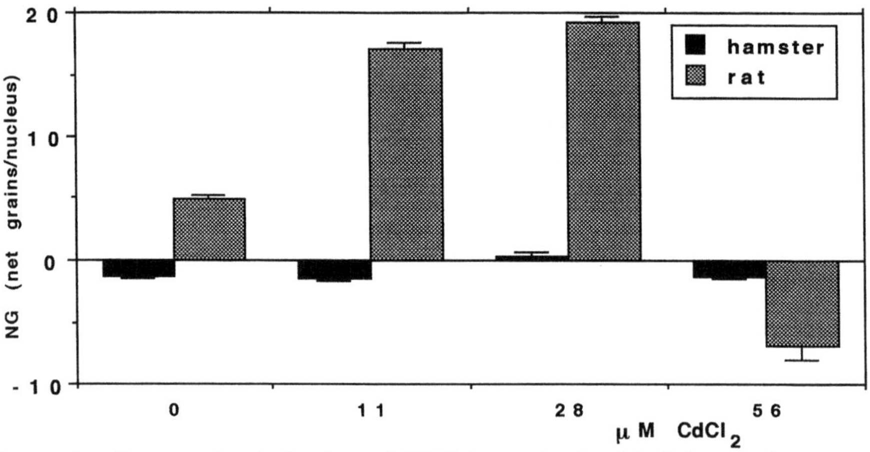

Figure 1 : Comparative induction of UDS in tracheal epithelial cells from rat and hamster, exposed to Cd Cl_2 *in vitro*.

A dose-response relationship of UDS induction with increasing doses of cadmium is shown on figure 1. The response decreases for dose levels of 56 µM presumably in relation with toxic effect.

A non significant increase of the grain count is observed within the hamster cell population treated by 28 µM of $CdCl_2$ i.e. a dose level which induced the maximum of response in rat tracheal cell.

Robison et al. (1984) have observed an unscheduled DNA synthesis induced by $CdCl_2$ in Syrian hamster embryo cells at concentrations as low as 10 µM. One study (Kuper et Benford, 1991) demonstrated induction of DNA repair in tracheal epithelial cells isolated from hamsters and treated *in vitro* by methylmethanesulphonate and benzo(a)pyrene (BP) whereas only methylmethanesulfonate induced UDS in rat epithelial cells. BP is known to be a carcinogen and genotoxic agent after enzymatic activation and the results indicated the lack of appropriate enzyme systems in rats for producing active metabolite. Methylmethanesulfonate as direct genotoxic agent, was effective in both species. It was reasonable to consider that either metal act directly to induce DNA damage or metal-induced genotoxicity was mediated via the formation of active oxygen species (Snyder, 1988). The sensitiveness of the tracheal epithelial cells to Cd, species dependent, might be related to the difference in the intracellular cadmium binding structures and their localization in the cell. Active oxygen species, for example, could produce DNA lesions if they were generated near DNA strand. Further work is in progress to elucidate these interspecies differences.

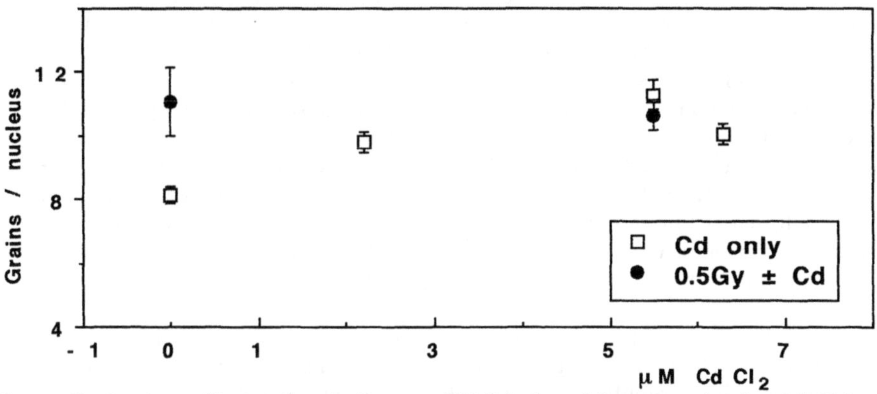

Figure 2 : *In vitro* effects of cadmium on UDS induced in rat tracheal epithelial cells by gamma irradiation. (The values of grains/nucleus were not corrected by cytoplasmic background count)

The multistage model for carcinogenesis discreminates between initiation (first event corresponding to DNA damage) and promotion or progression.
Cd compounds might act as direct carcinogen in epithelial cells by involving DNA damage but also as complete carcinogen by interacting with several different mechanism like the repair capacity of the cells. A dose of 5.5 μM of $CdCl_2$ which induced alone a significant increase of UDS in rat epithelial cells did not influence the repair process induced after gamma irradiation as shown in figure 2. This result might be confirmed under other experimental protocols nevertheless Cd did not seem to affect the repair process induced in the rat tracheal epithelial cells by a low dose of ionizing radiation.

This work was supported in part by a grant in aid (n° 92.C.0804) from the Ministére de la Recherche

References

Coogan T.P., Bare R.M., and Waalkes M.P. (1992). Cadmium-induced damage in cultured liver cells: Reduction in cadmium genotoxicity following zinc pretreatment. Toxicol. Appl. Pharmacol. 113, 237-246.

Foulkes E.C. (1986) Cadmium, pp. 135-178. Springer-Verlag, Berlin.

Friberg L., Elinder C.G., Kjeelström T. and Nordberg G.F. (1986). Cadmium and Health: A Toxicological and Epidemiological Appraisal, vol. II, Effects and Response, pp. 21-204. CRC Press, Boca Raton, FL.

Glaser U. (1989). Lung cancer in rodents after long term inhalation of cadmium compounds. In: Hiscock S.A., Volpe R.A. (eds) Proc. 6th Intern. Cadmium Conference, Paris, pp. 143-146.

Hadley J.G., Conklin A.W., Sanders C.L. (1979). Systemic toxicity of inhaled cadmium oxyde. Toxicol. Letters 1, 107-111.

Ochi T., Ishiguro T. and Ohsawa M. (1983). Participation of active oxygen species in the induction of single-strand scissions by cadmium chloride in cultured Chinese hamster cells. Mutation. Res. 122, 169-175.

Oldiges H., Hochrainer D., Glaser U. (1989). Long term inhalation study in Wistar rats and four cadmium compounds. Toxicol. Environ. Chem. 19, 217-222.

Maximilien R., Poncy J.L., Monchaux R., Masse R. (1992) Validity and limitations of animal experiment assesing lung carcinogenicity of cadmium. In: Cadmium in the Human Environment: Toxicity and Carcinogenesis, Nordberg G.F., Herber R.F.M. and Alessio L. (eds). pp 415-424. IARC Lyon.

Paterson J.C. (1947). Studies on the toxicity of inhaled cadmium. III. The pathology of cadmium smoke poisoning in man and experimental animals. J. Ind. Hyg. Toxicol. 29, 315-320.

Robison, S.T., Cantoni O.C. and Costa M. (1984) Analysis of metal-induced DNA lesions and DNA-repair replication in mamallian cells. Mutat. Res. 131, 173-181.

Snider R.D. (1988). Role of active oxygen species in metal-induced DNA strand breakage in human diploid fibroblasts. Mutation Res. 193, 237-246.

Takenaka S., Oldiges H., Konig H., Hochreiner D. and Oberdorster G. (1983). Carcinogenicity of cadmium chloride aerosols in Wistar rats. J. Natl. Cancer Inst. 72, 929-940.

Pulmonary toxicity of cobalt-containing dusts.

D. Lison and R. Lauwerys
Unité de Toxicologie Industrielle et de Médecine du Travail
Catholic University of Louvain
Clos Chapelle aux Champs 30.54
1200 Brussels
Belgium

INTRODUCTION

Cobalt is a hard metal with magnetic properties, resembling iron or nickel. It is practically insoluble in water but dissolves readily in biological fluids like plasma. Ionic cobalt exists in either a bivalent or a trivalent form and different oxides and salts are encountered in the industry. Cobalt is an essential nutrient to humans with vitamin B12 representing the major physiological form in which cobalt functions in the body. Cobalt is mostly produced as a secondary product during the processing of other metals like copper, nickel and lead. The major sources of minerals are Zaire, Zambia, Canada and ex-USSR while the principal consumer of cobalt is the US which uses about 50% of the world production (Lison 1993).
Because of its interesting properties of high melting point (1500°C), strength and resistance to oxidation, cobalt is widely used alone or in association with other compounds for a number of industrial applications. Large amounts of cobalt are handled in production plants but the most important application is in the manufacture of alloys and cemented tungsten carbide where it is used as a binder of the different powder components. Hard metals are composed of about 90 % tungsten carbide, less than 10 % of cobalt and small amounts of a few other metallic compounds. These metals possess extraordinary properties of hardness, equivalent to about 90 % that of diamond, and remarkably resist heat and wear. These characteristics are used in the manufacturing of saws, cutters, drill bits, armaments, tire studs and some

NATO ASI Series, Vol. H 85
Cellular and Molecular Effects
of Mineral and Synthetic Dusts and Fibres
Edited by J. M. G. Davis and M.-C. Jaurand

ballpoint pens. Cobalt exposure has also been reported among diamond polishers using certain polishing disks consisting of microdiamonds cemented into ultrafine cobalt metal powder (Demedts 1984). Additional uses are for pigments, drying agents and catalyst in the oil industry. Trace amounts have also been used in agriculture and medicine, and cobalt may be present in certain types of cement.

Industrial exposure to cobalt may cause allergic contact dermatitis. The inhalation of cobalt dust can cause irritative or allergic upper respiratory tract manifestations, asthma and a parenchymal disease called hard metal disease which is a subacute or chronic condition characterized by the presence of alveolitis and in some instances by an evolution to interstitial fibrosis. The responsibility of cobalt in the development of asthma is well established (Roto 1980; Shirikawa et al. 1989) but its exact role in alveolitis and fibrosis is still debated.
It has frequently been stated that cobalt alone is the causal agent for the development of hard metal disease, but this conclusion has been questioned since most cases have been observed in the hard metal and in the diamond polishing industries where workers are usually exposed to particles not only containing cobalt but also many other components such as tungsten carbide. On the contrary, no similar cases have been reported among workers exposed to pure cobalt powders.
The present paper summarizes some of the studies from our laboratory trying to clarify the exact role of cobalt in hard metal disease and its interaction with carbides.

EPIDEMIOLOGICAL STUDY.

A cross sectional study has been performed among workers occupied in a Belgian cobalt refinery (Swennen et al. 1993). The exposed workers were compared with a group of control subjects matched for age and smoking habits in order to assess whether exposure to pure cobalt dust may lead to adverse pulmonary effects. The mean exposure duration was of 8 years with very high levels of airborne cobalt. The mean time weighted average exposure assessed with personal samplers was about 125 µg/m3 (inhalable dust)and 25% of the values were higher than 500 µg/m3 which is ten times the current TLV-TWA.

No difference between lung volumes, ventilatory performances and diffusing capacity were found between cobalt and control groups and no lung abnormalities were detected on chest radiographs in both groups. These results suggest that exposure to high airborne concentrations of cobalt alone is not sufficient to cause a parenchymal pulmonary lesion and supports the hypothesis that hard metal disease results from an interaction between inhaled carbide and cobalt particles.

EXPERIMENTAL STUDIES.

To assess this hypothesis, we have performed a number of studies in the rat. We have compared the pulmonary responses obtained with pure cobalt and cobalt-carbide mixtures. The mixture used in all these studies was obtained from a hard metal producing factory and contained 6 % of cobalt metal. Its toxicity was compared with that of equivalent doses of cobalt and WC. Crystalline silica was included in the protocol as a positive control substance.

Acute toxicity study.

The acute response has been studied 24 h after a single intratracheal administration of the different preparations (Lasfargues et al. 1992).

The lung weights of the animals treated with tungsten carbide alone (15.7 mg/100g bw) were not different from those of control rats, but significant increases were noted in the Co (1 mg/100g bw) and mixture groups (15.7 mg/100g bw). The lung weight increase was however much more important in the mixture group than in the group instilled with an equivalent dose of pure cobalt.

The analysis of the cellular fraction of the bronchoalveolar lavage showed an important increase of the total cell number, similar to the response obtained with the same dose of silica. The increase of the neutrophil fraction was however more pronounced than in the silica group.
Similarly, the biochemical analyses showed the same pattern with an increase of LDH, total protein and albumin in the mixture group while the isolated components did not produce any modification.
Histological sections from rats instilled with pure cobalt were almost normal but an intense alveolitis was observed with an equivalent dose of cobalt mixed with tungsten carbide.
An interesting finding in this study was that, even at a dose that did not produce any pulmonary damage, the urinary excretion of cobalt during the 24 h following the instillation was about 4 times higher in the mixture group compared to rats instilled with an equivalent dose of pure cobalt.

Delayed toxicity.
We have also examined the delayed response after a single administration of the same particles.
The lung responses to WC-Co (1, 5 and 10 mg/100 g bw) was characterized by an immediate inflammatory reactgion followed by a regression at day 7 and a progressive flare up at later time points. The effect of cobalt (0.06, 0.3 and 0.6 mg/100g bw) or tungsten carbide alone (1, 5 and 10 mg/100G bw) were very modest if any. The biphasic response was very similar to that obtained with crystalline silica (1 mg/100g bw). However, 4 months after the instillation we could not identify clear histological fibrosis in the lungs of the animals treated with the mixture suggesting a reversibility of the lesions.
We have therefore examined the pulmonary response following the administration of 4 monthly doses of the different preparations. This regimen did not produce any fibrotic reaction in the group treated with pure cobalt (0.06 mg/100 g bw) but in contrast, clear fibrotic lesions were observed in the group that had been instilled with an equivalent dose of cobalt associated with tungsten carbide (1 mg/100g bw). No effect of pure tungsten carbide was noted (1 mg/100g bw).

These descriptive studies clearly confirm that the pulmonary response produced by cobalt is markedly more pronounced when it is associated with tungsten carbide.

In vitro studies.

For the study of the mechanism of the interaction between cobalt and tungsten carbide, an in vitro approach was also used.
We have first demonstrated that the interaction could also be evidenced in a macrophage culture model (Lison and Lauwerys 1990). Using for instance LDH release as a marker of biological reactivity, we found that the mixture was almost as reactive as crystalline silica whereas, when tested separately, tungsten carbide had no effect and pure cobalt only moderately impaired the cell viability.
It is also important to mention that this elective toxicity is not due to a particular processing of the hard metal powder since it can be reproduced by direct mixing, without extensive milling or any addition, pure tungsten carbide with 6 % of cobalt.
Using the same system, exposure of the macrophages to cobalt first and then to tungsten carbide or conversely, did not result in any cytotoxic effect. Clearly the two particles have to be present together to produce a toxic effect.

Since the response observed in this macrophage culture model was similar to the in vivo response, we have used this in vitro system as a tool to investigate the mechanism of the elective toxicity of the cobalt-carbide mixture.

First, we assessed whether this interaction could also be produced with other carbides. We have tested a number of different carbides alone or mixed with 6 % of cobalt and found that some of them like titanium carbide, niobium carbide, chromium carbide and tantalum carbide also interact with cobalt in vitro. With other carbides such as molybdenum and silicon carbide we did not observed such an effect. However, since the size of these latter carbide particles was much more important (>10 µm) than that of the four others (2µm), it may be possible that the interaction with cobalt is not dependent on the nature but rather on the size of the particles. This has therefore been

repeated with fine silicon carbide particles (<2 µm) and an interaction with cobalt was found. Additional experiments with a range of molybdenum carbides with different sizes are to be performed to confirm this finding.
Other particles such as crystalline silica, titanium dioxide, diamond, iron, latex beads were tested and none of them did interact with cobalt. We have therefore concluded that the interaction with cobalt occurs with several carbides but is not simply the result of an increased uptake of the cobalt particles due to an aspecific stimulation of the phagocytosis process caused by the presence of other particles in the extracellular fluid (Lison and Lauwerys 1992).

It has previously been suggested that the enhanced solubility of cobalt in the presence of a carbide might be responsible for the toxicity of hard metal particles (Schepers 1955). To test this possibility, macrophages were cultured in classical culture wells and a toxic dose of WC-Co mixture (150 µg/ml) was added either directly to the monolayer or in a Millicel insert which allowed solubilized material to come in contact with the cells but retained particulate materials (Lison and Lauwerys 1992). Soluble cobalt concentrations were identical whether the particles were introduced directly in contact with the cells or in the Millicel but the toxic effect was only observed in the first situation. This clearly demonstrated that cobalt solubilization is not the critical factor explaining the toxicity of the WC-Co mixture since the amount of cobalt solubilized from a toxic dose of mixture is insufficient to affect by itself macrophage viability. Furthermore, cobalt chloride which is completely soluble, did not induce any toxic effect in this model up to millimolar concentrations.
Another hypothesis that has been examined was the possible production of reactive oxygen species in the presence of cobalt (Lison and Lauwerys 1993). Some recent publications had suggested that the toxicity of cobalt containing powders might be mediated by the production of hydroxyl radicals through a Fenton reaction where cobalt replaces ferrous ions (Moorhouse et al. 1985; Kadiiska et al. 1989; Kennedy et al. 1989).
Using the deoxyribose assay to measure hydroxyl radical

production, we have indeed verified that cobalt ions and cobalt metal produce hydroxyl radicals in the presence of hydrogen peroxide; the degradation of deoxyribose was higher with an equivalent concentration of WC-Co mixture but this was due to a direct reaction of the carbide with the substrate and hydroxyl radical scavengers such as mannitol and benzoate were ineffective in preventing this reaction. In another series of experiments, we have used different scavengers and enzymes to try to reduce the toxicity of the mixture towards macrophages but none of them were effective. These results led us to conclude that the production of reactive oxygen species does not contribute to the elective toxicity of the carbide cobalt mixture.

The last series of experiments has examined the production of cytokines and growth factors by macrophages exposed to these particles. Il-1 and TNF have been shown to play an important role in the pathogenesis of alveolitis and pulmonary fibrosis (Driscoll and Maurer 1991). In vitro, macrophages exposed to a non-cytotoxic dose of silica (20 µg/ml) produce increased amounts of TNF alpha as compared to controls, but no difference was observed with the cobalt dusts (WC-Co 50 µg/ml).

Similarly, silica increases the production of Il-1 in vitro after LPS stimulation but no influence of the cobalt powders could be evidenced. We did not find any modification of the production of fibronectin after in vitro exposure to either silica or cobalt containing particles.

To summarize, a series of epidemiological and experimental arguments have been presented leading to the conclusion that hard metal disease is not only caused by the inhalation of cobalt but result from an interaction between cobalt and carbide particles. This suggests that hard metal dust behaves as a specific toxic entity. The mechanism of this interaction remains to be clarified but these observations may have some practical implications in industrial hygiene since permissible exposure levels to cobalt metal dust might have to be different when exposure is to pure cobalt particles or to cobalt metal associated with carbides.

REFERENCES

Demedts M (1984) Case reports : cobalt lung in diamond polishers. Am Rev Respir Dis 130:130-135

Driscoll K, Maurer JK (1991) Cytokine and growth factor release by alveolar macrophages-Potential biomarkers of pulmonary toxicity. Toxicol Pathol 19:398-405

Kadiiska MB, Maples KR, Mason RP (1989) A comparison of cobalt(II) and iron(II) hydroxyl and superoxide free radical formation;. Arch Biochem Biophys 275:98-111

Kennedy TP, Dodson R, Rao NV, Ky H, Hopkins C, Baser M, Tolley E, Hoidal JR (1989) Dusts causing pneumoconiosis generate °OH and produce hemolysis by acting as Fenton catalysts. Arch Biochem Biophys 269:359-364

Lasfargues G, Lison D, Maldague P, Lauwerys R (1992) Comparative study of the acute lung toxicity of pure cobalt powder and cobalt-tungsten carbide mixture in the rat. Toxicol Appl Pharmacol 112:41-50

Lison D (1993) Cobalt dusts. In: Corn M (ed) Selected topics in hazardous materials. Academic Press,

Lison D, Lauwerys R (1990) In vitro cytotoxic effects of cobalt containing dusts on mouse peritoneal and rat alveolar macrophages. Environ Res 52:187-198

Lison D, Lauwerys R (1992) Study of the mechanism responsible for the elective toxicity of tungsten carbide-cobalt powder toward macrophages. Toxicol Letters 60:203-210

Lison D, Lauwerys R (1993) Evaluation of the role of reactive oxygen species in the interactive toxicity of carbide-cobalt mixture in macrophages in culture. Arch Toxicol 67:347-351

Moorhouse CP, Halliwell B, Grootveld M, Gutteridge JMC (1985) Cobalt(II) ion as a promoter of hydroxyl radical and possible'crypto-hydroxyl' radical formation under physiological conditions. Differential effects of hydroxyl radical scavengers. Biochim Biophys Acta 843:261-268

Roto P (1980) Asthma, symptoms of chronic bronchitis and ventilatory capacity among cobalt and zinc production workers. Scand J Work Environ Health 6:1-49

Schepers G (1955) The biological action of particulate cobalt metal. Arch Ind Health 12:127-133

Shirikawa T, Kusaka Y, Fujimura N, Goto S, Kato M, Heki S (1989) Occupational asthma from cobalt sensitivity in workers exposed to hard metal dust. Chest 95:29-37

Swennen B, Buchet JP, Stanescu D, Lison D, Lauwerys R (1993) Epidemiologic survey on workers exposed to cobalt oxides, cobalt salts and cobalt metal. Br J Ind Med 50:835-842

Chronic Inhalation Toxicity of Fibrous Glass in Rats

Thomas W. Hesterberg, William C. Miiller and Philippe Thevenaz[1] and R. Anderson
Mountain Technical Center, Schuller International, Inc., Littleton, CO 80127

Abstract

Inhalation studies were conducted to determine the chronic biological effects in rodents of respirable fractions of fibrous glasses having compositions representative of commercial insulation products. Rats were exposed nose-only, 6 hrs/day, 5 days/week, for 24 months to several concentrations (3, 16 and 30 mg/m^3) of fibrous glass (FG). Positive control rats were exposed to chrysotile asbestos (10 mg/m^3) and negative controls to filtered air. At 3 to 6 month intervals, interim sacrifices took place to monitor progression of pulmonary changes and to analyze lung fiber burden. Lung fibrosis was evident within 3 months of initial exposure to chrysotile asbestos. Asbestos also induced an elevation in lung tumors and pleural mesotheliomas by the end of the study. Lung burdens of FG increased dramatically after 78 weeks of exposure in the 16 and 30 mg/m^3 exposure groups. Further, FG exposure resulted in a dose-related increase in inflammation and an increase in lung weights at the highest concentration (30 mg/m^3). These effects indicate that the lung clearance mechanisms were being impaired at the highest doses of FG and provide evidence that the maximum aerosol concentration (MAC) had been reached or exceeded at the highest dose. No lung fibrosis or mesotheliomas were observed in the FG exposed animals. When FG exposed groups were compared to negative controls, there was also no statistically significant increase in lung tumor incidence.

Introduction

The present study is part of a series of investigations into the chronic inhalation effects of the major categories of man-made vitreous fibers (MMVFs) using state-of-the-art inhalation techniques that have been described in detail elsewhere (Hesterberg et al., 1991; Hesterberg et. al., 1993; Mast et al., 1994).

A variety of fibrous glass compositions have previously been evaluated using animal inhalation models (Gross et al., 1970; Lee et al., 1981; Wagner et al., 1984; McConnell et al., 1984; Mitchell et al., 1986; Muhle et al., 1987; Le Bouffant et al., 1987; Smith et al., 1987). None of these studies identified a significant increase in either fibrosis or neoplasms following glass fiber inhalation in spite of FG lung burdens in excess of several hundred thousand fibers/mg dry lung tissue. While none of these studies is considered completely adequate for health risk classification, considered as a group, these studies provide strong evidence for a lack of significant adverse effects following glass fiber inhalation.

Recent development of a new fiber aerosol generation and a nose-only exposure system using specially size selected fibers has led to the design and conduct of a chronic multidose

---

[1] Research and Consulting Company, Geneva, Switzerland

NATO ASI Series, Vol. H 85
Cellular and Molecular Effects
of Mineral and Synthetic Dusts and Fibres
Edited by J. M. G. Davis and M.-C. Jaurand

inhalation study of a variety of MMVFs (Hesterberg et al., 1991). The present study reports the chronic biological effects of respirable fractions of two fibrous glasses having compositions representative of commercial insulation products.

Materials and Methods

Fibers Two fibrous glass compositions (MMVF 10 and 11), which represent over 90% of the commercially available insulation wool, were presized so that they would be comparable to the dimensions of fibers found in workplace air and also rat respirable. Positive control animals were exposed to intermediate length NIEHS chrysotile asbestos (Hesterberg et al., 1993).

Fiber Aerosol Exposure Rats were exposed in nose-only inhalation chambers, 6 hrs/day, 5 days/week, for 24 months to various concentrations of test fiber. Negative control rats were exposed similarly to filtered air. Positive controls were exposed to 10 mg/m^3 of chrysotile asbestos. Fiber size distributions were determined on a quarterly basis using scanning electron microscopy. Aerosol concentrations and fiber size distributions of the MMVF aerosols are shown below in Table 1.

Table 1. Fibrous Glass Aerosol Characteristics

Means±Std Dev

Exp. Grp.	Average Aerosol mg/m3	Average Aerosol WHO fibers/cc	Means Diameter (µm)	Means Length (µm)
MMVF 10				
Geometric			1.26 ± 1.81	13.1 ± 2.0
Arithmetic			1.43 ± 0.75	16.8 ± 12.8
Low	3.1 ± 0.1	29 ± 8		
Medium	17.1 ± 0.5	145 ± 35		
High	27.8 ± 1.8	232 ± 56		
MMVF 11				
Geometric			0.69 ± 2.10	13.7 ± 2.2
Arithmetic			0.90 ± 0.73	18.3 ± 14.6
Low	4.8 ± 1.4	41 ± 29		
Medium	15.8 ± 0.8	153 ± 69		
High	28.3 ± 0.4	246 ± 76		

Animals Weanling Fischer 344 male rats, obtained from Charles River were randomly distributed into the exposure groups. Following the 24 month exposure, the animals were held for lifetime observation (until ~20% survival), and were then sacrificed and examined. Rats were euthanized via intraperitoneal injection of pentobarbital sodium.

Pathology Three to six rats were randomly selected from each exposure group and killed at 3, 6, 12, 18, and 24 months. Lungs were removed in toto, weighed, sectioned and stained with hematoxylin and eosin (H&E) and Masson's trichrome stain for collagen deposition. Histopathology of the lungs was examined and each lung was given a Wagner Pathology Grading Score (Table 2) in accordance with the guidelines presented at the WHO conference on

"Biologic Effects of Man-made Mineral Fibres" in 1982 (McConnell et al., 1984). A normal appearing lung is given a Wagner score of 1, while lungs showing increased inflammatory response are given scores of 2 and 3. A Wagner score of 4 or greater indicates that interstitial fibrosis has occurred in the lung. The lungs and thoracic cavity were examined grossly and mircoscopically for pulmonary adenoma, carcinoma and mesothelioma. Pairwise comparisons of tumor incidence between exposure groups were made using Fisher's exact test.

Lung Burden Analysis Immediately after necropsy, the infracardiac lobe of each animal's lung was removed and frozen for later analysis of lung fiber burden. Lung tissue was rapidly dehydrated with acetone and ashed using a low-temperature process. Recovered fibers were dispersed in distilled water, filtered onto membranes and examined using scanning electron microscopy. Number, dimensions and other characteristics of the lung fibers were determined.

Results

Histopathology The first evidence (3 mo.) of fiber-induced, microscopic changes in the lung consisted of a dose related inflammatory response, including influx of pulmonary macrophages. Occasional microgranulomas were noted along the walls of the alveolar duct in the high dose groups of both FGs. A minimal amount of alveolar bronchiolization was observed in the high dose MMVF 11 group. A Wagner grade of 2.0 was given to each of the groups except the high dose MMVF 11, which was given a grade of 3.0 (Table 2).

Table 2. Pathology Scores (Wagner Scale) After Fibrous Glass Exposure

	MMVF 10 Concentration (mg/m^3)				MMVF 11 Concentration (mg/m^3)			Chrysotile Asbestos
Time	3	16	30	Air	3	16	30	10 mg/m^3
3 Mo	1.0	2.0	2.0	1.0	1.0	2.0	3.0	4.0
6 Mo	1.0	2.7	3.0	1.0	1.3	2.3	3.0	4.0
12 Mo	2.2	3.0	3.0	1.0	2.2	3.0	3.0	4.0
18 Mo	2.7	3.0	3.0	1.0	2.5	3.0	3.0	4.0
24 Mo	2.2	2.7	3.0	1.0	2.5	2.7	2.5	4.0
Term.	2.2	2.8	2.7	1.0	1.9	2.1	2.4	4.0

After 6 months exposure there was a slight increase in the inflammatory response and number of micrograpulomas. Bronchiolization was noted in both the MMVF 10 and 11 high dose rats. Fibers were present in many macrophages and within the interstitium. Short fibers were also noted in macrophages within the peribronchial lymphoid sheaths. At 6 months the overall lung response was similar for both FG types, ranging from Wagner grade 2.0 for the low dose to 3.0 for the high dose (Table 2). At 12 months the macrophage and micrognanuloma responses were slightly more intense. Bronchiolization was noted in the mid dose as well as the high dose lungs of both MMVFs 10 and 11. Average Wagner grades were 2.5 for low-dose of both types of fibers and 3.0 for the mid- and high-dose groups. At 18 and 24 months the pulmonary changes observed were comparable to those observed at 12 months for all dose groups (Table 2). In addition to the finding of no interstitial lung fibrosis in the FG exposure groups, there were no mesotheliomas, nor was there a statistically significant elevation of lung tumor incidence (Table 3). Chrysotile asbestos induced a significant increase in lung tumors and a single pleural mesothelioma.

Table 3. Incidence of Thoracic Tumors in Fibrous Glass Exposed Animals.

Exposure Group	Animals	Adenomas	Carcinomas	Total Lung Tumors	Mesotheliomas
Air Controls	123	3 (2.4%)	1 (0.8%)	4 (3.3%)	0
Chrysotile	63	6 (8.7%)	6 (8.7%)	12 (17.4%)	1 (1.4%)
MMVF 10					
3 mg/m^3	117	0	0	0	0
16 mg/m^3	118	1 (0.8%)	0	1 (0.8%)	0
30 mg/m^3	119	6 (5.0%)	1 (0.8%)	7 (5.9%)	0
MMVF 11					
3 mg/m^3	118	3 (2.5%)	1 (0.8%)	4 (3.4%)	0
16 mg/m^3	120	6 (5.0%)	3 (2.5%)	9 (7.5%)	0
30 mg/m^3	112	3 (2.7%)	0	3 (2.7%)	0

FG Lung Burden Analyses The FG lung fiber burden data is expressed as the total number of fibers per lung and is graphed in Figures 1 and 2. The number of fibers/lung increased dramatically after 78 weeks in the 16 mg/m^3 (MMVF 10) and 30 mg/m^3 (MMVFs 10 and 11) exposure groups, suggesting that the lung clearance mechanisms were impaired at these doses.

Figure 1. Lung Burden of MMVF 10

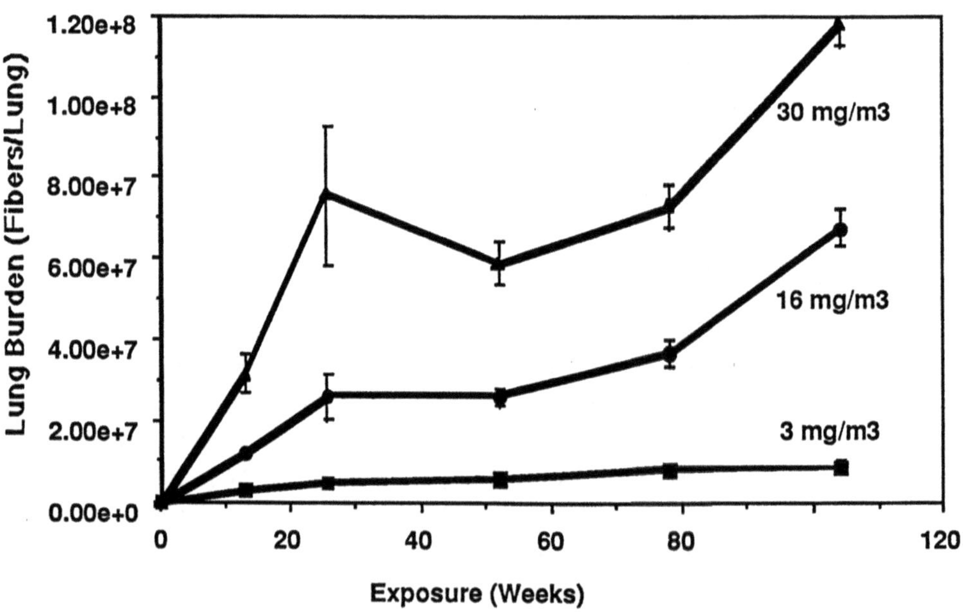

Lung burden for MMVF 11 was consistently higher than for MMVF 10, for each dose group (low, medium and high) and at each exposure time.

Figure 2. Lung Burden of MMVF 11

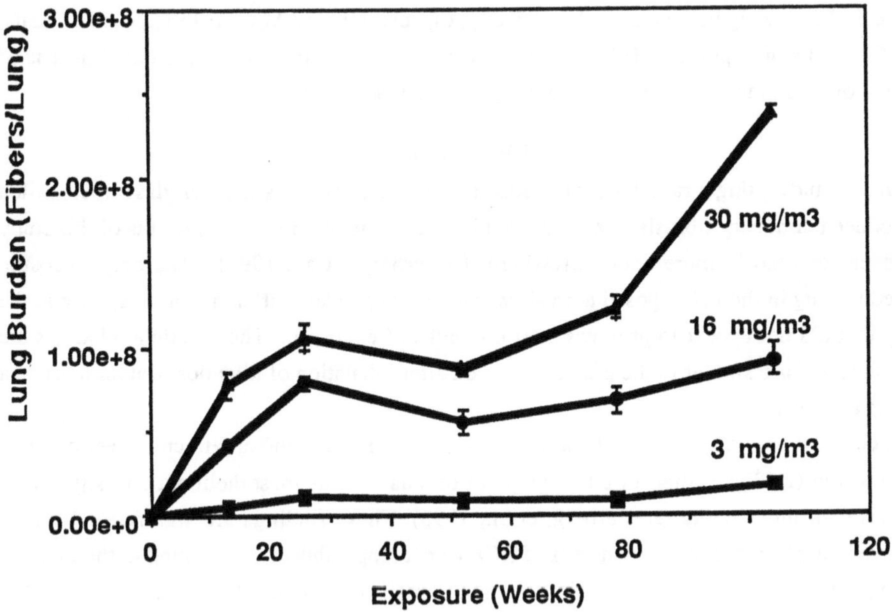

Figure 3. MMVF 10 and MMVF 11 Lung Weights

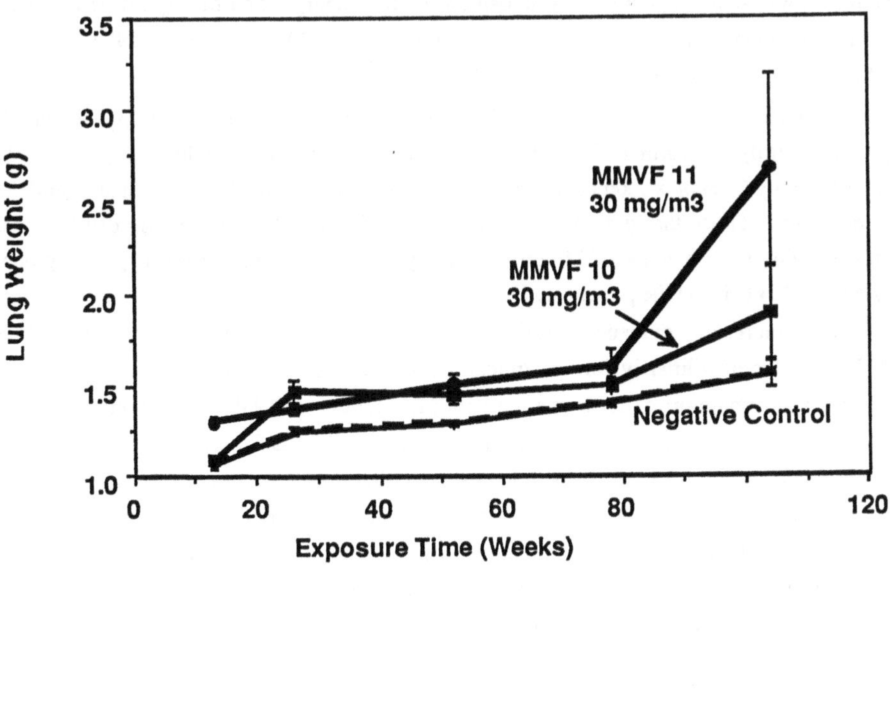

Lung Weights Lung weights of animals exposed to the highest dose of MMVFs 10 and 11 were significantly elevated over negative control lungs at all times during exposure as shown in Figure 3. Lung weights increased dramatically, especially in the MMVF 11 exposed animals, after 78 weeks of exposure. This was the same time frame during which the lung burdens of fibers showed a dramatic increase, suggesting that the two were related.

Discussion

In this study using a rat inhalation model, two different compositions of glass fibers failed to produce pulmonary fibrosis or a significant increase in lung tumors. The results of this study has been reported in more detail elsewhere (Hesterberg, et al., 1993). The only exposure related finding in the FG-exposed animals was a dose-dependent inflammatory response in the lungs that did not appear to progress after 6 months of exposure. These cellular changes are reversible and are similar to the effects observed after inhalation of high doses of an inert dust (Lee et al., 1981).

It has been previously reported that exposure of rats to 30 mg/m^3 of refractory ceramic fiber composition (RCF) resulted in pulmonary fibrosis as well as mesothelioma and significant increases in lung tumors (Hesterberg, et al., 1993). These findings are significant because observed lung burdens and dimensions of RCF were comparable to those found in the lungs of FG exposed animals yet FG did not induce lung fibrosis or tumors. This suggests that internal dose and dimension are not sufficient to explain the toxic potential of two chemically different types of fibers. Chemical composition and surface physicochemical properties are also important determinants of fiber toxicity for similarly sized fibers. The importance of other fiber characteristics, such as chemical composition, surface charge, and biological persistence as critical determinants of fiber toxicity is now well recognized (McClellan et al., 1992; Hesterberg et al., 1993; Morgan and Holmes, 1986; Bellman et al., 1987).

It is significant that lung burdens in the exposed animals are very much higher than those reported recently for former glass fiber production workers (McDonald et al., 1990). Lung samples of only 26% of workers contained any MMVF, almost all siliceous in nature and in low concentration. Those that did contain fibers averaged only 70 fibers per mg dry lung tissue, compared with more than 600,000 fibers per mg dry lung tissue in the highest exposure group (30 mg/m^3 of MMVF 11) in the present study.

There is increasing evidence that the traditional concept of the Maximum Tolerated Dose (MTD) developed primarily for systemically toxic agents may not apply to agents that are insoluble and remain primarily in the lung after inhalation. In 1989, the National Toxicology Program (NTP) organized a working group to determine criteria for selection of the maximum aerosol concentrations (MAC) to use in inhalation toxicity studies. Special emphasis was placed on substances that were relatively insoluble and had a low order of systemic or respiratory toxicity. One conclusion from this meeting was "In general, the highest concentration of test material used should produce only minimal interference with the lung defense mechanisms as judged by impaired particulate clearance." (Lewis et al. 1989).

In a more recent paper by Mauderly et al., it was concluded that "it may be useful to include at least a "minimal overload" level in inhalation bioassays of poorly-soluble particles, and this approach might be a useful substitute for the classical maximum tolerated dose in setting exposure limits" (Mauderly et al., 1990). Levels of dust that "overload" the lung were defined as levels which produced slowed lung clearance, persistent inflammation, increased lung weight and epithelial cell proliferation. If these levels were exceeded, fibrosis and even lung tumors could be eventually induced by non specific mechanisms. This "minimal overload" exposure level appears to have the same criteria, i.e. impairment of clearance, as the MAC defined by the NTP working group (Lewis et al. 1989). Thus for relatively non toxic insoluble particulates a new concept has evolved for setting the maximum aerosol concentration to which animals should be exposed in a chronic inhalation study. This exposure concentration and the resulting lung burden should cause only minimal impairment of the normal clearance mechanisms of the lung.

The results presented in this paper provide evidence that the MAC was attained or possibly even exceeded at the highest concentrations of fibrous glass used (30 mg/m^3). As shown in Figures 1 and 2, the lung burdens increased dramatically after 78 weeks in the lungs of animals exposed to 30 mg/m^3 of fibrous glass. This is a clear indication that the clearance mechanisms of the lung were being impaired at this dose level. In addition the other manifestations of clearance overload described by Mauderly et al., 1990, such as persistent inflammation (Table 2) and increased lung weights (Figure 3), were observed.

In conclusion, these findings demonstrate that this inhalation model provides a sound basis to identify the potential hazards of fibrous materials in man. Even at concentrations of fibrous glass that were at or above the maximum aerosol concentration (MAC), no lung fibrosis, mesotheliomas or significant increase in lung tumors was observed. These results suggest that respirable fibrous glass represents no significant hazard for fibrotic or neoplastic disease in humans.

References

Bellmann, B., Muhle, H., Pott, F., Konig, H., Kloppel, H. and Spurny, K. (1987) Persistence of man-made mineral fibres (MMMF) and asbestos in rat lungs. Ann. Occup. Hyg. 31: 693-709.

Gross, P., Kaschak, M., Tolker, E.B., Babyak, M.A., and De Treville, R.T.P. (1970) The pulmonary reaction to high concentrations of fibrous glass dust. Arch. Environ. Health 23: 67-76.

Hesterberg, T.W., Miiller, W.C., McConnell, E.E., Chevalier, J., Hadley, J., Bernstein, D.M., Thevenaz, P. and Anderson, R. (1993) Chronic inhalation toxicity of size-separated glass fibers in Fischer 344 rats. Fundamental and Applied Toxicology 20: 464-476.

Hesterberg, T.W., R. Mast, E.E. McConnell, J. Chevalier, D.M. Bernstein, W. B. Bunn and R. Anderson. (1991a) Chronic inhalation toxicity of refractory ceramic fibers in Syrian hamsters. In: "Mechanisms in Fibre Carcinogenesis" (eds. R.C. Brown, J.A. Hoskins and N.F. Johnson), Plenum Press, pp. 531-538.

LeBouffant, L., Daniel, H., Henin, J.P., Martin, J.C., Normand, C., Thichoux, G. & Trolard, F. (1987) Experimental study on long-term effects of inhaled MMMF on the lung of rats. Ann. occup. Hyg., 31: 765-790

Lee, K.P., Barras, E., Griffith, F.D., Waritz, R.S., and Lapin, C.A. (1981). Comparative pulmonary responses to inhaled inorganic fibers with asbestos and fiber glass. Environ. Res. 24: 167-191.
Lewis, T.R., Morrow, P.E., McClellan, R.O., Raabe, O.G., Kennedy, G.L., Schwetz, B.A., Goehl, T.J., Roycroft, J.H., and Chhabra, R.S. (1989) Contemporary Issues in Toxicology: Establishing aerosol exposure concentrations for inhalation toxicity studies. Tox. and Appl. Pharm. 99: 377-383.
Mast, R.W., McConnell, E.E., Anderson, R., Chevalier, J., Kotin, P., Bernstein, D.M., Thevenaz, P., Glass, L.R., Miller, W.C., and Hesterberg, T.W. (1994) Chronic toxicity (inhalation) of refractory ceramic fiber in male Fischer 344 rats. Inhalation Toxicology (in press).
Mauderly, J.L., Chang, Y.S. and Slipes, M.B. Particle overload in toxicological studies: friend or foe? (1990) J. Aerosol Med. 3: 169-187.
McClellan, R.O., Miller, F.J., Hesterberg, T.W., Warheit, D.B., Bunn, W.B., *et al.* (1992) Approaches to Evaluating the Toxicity and Carcinogenicity of Man-Made Fibers. Regulatory Toxicology and Pharmacology 16: 321-364.
McConnell, E.E., Wagner, J.C., Skidmore, J.W. and Moore, J.A. (1984). A comparative study of the fibrogenic and carcinogenic effects of UICC Canadian chrysotile B asbestos and glass microfibre (JM 100). In Biological Effects of Man-made Mineral Fibres (Proceed. WHO/IARC Conference, 20-22 April, 1982). World Health Organization, Copenhagen. Vol. 2, pp. 234-252.
McDonald, J.C., Case, B.W., Enterline, P.E., Hendersen, V., McDonald, A.D., Plourde, M., and Sebastien, P. (1990). Lung dust analysis in the assessment of past exposure of man-made mineral fibre workers. Ann. Occup. Hyg. 34(5): 427-441.
Mitchell, R.I., Donofrio, D.J., Moorman,W.J. (1986) Chronic inhalation toxicity of fibrous glass in rats and monkeys. J. Amer. Coll. Tox. 5: 545-575.
Morgan, A. and Holmes, A. (1986) Solubility of asbestos and man-made mineral fibers in vitro and in vivo: Its significance in lung disease. Environ. Res. 39: 475-484.
Muhle, H., Pott, F., Bellmann, B., Takenaka, S. & Ziem, U. (1987) Inhalation and injection experiments in rats to test the carcinogenicity of MMMF. Ann. occup. Hyg., 31: 755-764.
Smith, D.M., Ortiz, L.W., Archuleta, R.F. and Johnson, N.F. (1987). Long-term health effects in hamsters and rats exposed chronically to man-made vitreous fibres. Ann. occup. Hyg. 31: 731-754.
Wagner, J. C., Berry, G. B., Hull, R. J., Munday, D. E. and Skidmore, J. W. (1984). Animal experiments with MMM(V)F- Effects of inhalation and intrapleural inoculation in rats. In: "Biological Effects of Man-made Mineral Fibres" (Proceedings of WHO/IARC Conference, 20-22 April, 1982). World Health Organization, Copenhagen. Vol. 2, pp. 209-234.
WHO (1985). Reference methods for measuring man-made mineral fibers (MMMF). Prepared by WHO/EURO Technical Committee for evaluating MMMF, WHO Regional Office, Copenhagen, Denmark.

A comparison of alveolar macrophage cytotoxicity and ability to cause inflammation in the mouse peritoneal cavity for a range of different fibre types at equal fibre number.

Kenneth Donaldson and Brian G. Miller,
Napier University and The Institute of Occupational Medicine, Edinburgh Scotland.

INTRODUCTION

We have previously reported on the ability of fibres to damage epithelial cells and fibroblasts and described a relationship between increasing toxicity and increasing fibre length for a range of asbestos samples (Brown et al 1986). We have also utilised the mouse peritoneal cavity as site to study the inflammogenic potential of fibres (Donaldson et al 1989). As part of a programme aimed at examining the relationship between the biological activity of a range of non-asbestos fibres and aspects of their structure, we examined the ability of these vitreous and composite fibres to damage alveolar macrophages and to cause peritoneal inflammation. The study was carried out at equal fibre number in both studies so that a direct comparison could be made for fibres of varying density where matching by mass would have resulted in dramatic differences in fibre number.

MATERIALS AND METHODS

Fibres. The fibres used in the study are shown in the table below. All fibres were sized for length by counting 400 fibres by phase contrast optical microscopy according to WHO rules and the % of fibres longer than 10 and 20μm are shown in Figure . It shows that except for Short fibre amosite, all of the samples had a large proportion of fibres in the Stanton range (i.e. > 8μm) but there was variation in the proportions of longer (>20μm) fibres.

Cytotoxicity assay. This assay of cytotoxicity was as described in Donaldson et al (1988). In brief normal rat alveolar macrophages were labelled overnight with ^{51}Cr and then incubated with equal numbers (8.2×10^6 fibres/ml) of the various fibre types. Cytotoxicity was expressed as the

NATO ASI Series, Vol. H 85
Cellular and Molecular Effects
of Mineral and Synthetic Dusts and Fibres
Edited by J. M. G. Davis and M.-C. Jaurand

amount of free radioactive label in the supernatant after 24 hours of incubation. Experiments were carried out on 3 separate occasions, in triplicate, and the data were log transformed and averaged.

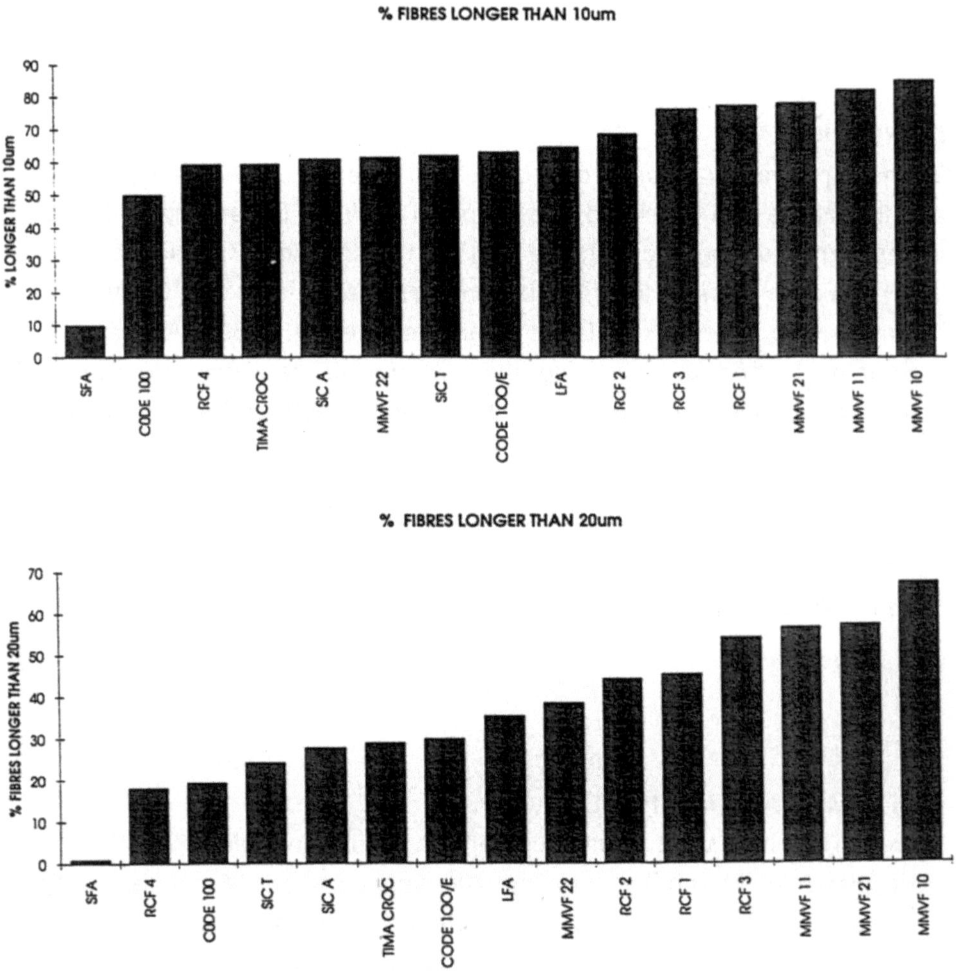

Figure 1 Percentage of fibres of each type used in the study that were longer than 10µm (upper panel) and 20µm (lower panel).

Inflammation in the mouse peritoneal cavity. This was assessed as described previously (Donaldson et al 1989). Briefly groups of 3 or 4 C57 Bl6 mice aged > 6 weeks were instilled with 0.5 ml of saline containing a mass of fibres that yielded 8.2 x 10^7 fibres. At 4 days the peritoneal cavity was lavaged with 3 x 2ml volumes of saline and the total and differential cell count determined. All data were expressed as mean millions of macrophages or granulocytes for 3-4 mice.

TABLE The fibre samples used in the study

Fibre	**Abbreviation**	**Source/reference**
Long fibre amosite	LFA	Davis et al 1988
Short fibre amosite	SFA	Davis et al 1988
Crocidolite	TIMA Croc	TIMA repository
Code 100/475 vitreous fibre	Code 100/475	Shuller (Manville)
Code 104/E vitreous fibre	Cide 100/E	ditto
Silicon carbide A	SiC A	Industrial sample
Silicon carbide T	SiC T	ditto
Refractory ceramic fibre 1	RCF 1	TIMA repository
Refractory ceramic fibre 2	RCF 2	ditto
Refractory ceramic fibre 3	RCF 3	ditto
Refractory ceramic fibre 4	RCF 4	ditto
Man-made vitreous fibre 10	MMVF 10	ditto
Man-made vitreous fibre 11	MMVF 11	ditto
Man-made vitreous fibre 21	MMVF 21	ditto
Man-made vitreous fibre 22	MMVF 22	ditto

RESULTS

Peritoneal inflammation assay.

Despite differences in chemical composition, at equal fibre number the different fibre types produced inflammation that was essentially similar (Figure 2). There were exceptions, principally RCF 3 and RCF 4. RCF 4 was, in general much shorter than any of the other samples and so the low response it caused was not surprising but there was no obvious reason for the reduced inflammogenicity of RCF 3. When millions of granulocytes, taken as a general measure of inflammation was plotted against % of fibres longer than 20μm for all of the fibres types, there was no obvious relationship (Figure 3).

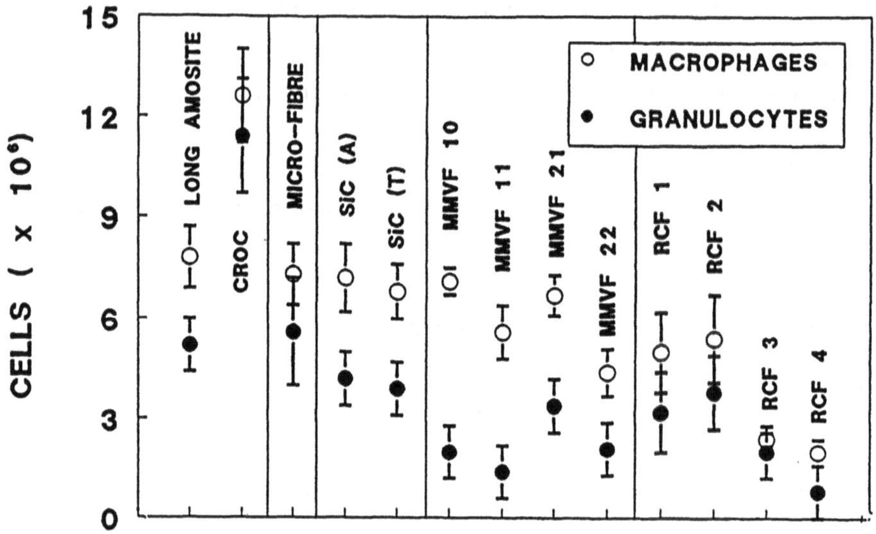

Figure 2 Number of macrophages (open circles) and granulocytes (closed circles) lavaged from the peritoneal cavity of mice instilled with equal numbers of the indicated fibre types. Data given as mean ± SD millions.

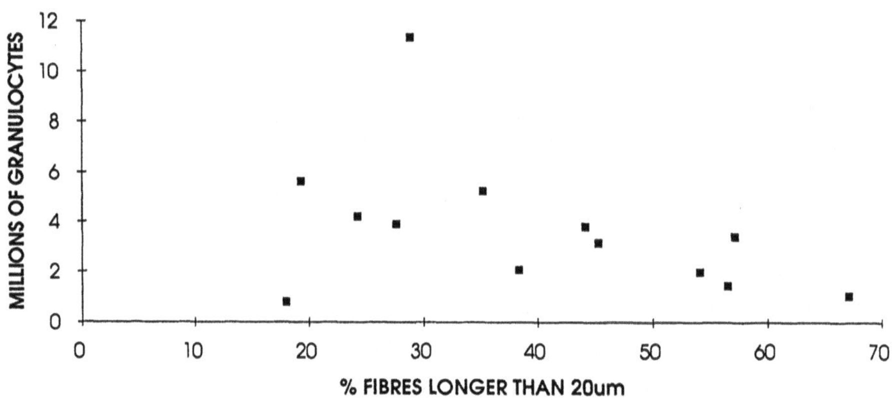

Figure 3 Peritoneal inflammation data of mean millions of granulocytes plotted against percentage of fibres longer than 20μm.

Macrophage cytotoxicity assay. There were differences in the ability of different fibre populations to cause cytotoxic effects to alveolar macrophages at equal fibre number (Figure 4). When toxicity to macrophages was plotted against the % of fibres longer than 20µm in each fibre sample no relationship was found (data not shown) .

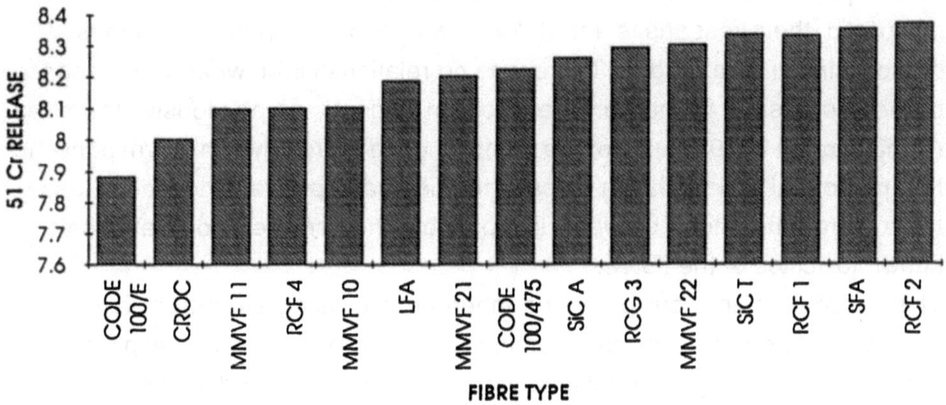

Figure 4 Cytotoxicity of different fibres at equal fibre number; cytotoxicity data log transformed and expressed as the average of triplicate wells in 3 separate experiments.

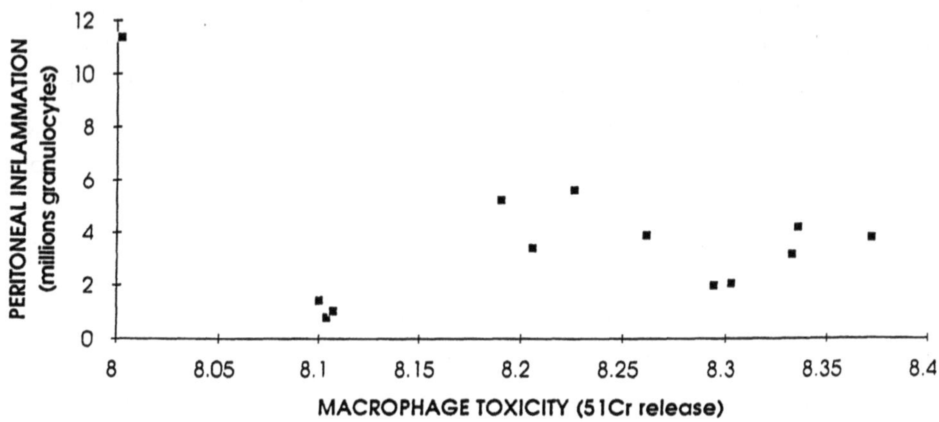

Figure 5 Macrophage toxicity versus peritoneal inflammation for the different fibres. Data expressed as in the above figures.

Comparing peritoneal inflammation with cytotoxicity to alveolar macrophages. When ability to cause inflammation was compared with ability to damage alveolar macrophages, no relationship was found (Figure 5).

DISCUSSION

In this paper we have utilised two short-term assays of fibre toxicity and compared their responses for a large range of different fibre types with broadly similar size ranges. There was no relationship between the response of the two assays to the fibre samples. In general, as previously described (Donaldson et al 1993a) all of the samples gave a broadly similar response in the peritoneal inflammation assay; in the macrophage cytotoxicity assay there were differences between samples but they showed no relationship to known toxicities of the fibres.

This exercise points out problems inherent in utilising short-term assays of toxicity for fibres but the shortcomings of the two assays are explicable in different ways. The peritoneal inflammation assay was found to respond more or less equally to all of the samples. This was not surprising since all of the fibre preparations contained a large proportion of fibres that were longer than 8-10μm. This assay responds to 'Stanton' fibres (Donaldson et al 1989) but does not apparently discriminate between fibres longer than 20μm as shown by the lack of correlation.

The likely importance of durability in determining the ultimate pathogenicity of fibres has led us to suggest that the peritoneal inflammation assay will be of value only if the dissolution potential of fibres is taken into consideration (Donaldson, in press). We suggest that if a fibre is pre-treated to mimic residence in the lung (eg immersion in mild acid for a time) before it is tested in this assay there will be more likelihood that the true toxicity of the fibre sample will be demonstrated. When this pre-treatment was carried out for a soluble vitreous fibre and a more durable ceramic fibre we demonstrated loss of inflammogenic potential in the case of the soluble fibres but little change in the case of the ceramic fibre (Donaldson, in press).

With the macrophage toxicity assay there were differences between fibre samples but these bore no relationship to the known pathogenIcities of the samples. For instance the two least cytotoxic samples were Code 100/E glass and TIMA crocidolite; whilst the glass fibre could be low in toxicity the crocidolite sample is known to be highly pathogenic. Also, long fibre amosite, known to be highly pathogenic, was around the middle of the ranking, being similar to Code 100/475. We have previously reported the Code 100/475 sample to be substantially less toxic than long amosite (Donaldson et al 1993b).

Cytotoxicity to alveolar macrophages was considered at one time to be a major descriptor of the pathogenenicity of dusts, as exemplified by quartz. However, it is now clear that to be fibrogenic in the lung the ability of any dust to stimulate the release of key macrophage products is just as likely to be a determining factor as the ability to kill macrophages. These products, which tend to enhance inflammation and promote pathological change include chemotactic factors, histotoxic agents such as proteases and oxidants, and also cytokines. Thus more subtle measures of 'toxicity' such as activation of cell secretion are becoming the markers of choice for predicting/explaining the toxicity of respirable dusts and these may bear no relation to ability to cause cell death. In addition the same arguments that are used above for the peritoneal inflammation assay are applicable to the toxicity assay and fibres pre-treated with mild acid would be more likely to reflect the long term toxicity of these samples in the lung.

REFERENCES

Brown GM, Cowie H, Davis JMG, Donaldson K. In vitro assays for detecting carcinogenic mineral fibres: a comparison of two assays and the role of fibre size. Carcinogenesis 1986; 7: 1971-1974.

Donaldson K, Brown GM. Assessment of mineral dust cytotoxicity towards rat alveolar macrophages using a ^{51}Cr release assay. Fundamental and Applied Toxicology 1988; 10: 365-366.

Donaldson K, Brown GM, Brown DM, Bolton RE, Davis JMG. The inflammation-generating potential of long and short fibre amosite asbestos samples. British Journal of Industrial Medicine 1989; 46: 271-276.

Donaldson K, Brown GM, Slight J, Addison J. Inflammation in the mouse peritoneal cavity in the investigation of factors determining the biological activity of respirable industrial fibres. Proceedings of the 8th International Conference on Occupational Lung Diseases. 1993 Vol 1 531-539. ILO.

Donaldson K. Addison J. Brown D.M. Brown G.M/ Jones A.D. Miller B.G. Slight J. Davis J.M.G. Short-term bronchoalveolar leukocyte responses in rats inhaling glass micro-fibre and amosite asbestos at equal airborne fibre number. Proceedings of the 8th International Conference on Occupational Lung Diseases. 1993 (Vol. 1 515 - 524. ILO.

Donaldson K. Biological activity of fibres treated to mimic residence in the lung. Toxicol. Letters, in press.

CORRELATION OF IN VITRO AND IN VIVO STUDIES ON THE BIOEFFECTS OF MINERAL PARTICLES

J. Bruch, B. Rehn
Institute of Hygiene and Occupational Medicine, University Clinicum
Essen University, D 45122 ESSEN, Germany

INTRODUCTION

The aim of testing mineral dusts on alveolar macrophages in vitro is to identify potentially health damaging bioeffects of the particles on cells. Such studies could be applied to

- the identification of endangerments to health through emission of new technologic materials
- bio-monitoring of production procedures involving materials with labile physico-chemical constitution
- knowledge on the molecular mechanisms of dust-cell interaction and the effectual routes leading to disease

The critical question is the extent to which the observed cytopathogenic effects are relevant in relation to human disease. In principle, the prerequisites for the extrapolation of cytopathogenic test results to potential disease effects are relatively favourable, since physiologicaly the macrophages are the primary target cells following the deposition of dusts in the alveoli.

Depending on the type of dust, macrophages will follow different specific pathways resulting in pulmonary disease as, for example, fibrosis. Numerous investigations with strongly acting dusts such as crystalline silica or asbestos dusts were able to initially confirm this working principle. Hence, silica dusts are cytotoxic (Allison AC et al. 1966; Marks J et al. 1956) and induce the release of mediators from macrophages at lower doses, stimulating connective tissue formation in neighbouring fibroblasts (Bruch J et al. 1993b; Dubois C et al. 1989). Similar in vitro results were encountered with respect to proliferation/fibrogenicity (Bonner JC et al. 1993; Dubois C et al. 1989) as well as carcinogenicity (Heintz NH et al. 1993) in the case of asbestos dusts.

The following questions need to be clarified for a further validation of the concept of in vitro evaluation of dust toxicity.

1. How do the tests in vitro and the in vivo changes referred to respond to gradated toxicities, particularly with respect to dose/effect assessments

2. Dusts exert different effects on cells, for example, in vitro toxicity, mediator release, formation of reactive oxygen (Donaldson K et al. 1984; Goodglick LA et al. 1986; Goodglick LA et al. 1989; Hansen K et al. 1987) and nitrogen species. Which of the in vitro effects are relevant for the in vivo extrapolation? Are the effects similarly directed or graded?

3. Which of the in vivo changes observed in animals are representative of human diseases, for example, connective tissue formation, weight increase, changes in cellular composition in the alveoli (granulocyte recruitment). What are the species specific peculiarities which cannot refered to human disease.

In the past, there have been relatively few studies addressing these questions in comparison with the vast number of publications on the in vitro effects of quartz or asbestos. The experimental animal and cell toxicological studies involving more than 10 different colliery fine dusts with differing quartz and mineral contents have shown very finely gradated effects . However, the substantial result was the definition of the multivectorial effects of dusts on tissue (Bruch J et al. 1975).

By means of multivariant analyses, vectors were able to be established which exercised a) different and

NATO ASI Series, Vol. H 85
Cellular and Molecular Effects
of Mineral and Synthetic Dusts and Fibres
Edited by J. M. G. Davis and M.-C. Jaurand

b) not always concordant effects along the different sections of the pathogenic pathway. The decisive sections of this pathway have been subdivided into (Bruch J 1987):

1. Mechanisms of deposition and elimination

2. Biopathological effects on the dose translocation from the alveolar region to the inside (lymphtropism of dusts)

3. The specific pathogenic effects of dusts translocated into tissues

The cytotoxicity, parameterised according to the TTC test, shows a highly significant regression for the lymphotropism of dusts, i.e. a vector at the level of tranlocation (Bruch J et al. 1985). By contrast, the assignment of cytotoxicity to fibrogenicity in animal lungs was not very conspicuous, but so for the quartz content of dusts. From this it can be concluded that the specific pathogenic effects with respect to the induction of fibrosis cannot be directly attributed to toxicity.

The following studies are focused in particular on the questions

- Is it possible to make predictions from dusts with gradated in vitro pathogenicity for further in vivo reactions ?
- Are different in vitro detectable effects on cells similarly directed in principle or are there signs for dust specific patterns of cellular effects which can be brought in line with the above mentioned vectors

MATERIAL AND METHOD

Series 1: In vitro toxicity and referencing to in vivo pulmonary reactions

A total of 4 coal mine dusts from a seam with varying quartz and mineral contents (Table 1) were studied for cell damage, lymph node weight and composition of lung surfactant. Dusts were injected into female Wistar rasts (Charles River; 180 g) intratracheally at a dose of 50 mg dust/animal. They were killed 12 months later and the following studies carried out:

- Determination of lymph node weights as an integral measure of tissue effects of the penetrating dusts
- Lavage of the lungs and determination of the phospholipid composition of lung surfactant (LS-PL), parameterisation of the ratio phosphatidylglycerol (PG) and phosphatidylinositol (PI).
- Determination of the TTC-value as a parameter of in vitro toxicity of dusts

TABLE 1

MINERAL- AND QUARTZ CONTENT OF THE MINE DUSTS

Dust Sample	Mineral Content (%)	Quartz Content (%)
Q-KG-01	71.0	15.6
Q-KG-02	57.0	10.2
Q-KG-04	45.0	8.1
Q-KG-05	39.0	4.1

The LS-PL are a metabolic product of the type II pneumocytes and represent an important factor for the alveolar microenvironment. Thus, these cells have a central function in the fibrotic reactions of the lung.

Series 2: Comparative testing of different in vitro effects of dusts of different origin.

The rationale for this section is that different dusts each exert specific effects on the cells with respect to free radical formation (a), release of the mediator TNF (b) and lysosomal damage (glucuronidase) as well as general cell damage by TTC. For these studies, dusts were explicitly chosen from various areas, namely quartz, coal mine dust, silicon carbide and asbestos (chrysotile). All dusts were of uniform grain size distribution and the samples stemmed from the following sources: quartz: Dörentruper quartz DQ 12; coal mine dust gathered as authentic work place dust, quartz content 0.9%; silicon carbide: standard product F1200 of FEPA-Norm; chrysotile: UICC sample.

IN VITRO METHODS

For the cell assays alveolar macrophages of male guinea pigs (strain: PBW; b.w 400-600 g) were harvested by routine method of bronchoalveolar lavage.

TTC (triphenyl-tetrazolium-chloride) reduction activity of macrophages was measured using a modified method described by Absolom (Absolom DR 1986). The test was performed in quadruplicate.

Hydrogen peroxide release of alveolar macrophages was measured in an assay based on the method of De la Harpe and Nathan (De la Harpe J et al. 1985); H_2O_2 is taken as a parameter for the release of reactive oxygene species (ROS) from cells. Briefly, microtiter plates were incubated with macrophages (3×10^5 cells/well); after a 16 hour period of settlement the dust samples suspended in minimum essential medium (MEM) were added at doses of 60, 120, 180 and 240 µg per 1×10^6 cells respectively and incubated for 90 minutes. During the dust incubation period the hydrogen peroxide secretion of macrophages was monitored (fluorescence plate reader) for 90 minutes by the horseradish peroxidase-catalyzed oxidation of fluorescent scopoletin. The test was performed in quadruplicate.

The release of glucuronidase was measured enzymatically based on the method of Absolom (Absolom 1986). Briefly, microtiter plates were incubated with macrophages (3×10^5 cells/well); after a two hour period of settlement the dust samples suspended in minimum essential medium (MEM) were added at doses of 60, 120, 180 and 240 µg per 1×10^6 cells respectively and incubated for 16 hours. After the dust incubation 50 µl of the supernatant per well were incubated for 2 hours at a temperature of 37°C with 50 µl of the assay mix (10 ml Na-acetate buffer ph 5; 2ml 1% Triton in a.bidest; 2 ml 0.0133 M p-nitro-ß-D-glucorunide). Afterwards the reaction was stopped by adding 0.2 M NaOH (100 µl per well) and the extinction was was measured at 405 nm in a plate reader. The test was performed in quadruplicate.

Cell viabiltity: Following the H_20_2 mesurement, the cells were washed and incubated with a mixture of ethidium bromide (non viable cells) and fluorescein diacetate (viable cells) and fluorescence was quantified using the fluorescence plate reader.

TNF-alpha : Microtiter plates were incubated with macrophages (3×10^5 cells/well); after a two hour period of settlement the dust samples suspended in Ham´s F 10 were added at doses of 25 and 100 µg per 1 x 106 cells and incubated for 24 hours. Additionally protease inhibitor (PMSF 1 mM) was added to stop possible proteolytic activity in the culture fluid

The activity of TNF-alpha was monitored using a cell lytic assay based on the method of Aggarwal et al. (Aggarwal BB et al. 1985). Briefly mouse L-929 fibroblasts were grown in 96-well flat-bottom trays at 50000 cells (100µl volume) per well and incubated with the macrophages supernatant. Following the18 hours incubation period, the the culture fluid was removed, the fibroblasts washed and stained with a 0.5 % solution of crystal violet. Afterwards cell lysis was measured colorimetric in an plate reader. Untreated cells and cells treated with known concentrations of recombinant murineTNF alpha (Boehringer, FRG) served as controls. As a additional control one part of the samples were neutralized using polyclonal rabbit anti-murine TNF alpha (B Dunn. Alsbach FRG) One unit of TNF-alpha is defined as the amount of TNF alpha required for 50 percent cell lysis. The test was performed quadruplicate.

Statistical analysis was performed using the one sided Student's t-Test and the Mann Whitney-U Test. An alpha level of less than 0.05 was considered significant. Statistics and correlations were computed with a statistic programme (Statview, Version 1.03, Acabus Concepts Inc., USA).

Series 3: Effects of ROS release and cytotoxicity of three dust samples on lung clearance.

For these studies, the results of the above investigations on in vitro damage of dusts were linked to those of inhalation studies using the same samples. The inhalation studies were carried out according to the standard Essen procedure (Bruch J et al. 1993c). Female rats weighing 170 g, 2-3 months old, inhaled various dust samples under standardised conditions (OECD-Norm), whose particle distribution was made uniform by previous sifting. The exposure times were 2 x 5 days, 2 weeks overall, 5 hours per day; in the standard procedure, animals are killed 3, 21 and 90 days after completion of inhalation. In the present study, short term elimination is of interest. The values for 3 day retention were compared with the in vitro parameters 'ROS release' and 'cytotoxicity'.

RESULTS

Series 1: The question was to what extent the cell damaging effect of dusts from work places affects different lung reactions. In this study, lymph node weight was assessed as an expression of long term tissue reaction and the ratio of PG to PI as that of pneumocyte type II activity. The coal mine dusts stemmed from a defined work area underground. These differed merely in their quartz content, ranging from 4.1 % to 15%. Figure 1a shows that the two samples with the highest quartz content of about 30% result in similar cell damage in the TTC test. By contrast, the two samples with the lower quartz contents show correspondingly lower cell damage. When one assesses the reaction in the experimental animals 12 months after intratracheal injection, it turns out that the gradation corresponding to the varying mineralogical composition of the dusts becomes very much clearer. The lymph node weights are completely in agreement with the quartz content, whereby dusts with the highest quartz content give rise to the largest lymph node reaction (fig. 1b). A similarly distinct gradation is seen in the ratio PG/PI as an expression of the pneumocyte type II reaction to chronic dust load. The dust with the highest quartz content induces the strongest decrease in the ratio to a value slightly below 1.0, while those for the intermediate samples (KG02, KG03) are correspondingly higher whereas the sample KG04 results in a ratio of 2.0 close to the controls (fig 1c).

In summary it can be stated that the cytotoxicity assay approximately agrees with the gradation of samples with respect to their differences in mineral contents. By contrast, the lymph node reaction appears to be closely correlated to the different compositions of coal mine dusts. This result agrees with earlier studies on coal mine dusts (Le Bouffant L 1985). The results for the composition of LSF are new. The previously mentioned assumption from short term experiments that the ratio of PG/PI possibly indicates a pathological situation in the lungs and in particular reflects the reaction of the pneumocyte type II to dust load could be entirely confirmed(Bruch, Rehn et al. 1993c; Honda Y et al. 1988; Horiuchi T et al. 1990; Robinson PC 1988). This also substantiates previous investigations which emphasize the significance of the pneumocyte type II reaction in the pathological lung reaction (Bruch J et al. 1993a; Crouch C1990).

Series 2.:The hypothesis underlying this study is that with dusts of different origin, the effects of the particles on the different cell functions and their restrictions may possibly not be concordant. The dust samples used here reflect wide ranges of industrial production. The parameters studied can be considered to be key parameters of dust effects: the formation of reactive oxygen species (ROS), the release of mediators (parameterised by TNF) and the extracellular glucuronidase as a measure of lysosomal damage by dusts as well as the TTC values as a measure of general cellular damage. As the results in Tab. 2 show, the respective effects of dusts can be assigned very specifically to the special samples from the various industrial branches. Thus, for example, one finds a large release of glucuronidase activity in the case of quartz, corresponding to the well known damage to lysosomes by silica dusts. The release of ROS is not in accord with this parameter. Quartz induces a weakly enhanced release of ROS, while chrysotile in a significantly higher effect in agreement with the results from other studies. The extremely high release of ROS after incubating cells with silicon nitrite is striking. In turn, this dust exercises only a relatively slight

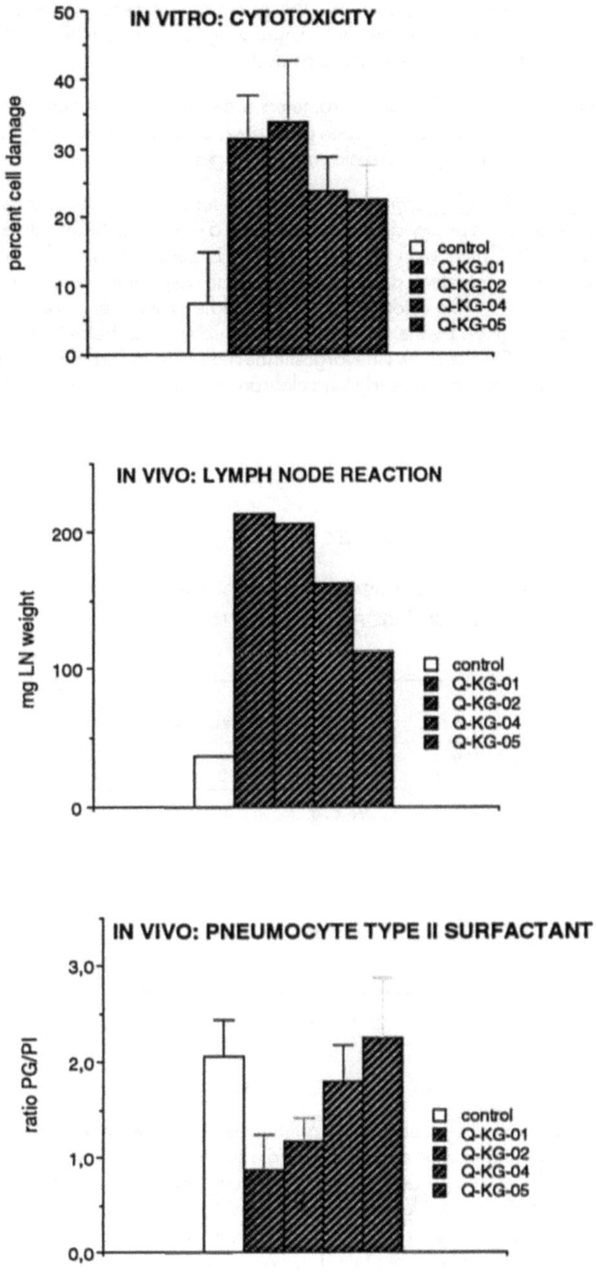

<u>Figure 1</u> in vitro cytotoxicity in relation to long term (12 months) in vivo changes as shown by lymph node changes and alteration of the ratio lung phospholipid exerted by four mine dusts

effect on the lysosomes (glucuronidase). The parameter TNF as expression of cell proliferative effects from the macrophages is particularly increased in the case of quartz and chrysotile : this also agrees with the known strong fibroblastic effects of quartz and asbestos dusts.

The original question to what extent the different parameters are stimulated in macrophages by dusts of different origin can be answered unequivocally: each dust shows a highly specific reaction pattern with respect to the different pathogenic effects in alveolar macrophages.

Series 3: Cytopathogenicity and lung clearance: The results on lung clearance for three dust samples studied show clear differences: coal is poorly cleared (approxx. 1.5 mg/lung), SiC is still retained to about 0.7 mg/lung while the quartz/SiC mixture in the ratio of 20/80 is cleared best of all. The three samples have different non concordant in vitro effects on macrophages (Table 2). Thus, the coal mine dust induces a relatively high secretion of TNF, while cytotoxicity is of a middle order. Very apparent is the almost complete inhibition of ROS formation by the coal mine dust, which is even below that of the untreated control or cells treated with SiC. Quartz shows the largest induction of ROS formation. The data indicate that ROS formation has a direct effect on the early lung clearance while none is seen for the other in vitro parameters studied (fig. 2).

TABLE 2

DUST DEPENDENT PATHOGENICTY MEASURED BY TNF-, ROS - STIMULATION, GLUCURONIDASE RELEASE AND TTC

Dust Sample	TNF a	ROS	Glucurondase	TTC
units dose per 10^6 cells	U/ml 25 mg	ng/ml 120 mg	ng/ml 120 mg	% 120 mg
quartz	20.3	97 ± 38	353	87
silicon carbide	7.4	45 ± 25	50	2
coal mine dust	19.0	2 ± 1	67	87
chrysotile	16.4	141 ± 21	141	90
cell control	7.9	64 ± 14	35	0

DISCUSSION

The present study was concerned with the possible quantitative and qualitative assessment of endangerment to health through dusts using in vitro methods. First of all, the question is whether samples from a qualitatively identical type of dust with gradated quantitative composition resulted in varied cell damage which can be compared with similar gradations in the tissue after being administered to animals. The colliery dusts from a distinct seam with quartz concentrations from 4.1% to 15.6% used by us fulfill these requirements. A further aspect of the experimental set-up was to compare the short term cell experiments with long term animal effects which in the end are relevant for possible harmful effects on humans. In the tests chosen, dusts were injected intratracheally and the lymph node weight determined 12 months later. These weights showed a very clearly graded development in relation to the different contents, where dusts with higher quartz content induced significantly more weight gain. The lymph node weights are an integral of the total tissue effect, whereby both the reticuloepithelial as well as the fibroblastic reactions are included (Le Bouffant L 1985). The short term tests with the aid of the TTC-test resulted in an approximately similar gradation.

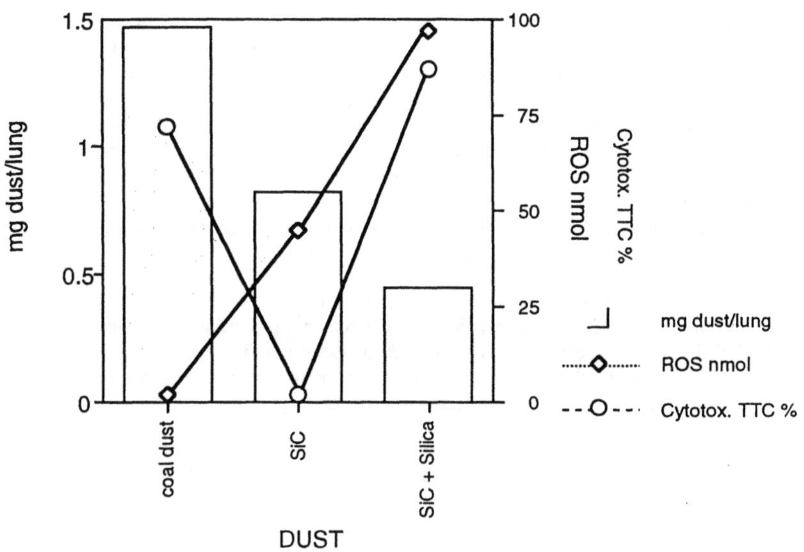

Figure 2 Short term elimination of three dust types (columns); related in vitro tests are ROS formation and TTC cytotoxicity (lines; values from table 2)

The ratio PG/PI from lung surfactant was determined as a further in vivo parameter. From our own previous studies (Bruch J et al. 1993c) as well as results from the literature (Honda Y et al. 1988; Horiuchi T et al. 1990), a PG/PI ratio < 2.0 is considered to be pathological. The results show that the ratios PG/PI completely parallel the weight development of lymph nodes and correlate with the quartz content in the samples. The alteration in the ratio PG/PI is due to a functional change in the pneumocyte type II in the alveolar membrane. In previous studies a decrease in the ratio could be seen only after a relatively short term burden. The results of this study show that the ratio remains lowered in the long term after chronic alterations to the lung. This indicates that the pneumocyte type II reaction apparently not only represents a short term inflammatory response to a lung damaging dust effect but is in principle incorporated in the pathological process of the long term dust effect. This is underlined by the results from other studies. From inhalative quartz burden tests using two different strains of mice, it was possible to define one strain (BALB/C) with more pronounced collagen development as a high responder in comparison with the co-studied DBA/2 strain with lower collagen formation (= low responder) (Bruch J et al. 1993a). Of all the other parameters studied, it could be shown that only the amount of surfactant as an expression of the pneumocyte type II reaction was clearly more expressed in the high responders. We are thus of the opinion that the pneumocyte type II alteration and, in agreement, the alteration in composition and amount of surfactant represent an important component of the pathological lung reaction to dusts.

The present investigation shows that a dust of gradated composition in conjunction with a harmful component (quartz) gives rise to a corresponding change in the composition of lung surfactant. In this connection, this finding lends significance to the question as to the extent to which in vitro experiments can predict the harmfulness of dusts for a further reason. Lung surfactant can be obtained from humans by means of bronchoalveolar lavage and analysed analogously to the animal experiment. We have studied such lavages from patients exposed to asbestos with respect to the composition of lung surfactant and could also observe a decrease in the PG/PI ratio (data to be published in 1994).

The question of the predictability of potential pulmonary harmfulness by dusts also applies to emissions from various industrial areas. The processing of work substances in principle always produces emissions whose harmfulness to tissues is not known a priori. In a second experimental test set-up, dusts from different areas were chosen with a qualitatively wide range of composition: fibre dusts (chrysotile) and mineral dusts (quartz, colliery mixed dust, silicon carbide). These dusts exert different pathogenic effects on alveolar macrophages: release of reactive oxygen radicals, mediators, lysosomal damage (glucuronidase). The results clearly show that for each of the pathogenic dust effects mentioned there is a specific response pattern for the respective dust. These findings are new since most publications to our knowledge have in principle reported redundant dust effects in cases where several parameters of cell pathogenicity were determined (Brown RC et al. 1980; Daniel FB et al. 1980).

The biopathological significance of this finding can only be assessed on a trial basis. In agreement with other studies, carcinogenically acting dusts show a corresponding gradation in free radical formation (Mossman BT 1993) which was also found in our study as exemplified by quartz and chrysotile . Similar findings could be confirmed for the induction of TNF formation in relation to the large values with quartz and chrysotile , since both dusts are known to be strongly fibrogenic.

In the context of this study, the question was whether and which of the observed pathogenic effects on cells correspond to an effect of dusts on animals. Of the three effect levels of dusts mentioned in the Introduction (1: Deposition/Elimination, 2: Translocation of dusts into the tissue, 3: Specific tissue pathogenic effect), the first level of deposition/elimination is, hierarchially, the highest and, in our opinion, also the most important in the pathogenesis of pneumoconioses. Three of the dusts mentioned could also be examined in parallel in inhalation studies enabling statements to be made as to the effect of dusts on short term elimination. The first result is the dependence of elimination on the type of dust inhaled. Since the dusts were screened for comparable particle size, this effect can not be attributed to a different deposition due to differing particle size distribution. The differences in dust retention can be assigned to biological effects on the elimination processes of dusts. The best agreements as to ROS formation are registered when comparing the cell pathological effect of the three dusts with the gradations of elimination. The highest initial elimination is seen for dusts with high ROS formation whereas those with poorer initial elimination (coal dust) are only eliminated to a very delayed extent. No correlation to cytotoxicity, which is decisive for the translocation of dusts from the alveoli into the interstitium (Bruch J et al. 1985), is seen. This applies particularly to the relatively strong cytotoxic colliery dusts with very delayed elimination. It remains to be discussed whether the differences in ROS formation by the different dusts, especially the large depression caused by coal-rich colliery dust, are due to a specific inhibition by the dust or an unspecific cell damage. This question can be answered unequivocally since other parameters such as, for example, TNF formation are enhanced by this colliery dust, thus excluding a general damage to cells. Overall, this study comes to the conclusion that there are specific relations between the different pathogenic effects on cells in vitro and changes in vivo in each case. With respect to lung clearance, it appears that harmfulness to cells, mediator release and glucuronidase activity are of a lower significance. The increased elimination may possibly be attributed to an enhanced cell recruitment, where other mediators not determined here may be of importance apart from ROS formation. (IL-8).

Conclusions

1. Estimates of the harmfulness of dusts of varied composition from a qualitatively similar work substance can be carried out using in vitro methods. However, it is necessary to reference the type of cell damage to relevant pathogenic reactions in the lung including, for example, effects on lung clearance, collagen formation, pneumocyte type II/ lung surfactant changes.

2. With dusts of varying origin with various work substances and unknown pulmonary effects, it is essential to parameterise a broad spectrum of pathogenic cell effects. In our opinion, the main cellular effects of dusts are:

- cellular toxicity
- mediator release
- ROS formation

The studies have clearly shown that the determination using one parameter does not provide sufficient information on potential harmfulness.

3. An estimation of the significance of pathogenic cellular effects should not only be carried out by means of referencing to corresponding in vivo animal systems. The present investigation shows one way, for example, via bronchoalveolar lavage, by which a referencing of animal experiments to long term results in humans could be carried out. An important part of the comparative assessment of dust effects is the alteration to lung surfactant; this parameter indicates the functional state of the type II cells, which play an important role in lung pathology, for example, fibroses according to the present results.

LITERATUR

Absolom DR (1986) Basic Methods for the study of Phagocytotsis. Immunochemical Techniques. Part I: Phagocytosis and Cell -mediated Cytotoxicity. Academic Press.

Aggarwal BB, Kohr WJ, Hass PE, Moffat B, Spencer SA, Henzel WJ, Bringman TS, Nedwin GE, Goedel DV and Harkons RN (1985) Human Tumor Nercrosis Factor. Production, Purification, and Characterization. J of Biological Chemistry. **260**: 2345-2354.

Allison AC, Harington JS and Birbeck M (1966) An examination of the cytotoxic effects of silica on macrophages. J Exper Med. **124**(2): 141-153.

Bonner JC, Goodell AL, Coin PG and Brody AR (1993) Chrysotile asbestos upregulates gene expression and production of alpha-receptors for platelet-derived growth factor (PDGF-AA) on rat lung fibroblasts. J Clin Invest. **92**(1): 425-30.

Brown RC, Chamberlain M, Davies R, Morgan D, Pooley FD and Richards RJ (1980) A Comparison Of 4 'In Vitro' Systems Applied To 21 Dusts. The In Vitro Effects of Mineral Dusts, Brown, R. C., M. Chamberlain, R. Davies, and I. P. Gormley, Editors; Academic Press, London, pages. :

Bruch J (1987) Value of in vivo and in vitro tests in regard to the evaluation of mixed dusts. Proc. Conf on: Silicose et pneumoconioses a poussieres mixtes a Chantilly 28-30 avril 1986. Colloque Inserm. **155**: 213-226.

Bruch J, Hilscher W and Krämer U (1975) Pathogenicity to animals of fine dusts from Ruhr mines. Inhaled Particles IV. Oxford/ New York , Pergamon Press.

Bruch J, Rehn B, Song H, Gono E and Gleichmann E (1993a) A new approach in silicosis research using balb/c (high responder and dba/2 (low reponder) mice as an animal model. Proceedings 8 th International Pneumoconioses Conference; Prag 1992. **in press:**

Bruch J, Rehn B, Song H, Gono E and Malkusch W (1993b) Toxicological investigations on silicon carbide. 2. In vitro cell tests and long term injection tests. Br J Ind Med. **50**: 807-813.

Bruch J, Rehn B, Song H, Gono E and Malkusch W (1993c) Toxicological investigations on silicon carbide.1. Inhalation studies. Br J Ind Med. **50**: 797-806.

Bruch J and Rosmanith J (1985) Properties of mixed dusts and their relationship to pneumoconiosis. In vitro effects of mineral dusts. 3rd International workshop Schluchsee Germany Oct1-4, 1984. Berlin , Springerverlag .

Crouch E (1990) Pathobiology of pulmonary fibrosis. Am J Physiol. **259**: L159-L184.

Daniel FB, Beach CA and Hart RW (1980) Asbestos-Induced Changes In The Metabolism Of Polycylic Aromatic Hydrocarbons In Human Fibroblast Cell Cultures. The In Vitro Effects of Mineral Dusts, Brown, R. C., M. Chamberlain, R. Davies, and I. P. Gormley, Editor; Academic Press, London, pages. :

De la Harpe J and Nathan CF (1985) A Semi-Automated Micro-Assay for H_2O_2 Release by Human Blood Monocytes and Mouse Peritoneal Macrophages. J. Immunological Methods,. **78**: 323-336.

Donaldson K and Cullen RT (1984) Chemiluminescence of asbestos-activated macrophages. Br J Exp Path. **65**: 81-90.

Dubois C, Bissonnette E and Rola-Plecszczynski MR (1989) Leukotriene B4 and Tumour Necrosis Factor Production after in vitro exposure of rat alveolar macrophages to mineral dust: Potential role in fibrogenesis. Effects of Mineral Dusts on Cells. Berlin, Heidelberg, Springer.

Goodglick LA and Kane AB (1986) Role of Reactive Oxygen Metabolites in Crocidolite Asbestos Toxicity to Mouse Macrophages. Cancer Research. **46**(11): 5558-5566.

Goodglick LA, Pietras LA and Kane AB (1989) Evaluation of the causal relationship between crocodolite asbestos-induced lipid peroxidation and toxicity to macrophages. Am Rev Respir Dis. **139**: 1265-1273.

Hansen K and Mossman BT (1987) Generation of superoxide (O2-.) from alveolar macrophages exposed to asbestiform and nonfibrous particles. Cancer Res. **47**(6): 1681-6.

Heintz NH, Janssen YM and Mossman BT (1993) Persistent induction of c-fos and c-jun expression by asbestos. Proc Natl Acad Sci U S A. **90**(8): 3299-303.

Honda Y, Tsunematsu K, Suzuki A and Akino T (1988) Changes in phospholipids in bronchoalveolar lavage fluid of patients with interstitial lung diseases. Lung. **166**(5): 293-301.

Horiuchi T, Mason RJ, Kuroki Y and Cherniack RM (1990) Surface and tissue forces, surfactant protein A, and the phospholipid components of pulmonary surfactant in bleomycin-induced pulmonary fibrosis in the rat. Am Rev Respir Dis. **141**: 1006-1013.

Le BL (1985) Relationships between In Vitro Toxicity, Tissue Effects and Mineralogical Composition of Coal Mine Dusts. In Vitro Effects of Mineral Dusts. Third International Workshop, E. G. Beck, J. Bignon, Editors; Berlin, Springer Verlag, pages. :

Marks J, Mason MA and Nagelschmidt G (1956) A Study Of Dust Toxicity Using A Quantitative Tissue Culture Technique. British Journal of Industrial Medicine. **13**(3): 187-191.

Mossman BT (1993) Mechanisms of asbestos carcinogenesis and toxicity: the amphibole hypothesis revisited. Br J Ind Med. **50**: 673-676.

Robinson PC (1988) Abnormalities in bronchoalveolar lavage fluid phospholipids. Am Rev Respir Dis. **137**: 585-591.

Comparative studies with natural and man-made mineral fibres in vitro and in vivo

Riebe-Imre, M., M. Aufderheide, M. Emura, M. Straub, M. Roller[1]
U. Mohr and F. Pott[1]

Margrit Riebe-Imre
Institut für Experimentelle Pathologie
Medizinische Hochschule Hannover
Konstanty-Gutschow-Str. 8
D-30625 Hannover
Germany

Abstract

Several types of fibrous dusts have caused lung tumours and mesotheliomas in humans and/or in laboratory animals. In order to study the effects of the fibrous dusts, cell toxicity and transformation were investigated. The cell systems used included epithelial lung cells from the Syrian golden hamster, human bronchoepithelial cells and mesothelial cells from the rat.

In general, the natural mineral fibres showed a higher cytotoxic and transforming potency per fibre than did the man-made fibres. Nickel powder and nickel oxide were used as well known chemically carcinogenic dusts and titanium dioxide as a negative non-fibrous dust control The extent of the biological fibre mass, whereas among the natural mineral fibres, crocidolite was more active on the basis of fibre numbers.

Corresponding with intraperitoneal carcinogenicity studies, the in vitro studies demonstrated a clear transforming capacity of the naturally occurring mineral fibres. This effect was much less for the glass microfibres, silicon carbide fibres and nickel dusts. In the reaction of the different cell systems to fibre treatment, no systematic differences could be observed.

---

[1]Medizinisches Institut für Umwelthygiene an der Heinrich-Heine-Universität Düsseldorf, D-40028 Düsseldorf, Germany

NATO ASI Series, Vol. H 85
Cellular and Molecular Effects
of Mineral and Synthetic Dusts and Fibres
Edited by J. M. G. Davis and M.-C. Jaurand

Introduction

The attempt to predict human cancer risk for mineral fibres offers the possibility to approach this problem by epidemiologic and experimental studies on animals. Adequate epidemiologic knowledge of the effects of mineral fibres is limited to asbestos and erionite fibres (Baris, 1987; Doll & Peto, 1985; WHO, 1986) whereas only little information is available for the man-made mineral fibre (IARC, 1988; IPCS, 1988). Therefore subchronic and chronic animal studies have been carried out to estimate the toxic and carcinogenic properties of these mineral dusts. In addition to these investigations, in vitro experiments have been performed to determine the toxic, genotoxic and even transforming capacity of different fibre types. The acceptance of the experimental data with regard to their relevance for the in vivo situation is always combined with the question whether the results of the in vitro studies can be compared to those of animal studies. To clarify this question, experiments using several fibrous and particulate dusts have been carried out with different in vitro systems, to test for their carcinogenicity in vivo. The cell systems represent possible targets for the interaction with fibres after inhalation and consist of epithelial lung cells from the Syrian golden hamster, human bronchoepithelial cells and mesothelial cells from the rat. In order to study the effects of natural and man-made fibres, the cytotoxicity and transformation based on anchorage-independent growth in soft agar were investigated. The data are described in more detail elsewhere (Riebe-Imre et al., 1992).

Materials and Methods

Cell lines and culture conditions
The Syrian hamster epithelial cell line (M3E3/C3) was derived

from the lung of a foetus on day 15 of gestation. Cells were grown in a medium consisting mainly of RPMI 1640 (80% Flow, Meckenheim) and foetal bovine serum (BSA; 20% Flow) as described previously (Emura et al., 1982). Cultures were maintained at 37°C in a humidified atmosphere with 10% CO_2 and 90% air. Confluent cultures were harvested by trypsinization (0.25% trypsin, 2% EDTA in PBS) and resuspended in the medium after low speed centrifugation. The epithelial character of the cell line and its application to pulmonary toxicology have been described elsewhere (Emura et al., 1989, 1990).

The human foetal bronchial epithelial cell line - HOK - is derived from a foetus artificially aborted at 18 weeks of gestation when the continuation of the pregnancy was contra-indicated for the mother. The establishment of the epithelial cell line was performed by applying replica explant techniques. After continuous cultivation, rapidly proliferating epithelial cell populations were stored frozen at passages 6-8.

The HOK cell line was cultured in plastic dishes (Falcon) containing a conventional medium (CM) consisting of RPMI 1640, 110μg/ml Na-pyruvate and 20% foetal bovine serum. The dishes were kept in a highly humidified atmosphere with 10% CO_2 and 90% air at 37°C. The cells were passaged using a mixture of 25% trypsin and 0.02% EDTA dissolved in a Ca^{2+}- and Mg^{2+}-free phosphate buffered saline (PBS). The epithelial character of the cells was confirmed by positive intraplasmatic PAS- and NTB-staining, immunoreactivity for cytokeratin (up to five passages), carcinoembryonic antigen (CEA) and epithelial membrane antigen (EMA). The usefulness of this system for toxicological studies has been demonstrated by Emura et al. (1989).

The mesothelial cells from the diaphragm of Han-Wistar rats RaD/C1, which showed the characteristic morphology of epithelial cells, were established from foetal rats at day 19 of gestation also by the replica explant techniques and cultured under the above-mentioned conditions. The subculti-vated cells were stored frozen at passages 3-8.

Mineral dusts

The characterization of the fibre samples was performed by

Dr. Bellmann, Fraunhofer Institute of Toxicology and Aerosol Research, and is given in Table 1.

Table 1: Characterization of fibre dusts

Fibre type	mass of critical fibres %	critical fibres/ng mean	fibre length (µm) 10%<	50%<	90%<	fibre diameter (µm) 10%<	50%<	90%<
Chrysotile, UICC-B	54	809	0.3	0.7	1.9	0.03	0.06	0.12
Crocidolite	42	99	0.9	1.8	4.7	0.11	0.19	0.32
Erionite, Oregon	28	40	0.8	1.8	4.6	0.08	0.16	0.41
Silicon carbide, Tateho	64	107	1.5	3.1	9.5	0.19	0.31	0.48
Glass:								
M 104/475[a]	88	158	0.7	2.3	8.2	0.06	0.14	0.38
B-3K[a]	54	57	1.6	3.3	7.7	0.19	0.37	0.70
B-1M[a]	15	3	6.7	10.7	21.1	0.88	1.68	2.49
RCF H II[b]	15	2	3.4	13.1	51.1	0.25	0.84	2.23

[a] chemical composition quoted by Pott et al., 1991a
[b] aluminium silicate refractory ceramic fibres

Results of long-term carcinogenicity studies were available for all dust samples used; they are summarized in Table 2.

Cell viability

Cells of the lines M3E3/C3, RaD/C1 and HOK21 (5×10^4 cells) were seeded in $25cm^2$ Falcon flasks with 5ml growth medium. 24h later the cells were treated for 96h with different doses of mineral fibres which were suspended in medium after being sterilized. At the end of exposure, cells were trypsinized and cell viability was determined by the trypan blue exclusion method in a haematocytometer.

Plating efficiency

From each treatment group 200 viable cells were plated in an $8cm^2$ Petri dish (Falcon) and cultured under the above-mentioned culture and test conditions for 8 days without medium change. The cells were fixed in methanol, stained with Giemsa and colonies were counted. Plating efficiency was calculated as the percentage of colony formation relative to the control groups.

Table 2: Results of intratracheal and intraperitoneal carcinogenicity studies with dusts which were also tested in cell culture experiments

Dust	Dose i.p. or i.tr. [mg]		Dose i.p. or i.tr. fibres* x 10^9	Fibres /ng	Fibre size median [µm] Length	Fibre size median [µm] Diam.	Tumour incid. [%]	Ref.: Pott et al.
Fibres								
Chrysotile	0.05	i.p.	0.04	809	0.7	0.06	33	1989
UICC-B	0.25	i.p.	0.2				68	1989
	1	i.p.	0.8				83	1989
Crocidolite	10	i.tr.	1.0	99	1.8	0.19	43	1987
	0.5	i.p.	0.05				56	1987
	2	i.p.	0.2				88	1987
Erionite	7.5	i.tr.	0.3	40	1.8	0.16	5	1991b
	0.5	i.p.	0.02				48	1987
	2	i.p.	0.08				90	1987
Glass:								
M-104-475	5	i.tr.	0.8	158	2.3	0.14	2	1991b
	10	i.tr.	1.6				13	1991b
	10	i.tr.	1.6				15	1987
	2	i.p.	0.32				26	1989
	5	i.p.	0.8				64	1989
B-1M	20	i.p.	0.05	3	10.5	1.7	2	1991a
	60	i.p.	0.16				2	1991a
B-3K	6.7	i.p.	0.38	57	3.3	0.37	21	1991a
	20	i.p.	1.14				64	1991a
Silicon	5	i.tr.	0.5	107	3.1	0.31	24	1991b
carbide	10	i.tr.	1.0				31	1991b
	1.25	i.p.	0.13				62	1991a
	6.25	i.p.	0.7				77	1991a
	25	i.p.	2.7				97	1991a
Non-fibrous Dusts								
Nickel	6	i.p.	-	-	-	-	26	1987
powder	9	i.tr.	-	-	-	-	25	1987
	6	i.p.	-	-	-	-	12	1991c
	12	i.p.	-	-	-	-	15	1991c
	25	i.p.	-	-	-	-	71	1991c
Nickel	50	i.tr.	-	-	-	-	27	1987
oxide	150	i.tr.	-	-	-	-	32	1987
"green"	25	i.p.	-	-	-	-	35	1991c
	100	i.p.	-	-	-	-	42	1991c
Titanium dioxide	45	i.tr.	-	-	-	-	5	1991b
NaCl		i.tr.					0	1991b
		i.p.					2	1989

* Fibre definition: Aspect ratio > 5/1; length > 5µm, diameter < 2µm

Proliferating activity

For the cell lines of rat and man, a proliferation factor (PF) was determined as a marker for chronic toxicity after toxic treatment. Following exposure, cells were trypsinized and seeded in defined numbers in several flasks and cultured under the above-mentioned conditions. After 8 days, cell viability was determined by observing the membrane integrity of the cells with the trypan blue exclusion method. The proliferation factor was calculated as follows:

$$PF = \frac{\text{viable cells after 8 days}}{\text{viable seeded cells}}$$

Transformation

The determination of transformation was based on anchorage-independent growth in soft agar. The treatment was carried out in the same way as for the cytotoxicity study. The cells of the different treatment groups were passaged 5 times by trypsinization at intervals depending on the growth speed in reaching about 80% confluency. The cells were then introduced into soft agar cultures to detect anchorage-independent growth indicating transformation. The plating efficiency of the cells for each dose group was determined as described above. Six weeks later, evaluation of cultures was performed by counting only colonies with a diameter greater than 30µm. Transformation frequency was calculated as the percentage of transformed colonies relative to the control cells unexposed to mineral fibres.

Statistics

The relationship between applied dust concentration and the logarithm of the effect variable (cell viability, colony formation, proliferation factor and transformation, respectively) was analysed by linear regression. The slope of the regression line was used as a measure of the potency of a dust. From the slope coefficient (k), the concentration that corresponds to a 50% effect (ED_{50}) can be calculated by 0.693/k.

Results and Discussion

The fibre samples tested in this study included naturally occurring dusts such as chrysotile, crocidolite and erionite, as well as man-made fibres, and moreover well-known carcinogenic nickel dusts with low and high solubility and "inert" titanium dioxide (see Table 2). For the comparison of carcinogenicity and cell transformation, results of the five fibre samples with relatively similar fibre size (crocidolite, erionite, silicon carbide, glass M-475-104, B-3k) and the non-fibrous dusts were used.
The application of increasing doses of crocidolite, chrysotile and erionite caused a significant decrease in the cell survival of the epithelial lung cells of the hamster (Figure 1).

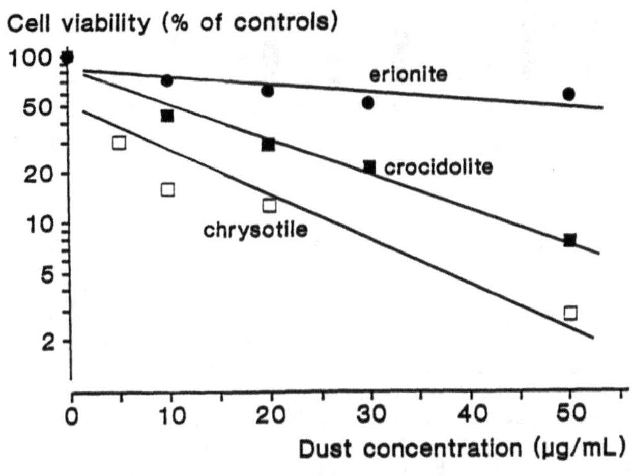

Figure 1: Results of the test of acute cytotoxicity of natural fibrous dusts on the Syrian hamster lung epithelial cell line M3E3/C3

All fibre types showed a clear cytotoxicity. Chrysotile has, on the basis of the dust mass, a much stronger effect than crocidolite and erionite. The effects differed in the three cell systems depending on the fibre dust applied. For comparison of the different dusts, the relative potency was calculated by relating the slope of the regression line or the reciprocal of the ED50-value of each experiment to the value obtained with crocidolite and the hamster lung epithelial cells. The effect of crocidolite on hamster epithelial lung cells is defined as 100.

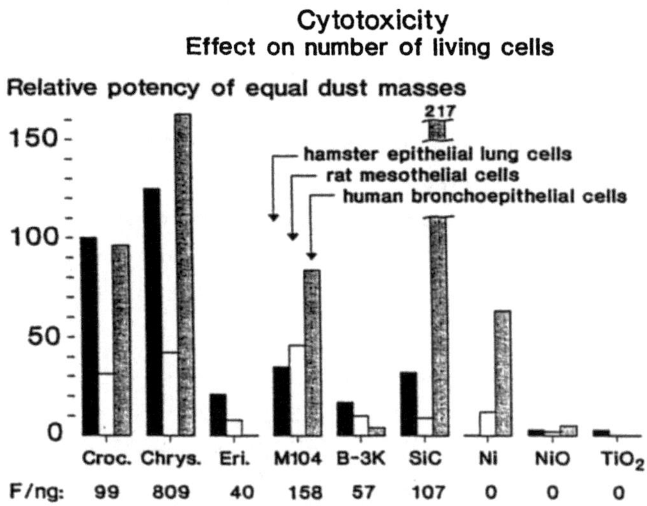

Figure 2: Relative cytotoxic potencies of equal dust masses of fibrous and control dusts. The cytotoxic potency of each dust calculated from the dose-effect relationship has been related to the value for crocidolite on hamster epithelial cells, which has been defined as 100. Abbreviations: Croc.: crocidolite; Chrys.: chrysotile: Eri.: erionite; M104: glass fibres M104/475; B-3K: glass fibres B-3K; SiC: silicon carbide fibres; Ni.: nickel powder; NiO: nickel oxide; TiO_2: titanium dioxide; F/ng: number of fibres with length > 5µm, diameter < 2µm and L./D. > 5/1 per nanogram dust

The relative potency of the same masses of the different dusts is shown in Figure 2. The effect on the rat mesothelial cells is altogether weaker than on the hamster epithelial lung cells and even on the human bronchoepithelial cells. Chrysotile revealed the strongest effect whereas the non-fibrous dusts showed only a weak reaction. The weak cytotoxic effect of erionite in comparison to the glass fibres and the other naturally occurring fibre types is remarkable, because erionite is considered to be the most active dust. This statement is true in the case of the dust mass administered. The circumstances change, however, according to the number of fibres administered.

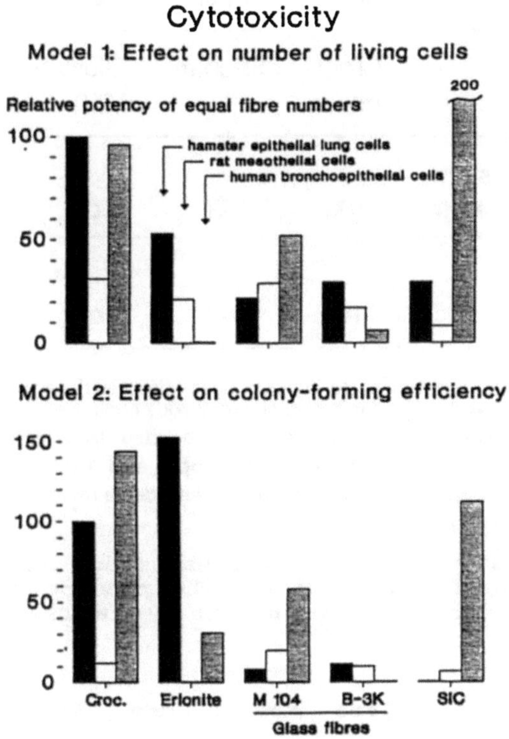

Figure 3: Relative cytotoxic potencies of equal fibre numbers of five fibrous dusts with similar specific fibre contents. The potencies have been calculated from the mass-related potencies (cf. Fig. 2) and from the specific fibre contents (fibre numbers per nanogram dust). All values have been related to crocidolite and hamster epithelial cells.

The results for cell viability and colony-forming efficiency are shown in Figure 3. In this parameter too, there is relatively little influence on the rat mesothelial cells and a variable strong influence on the hamster and human airway epithelial cells.

The cell transforming effect was studied in two cell systems, the hamster lung cells and the rat mesothelial cells. The results are summarized in Figure 4. The effect of crocidolite on hamster epithelial lung cells is again set at 100.

Cell transformation

Relative potency of equal fibre numbers

Cell type	Croci-dolite	Erionite	Glass fibres JM104	Glass fibres B-3K	Silicon carbide
Hamster lung epith.	100	48	n.d.	n.d.	pos.
Rat mesoth.	160	42	pos.	pos.	n.d.

n.d.: no effect detected (no clear dose-effect relationship or slope not clearly different from titanium dioxide controls)

pos.: effect detected (dose-effect relationship clearly different from titanium dioxide controls; but slope - equivalent to reciprocal of ED_{50}-value - less than 10 % of crocidolite on hamster epithelial cells)

Figure 4: Relative potency of equal fibre numbers in the cell transformation test. The potency of crocidolite on hamster lung epithelial cells has been defined as 100.

Depending on the fibre number, the rat mesothelial cells reacted more strongly. In this case there seems to be a correspondence between transformation in vitro and the carcinogenicity, since the hamster lung was not sensitive to asbestos in inhalation nor in instillation experiments. In general, it was noticeable that erionite demonstrated a weaker cell transforming effect than crocidolite, whilst glass fibres and silicon carbide fibres were either only weakly positive, or hardly differed from the titanium dioxide controls.

The results for the different parameters tested showed that there are differences in the test models implying a species-specific effect even in dependence on the fibre type administered. But there was no systematic differences observed. With regard to the cell transforming effect, crocidolite induced stronger effects than erionite, whilst glass fibres and silicon carbide showed only little or no activity. In comparison to in vivo carcinogenicity of the serosa, the in vitro cell transformation of crocidolite, erionite and titanium dioxide were in good agreement, whereas the association between the in vivo and in vitro results with glass fibres and silicon carbide fibres were insufficient (Figure 5).

Comparison:	in vivo /	in vitro
Dust	cancer, rat i.p.	cell transformation
crocidolite	+++	+++
erionite	+++	++
glass	++	+/-
silicon carbide	+++	+/-
nickel powder	++	+
nickel oxide	+	-
titanium dioxide	-	-

Figure 5: Comparison of the results of carcinogenicity tests in the rat and cell transformation tests in vitro

The reasons for these differences, especially in the case of silicon carbide fibres, which are unequivocally durable, are still unknown. Also the question as to whether surface properties are relevant for the carcinogenic or cell transforming effect is still open and needs further investigation.

References

Baris YI (1987) Asbestos and erionite related chest diseases. Semih Ofset Matbaacilik Ltd, Ankara

Doll R, Peto R (1985) Asbestos, effects on health of exposure to asbestos. Ed. by Health and Safety Commission. Her Majesty's Stationery Office, London

Emura M, Richter-Reichhelm H-B, Böning W, Eichinger R, Schoch C, Althoff J, Mohr U (1982) A fetal respiratory epithelial cell line for studying some problems of transplacental carcinogenesis in Syrian golden hamsters. J Cancer Res Clin Oncol 104:133-144

Emura M, Riebe M, Germann P, Brockmeyer C, Aufderheide M, Mohr U (1989) Functional culture of hamster and human airway epithelial cells and its application to pulmonary toxicology. Exp Pathol 37:224-227

Emura M, Riebe M, Ochiai A, Germann P, Mohr U (1990) New functional cell-culture approach to pulmonary carcinogenesis and toxicology. J Cancer Res Clin Oncol 116:557-562

IARC (International Agency for Research on Cancer, ed.) (1988) IARC Monographs on the Evaluation of the Carcinogenic Risk of Chemicals to Humans. Vol. 43, Man-Made Mineral Fibres and Radon. World Health Organization, International Agency for Research on Cancer, Lyon

Pott F, Ziem U, Reiffer FJ, Huth F, Ernst H, Mohr U (1987) Carcinogenicity studies on fibres, metal compounds, and some other dusts in rats. Exp Path 32:129-152

Pott F, Roller M, Ziem U, Reiffer FJ, Bellmann B, Rosenbruch M, Huth F (1989) Carcinogenicity studies on natural and man-made mineral fibres with the intraperitoneal test in rats. In: Bignon J, Peto J, Saracci R (eds) Non-occupational Exposure to Mineral Fibres. International Agency for Research on Cancer, Lyon (IARC Sci Publ No. 90, pp 173-197)

Pott F, Roller M, Rippe RM, Germann PG, Bellmann B (1991a) Tumours by the intraperitoneal and intrapleural routes and their significance for the classification of mineral fibres. In: Brown RC, Hoskins JA, Johnson NF (eds) Mechanisms in Fibre Carcinogenesis. Plenum Press, New York London (NATO ASI Series, Series A: Life Sciences, Vol. 223, pp 547-565)

Pott F, Dungworth DL, Heinrich U, Muhle H, Kamino K, Germann PG, Roller M, Rippe RM, Mohr U (1991b) Lung tumours in rats after intratracheal instillation of dusts. BOHS 7th Int. Symposium "Inhaled Particles" Edinburgh 16.-20.09.91. Ann Occup Hyg (Suppl) (In press)

Pott F, Rippe RM, Roller M, Rosenbruch M, Huth F (1991c) Vergleichende Untersuchungen über die Kanzerogenität verschiedener Nickelverbindungen und Nickellegierungen. Pub. by Bundesanstalt für Arbeitsschutz Dortmund. Wirtschaftsverl. NW, Bremerhaven (Schriftenreihe der Bundesanstalt für Arbeitsschutz. Forschung)

Riebe-Imre M, Emura M, Straub M, Roller M, Mohr U, Pott F (1992) Untersuchungen über Wirkungen von Mineralfaserstäuben auf Mesothelzellen und Lungenepithelzellen in vitro. Wissenschaftlicher Schlußbericht Mai 1992. Berufsgenossenschaft der keramischen und Glasindustrie, Würzburg

WHO (World Health Organization, ed) (1986) Asbestos and other natural mineral fibres, IPCS International Programme on Chemical Safety. World Health Organization, Geneva (Environmental Health Criteria 53)

WHO (World Health Organization, ed) (1988) Man-Made Mineral Fibres. IPCS International Programme on Chemical Safety. World Health Organization, Geneva (Environmental Health Criteria 77)

COMPARISONS OF PULMONARY EFFECTS IN RATS EXPOSED TO SIZE-SEPARATED PREPARATIONS OF PARA-ARAMID OR CHRYSOTILE ASBESTOS FIBRES AFTER 2-WEEK INHALATION EXPOSURES.

D.B. Warheit, M.A. Hartsky, S.R. Frame & C.J. Butterick*
DuPont Haskell Laboratory
P.O. Box 50, Elkton Rd
Newark, DE 19714 USA
and

*Texas Tech University Health Science Center
Lubbock, TX 79430 USA

INTRODUCTION

This multifunctional study was designed to compare the pulmonary toxic effects of inhaled, size-separated preparations of chrysotile asbestos fibres with Kevlar® aramid fibrils at similar aerosol fibre concentrations. One of the most important factors influencing the pathogenic development of fibre-related lung disease is fibre dimension (Davis et al., 1986). Although chrysotile asbestos fibres have previously produced fibrosis and tumors in the lungs of rats, the mean lengths of fibres used in those studies were short, i.e., in the range of 2 μm. Thus, for adequate comparisons, mean lengths of chrysotile asbestos fibres were size-separated, (i.e., increased) for this inhalation toxicity study.

Bulk Canadian chrysotile asbestos fibre samples were placed in a rotating sieve shaker and sieved through metal mesh screens. The longer fibres were collected and generated for inhalation

NATO ASI Series, Vol. H 85
Cellular and Molecular Effects
of Mineral and Synthetic Dusts and Fibres
Edited by J. M. G. Davis and M.-C. Jaurand

studies; fibres were collected on a filter and dimensional analysis was performed using scanning electron microscopy. The results showed that median and mean lengths of fibres were increased from 3 and 5 μm, respectively, in the original bulk sample to 6 and 9 μm in the generated sample preparation.

In the inhalation toxicity study, rats were exposed for 2 weeks to Kevlar® aramid fibrils or chrysotile asbestos fibres at design fibre concentrations of 750 and 400 f/cc. This study is still in progress; only interim results will be reported here. Following exposures, the postexposure recovery time periods used for evaluation were the following: immediately after 2-week exposure (i.e., time 0), as well as 5 days, 1, 3, 6, and 12 months postexposure. The major endpoints of this study are 1) pulmonary BrdU cell labeling and biochemical assessments of bronchoalveolar lavage fluids; and 2) durability/dimensional analysis of fibres recovered from the lungs of Kevlar® and chrysotile-exposed animals. Two-week exposures to Kevlar® aramid fibrils produced transient pulmonary inflammatory and terminal bronchiolar cell labeling responses. A transient increase in neutrophils was also measured in the lungs of rats exposed to chrysotile asbestos fibres for two weeks. However, substantial asbestos-induced increases compared to controls in pulmonary cell labeling indices were measured on terminal bronchiolar, parenchymal, and visceral and subpleural surfaces and these persisted through 1 month postexposure. Median lengths of fibres recovered from digested asbestos-exposed lung tissue were increased over time, suggesting that short asbestos fibres were selectively cleared from the lungs, with apparent insignificant pulmonary clearance of long fibres. The dimensional changes of asbestos fibres as well as the pulmonary cell labeling data indicate that chrysotile asbestos fibres may produce greater long-term pulmonary effects when compared to Kevlar® fibrils.

METHODS

General Experimental Design

Groups of male Crl:CDBR rats (7-8 weeks old, Charles River Breeding Laboratories, Kingston, New York) were used to assess the pulmonary effects of 2-week inhalation exposures to size-separated preparations of Kevlar® para-aramid fibrils or chrysotile asbestos fibres. For this study, Kevlar® was utilized as a representative para-aramid fibre. The two commercial types of para-aramid fibres are Twaron®, made by Akzo and Kevlar®, made by DuPont.

Following exposures, the lungs of Kevlar® or chrysotile-exposed animals and aged-matched sham controls were subsequently evaluated by bronchoalveolar lavage fluid analysis and/or BrdU pulmonary cell labeling assessments at several time periods postexposure (i.e., 0 hrs, 5 days, 1, 3, 6, and 12 months postexposure). Durability/dimensional analysis studies with fibre-exposed, digested lung tissue, and histopathological analysis will be carried out at postexposure time periods through 1 year.

Fibre Preparation and Inhalation Exposure

Ultrafine Kevlar® aramid fibrils were supplied by DuPont Fibres. A special preparation of respirable Kevlar® fibrils which had been prepared for the 2-year inhalation study (Lee et al., 1988) was utilized for this study. Bulk Canadian chrysotile asbestos fibres were obtained from John Addison of the Institute of Occupational Medicine in Edinburgh, Scotland. The bulk fibre preparation was size-separated (i.e., median fibre lengths were increased) by placing the fibres in a rotating sieve shaker and sieved through metal mesh screens. The longer fibres were collected and generated for inhalation studies; fibres were collected on a filter and dimensional analysis was performed using scanning electron microscopy. The results showed that median and mean lengths of fibres were increased from 3 and 5 μm, respectively, in the original bulk sample to 6 and 9 μm in the generated sample preparation. The median lengths and diameters of Kevlar® fibrils used in the study were 9 μm and 0.3 μm, respectively.

The methods for aerosol generation of Kevlar® fibrils have previously been reported (Warheit et al., 1992). Briefly, rats were placed in cylindrical polycarbonate or stainless steel holders equipped with conical nose pieces. The restrainers were inserted into face plates on the exposure chambers such that the nose of each animal protruded into the chamber. Atmospheres of Kevlar® fibrils were generated with a K-tron bin feeder equipped with twin screws. Baffles were inserted into the generation apparatus and served to increase the respirability of the samples. Kevlar® fibrils were metered into a plastic funnel connected to a cyclone where high pressure air transferred the test material into a microjet apparatus. This high pressure air-impingement device was utilized to separate Kevlar® fibrils from the larger fibre clumps. Aerosols of chrysotile asbestos fibres were generated in a similar manner, i.e., with a binfeeder and baffles, but without the microjet apparatus. Final mean fibre concentrations for the Kevlar® exposures were 772 and 419 f/cc. Final mean fibre concentrations for the chrysotile asbestos exposures were 782 and 458 f/cc.

Pulmonary Lavage and Preparation of Macrophage Monolayers

Bronchoalveolar lavage procedures were conducted according to methods previously described (Warheit et al., 1984a). The methods for quantifying cell numbers, differentials, and biochemical assays on bronchoalveolar lavage fluids have also been reported elsewhere (Warheit et al., 1991).

Lung Dissection, Tissue Preparation and BrdU Immunostaining

The lungs of rats exposed to Kevlar® and chrysotile asbestos fibres were prepared for light microscopy by airway infusion using methods previously reported (Warheit _et al_., 1984b, 1991). Sagittal sections of the left lung were made with a razor blade. Tissue blocks were dissected from upper, middle, and lower regions of the lung, and were subsequently prepared for light microscopy (paraffin embedded, sectioned, and hematoxylin-eosin stained). Pulmonary cell labeling experiments were designed to measure the effects of fibre inhalation exposure on terminal airway, lung parenchymal, subpleural, visceral pleural and mesothelial cell turnover in rats following 2-week exposures. Groups of fibre and sham-exposed rats were pulsed with an intraperitoneal injection of 5-bromo-2′-deoxyuridine (BrdU) dissolved in a 0.5 N sodium bicarbonate buffer solution at a dose of 100 mg/kg body weight as previously described (Warheit _et al_., 1992). For each treatment group, immunostained nuclei were counted by light microscopy at X1000 magnification.

Fibre Recovery from Lung Tissue

Kevlar® fibrils were recovered from the lungs of exposed rats using a dilute 2.6% hypochlorite (bleach) solution. The fixed lung tissue was incubated with a bleach solution for 10 minutes, further processed and deposited on filters. Chrysotile asbestos fibres were recovered from the lungs of exposed rats by incubating the lung tissue with a 5.25% hypochlorite solution for 3 hrs. Subsequently, the filters containing fibres recovered from lung tissue were mounted and prepared for phase-contrast light microscopy (for counting) and for scanning electron microscopy (for fibre dimensional analysis).

RESULTS AND DISCUSSION

Two-week exposures to Kevlar® aramid fibrils or chrysotile asbestos fibres produced transient pulmonary inflammatory responses as measured by bronchoalveolar lavage fluid analysis (data not shown). In this regard, small increases in lavaged neutrophils, lactate dehydrogenase and protein were measured immediately after and 5 days after 2-week exposure to chrysotile asbestos but returned to control levels by 1 month postexposure. Small but transient increases in lavaged neutrophils were also measured immediately after a 2-week exposure to Kevlar® fibrils.

Pulmonary cell labeling studies demonstrated that 2-week exposures to Kevlar® aramid fibrils produced a transient increase in terminal bronchiolar cell labeling responses (Fig. 1). No increases in lung parenchymal, subpleural or mesothelial cell labeling indices were measured relative to sham controls. In chrysotile asbestos-exposed rats, however, substantial increases compared to controls in pulmonary cell labeling indices were measured on terminal bronchiolar, parenchymal, mesothelial, and subpleural surfaces, and these were sustained through one month postexposure (Figs. 2-3).

Fibre clearance studies have not been completed through one year postexposure. However, preliminary dimensional analysis studies indicate that median lengths of fibres recovered from digested asbestos-exposed lung tissue were increased over time (Fig. 4), suggesting that short asbestos fibres were selectively cleared from the lungs, with apparent insignificant pulmonary clearance of long fibres. Preliminary studies with Kevlar® aramid fibrils are consistent with earlier data, suggesting biodegradability of inhaled Kevlar® fibrils (Warheit et al., 1992).

The BrdU pulmonary cell labeling results presented here are consistent with data from a number of studies reported by Brody and colleagues (Brody and Overby, 1984, McGavran et al., 1990; Coin et al., 1992a). In those studies, brief exposures to chrysotile asbestos fibres produced a biphasic cell labeling response in the lungs of exposed rats and mice. This was characterized by dramatic increases in epithelial cell DNA synthesis, followed several days later by enhanced labeling of interstitial cells. In follow-up studies, a three-day exposure prolonged the duration of labeled cells (Coin et al., 1992b). In another study, Coin et al., (1991) reported that a 5-hr exposure to chrysotile fibres in mice produced substantial increases in mesothelial and subpleural cell labeling indices at 2 and 8 days postexposure.

The biodurability data reported here are consistent with the results of previous studies by Roggli and colleagues (1984) and Bellmann et al., (1986, 1987). Roggli and coworkers exposed rats for brief periods to aerosols of chrysotile asbestos fibres and assessed fibre clearance and dimensional changes as a function of time using a lung digestion/fibre recovery technique. Their results demonstrated a progressive increase in mean fibre length of retained chrysotile fibres in the lung. Similar results have been reported by Bellmann and coworkers. In contrast, results with Kevlar® fibres have demonstrated biodegradability of the inhaled fibrils.

In conclusion, size separation techniques for chrysotile asbestos fibres increased median lengths from 3.3 μm to 6.2 μm. Both Kevlar® and chrysotile produced a transient pulmonary inflammatory response, as evidenced by the biochemical and cellular BAL fluid data. Pulmonary cell labeling studies demonstrated substantial increases in lung parenchymal, airway, and pleural/subpleural cell labeling indices following chrysotile exposures. Fibre biopersistence/durability results thusfar indicate that the long chrysotile fibres are cleared at a slow rate. In this regard, median lengths of chrysotile fibres were increased over time, while median lengths of Kevlar® fibrils were decreased over time. It is thus concluded that inhalation of chrysotile asbestos fibres is likely to produce greater long-term pulmonary toxic effects when compared to Kevlar® aramid fibrils.

MANUSCRIPT REFERENCES

Bellmann B, Konig H, Muhle H, and Pott F (1986) Chemical durability of asbestos and of man-made mineral fibres in vivo. J Aerosol Sci 17:341-345.

Bellmann B, Muhle H, Pott F, Konig H, Kloppel H, and Spurny K. (1987) Persistence of man-made mineral fibers (MMMF) and asbestos in rat lungs. Ann Occup Hyg 31 (4B) 693-709.
Brody AR, Overby LH (1989) Incorporation of tritiated thymidine by epithelial and interstitial cells in bronchiolar-alveolar regions of asbestos-exposed rats. Am J Pathol 134: 133-140.
Coin PG, Moore LB, Roggli V, and Brody AR (1991) Pleural incorporation of ^{3}H-TdR after inhalation of chrysotile asbestos in the mouse. Amer Rev Respir Dis 143:A604.
Coin PG, Roggli VL, Brody AR (1992a) Deposition, clearance and translocation of chrysotile asbestos from peripheral and central regions of the rat lung. Environ Res 58:97-116.
Coin PG, Roggli V, and Brody AR (1992b). Pulmonary fibrogenesis and BRDU incorporation after three consecutive inhalation exposures to chrysotile asbestos. Amer Rev Respir Dis 145:A328.
Davis JMG, Addison J, Bolton RE, Donaldson K, et al., (1986) The pathogenicity of long versus short fibre samples of amosite asbestos administered to rats by inhalation or intraperitoneal injection. Br J Exp Pathol 67:415-430.
Lee KP, Kelly DP, O'Neal FO, Stadler JC, and Kennedy GL, Jr. (1988) Lung response to ultrafine Kevlar aramid synthetic fibrils following 2-year inhalation exposure in rats. Fundam Appl Toxicol 11:1-20.
McGavran PD, Butterick CJ, Brody AR (1990) Tritiated thymidine incorporation and the development of an interstitial lesion in the bronchiolar alveolar regions of the lungs of normal and complement deficient mice after inhalation of chrysotile asbestos. J Environ Pathol Toxicol Oncol 9:377-92.
Roggli VL and Brody AR (1984) Changes in numbers and dimensions of chrysotile asbestos fibers in lungs of rats following short-term exposure. Exp Lung Res 7:133-147.
Warheit DB, Hill LH, and Brody AR (1984a) Surface morphology and correlated phagocytic capacity of pulmonary macrophages lavaged from the lungs of rats. Exp Lung Res 6: 71-82.
Warheit DB, Chang LY, Hill LH, Hook GER, Crapo JD, and Brody AR (1984b) Pulmonary macrophage accumulation and asbestos-induced lesions at sites of fiber deposition. Am Rev Respir Dis 129: 301-310.
Warheit DB, Carakostas MC, Hartsky MA, and Hansen JF (1991) Development of a short-term inhalation bioassay to assess pulmonary toxicity of inhaled particles: Comparisons of pulmonary responses to carbonyl iron and silica. Toxicol Appl Pharmacol 107: 350-368.
Warheit DB, Kellar KA, and Hartsky MA (1992) Pulmonary cellular effects in rats following aerosol exposures to ultrafine Kevlar® aramid fibrils: Evidence for biodegradability of inhaled fibrils. Toxicol Appl Pharmacol 116: 225-239.

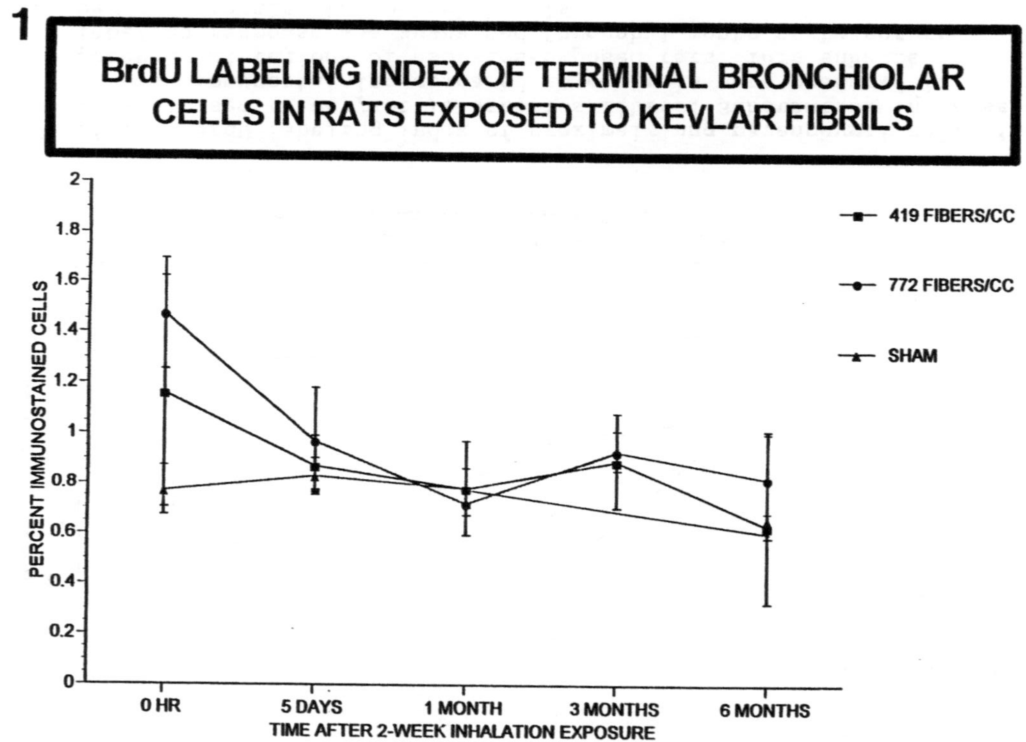

Fig. 1. BrdU labeling index of terminal bronchiolar cells in rats exposed to Kevlar® fibrils. A transient increase in terminal bronchiolar cell labeling was measured immediately after 2-week exposure but returned to normal levels after 5 days postexposure.

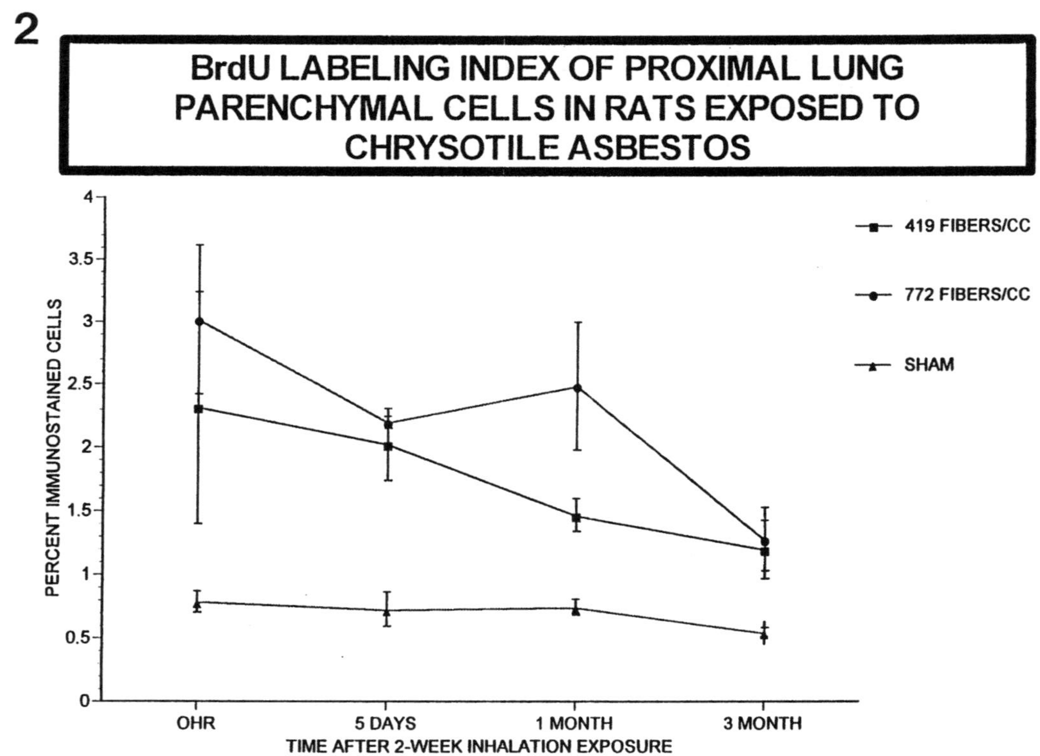

Fig. 2. BrdU labeling index of proximal lung parenchymal cells in rats exposed to chrysotile asbestos fibres. Substantial increases in BrdU immunostaining compared to controls were measured in parenchymal regions of asbestos-exposed rats. This enhanced labeling response was sustained through 1 month postexposure before returning to control levels.

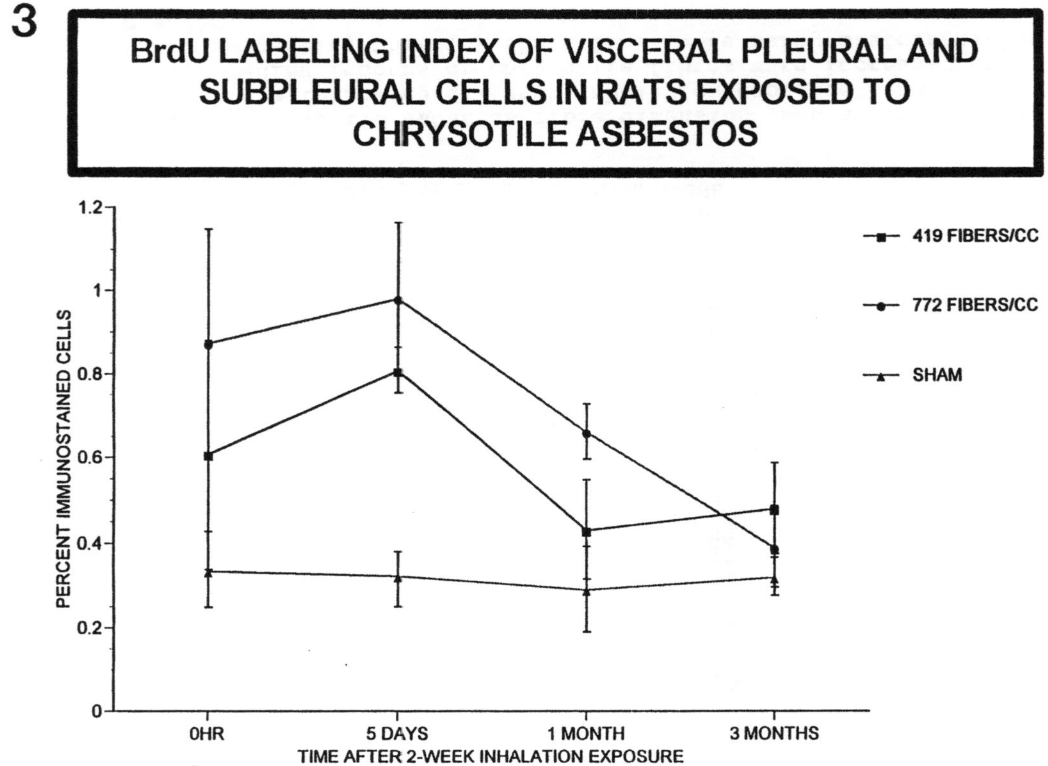

Fig. 3. BrdU labeling index of visceral pleural and subpleural cells in rats exposed to chrysotile asbestos fibres. Substantial increases in BrdU immunostaining compared to controls were measured in pleural and subpleural regions of asbestos-exposed rats. This enhanced labeling response was sustained through 1 month postexposure before returning to control levels.

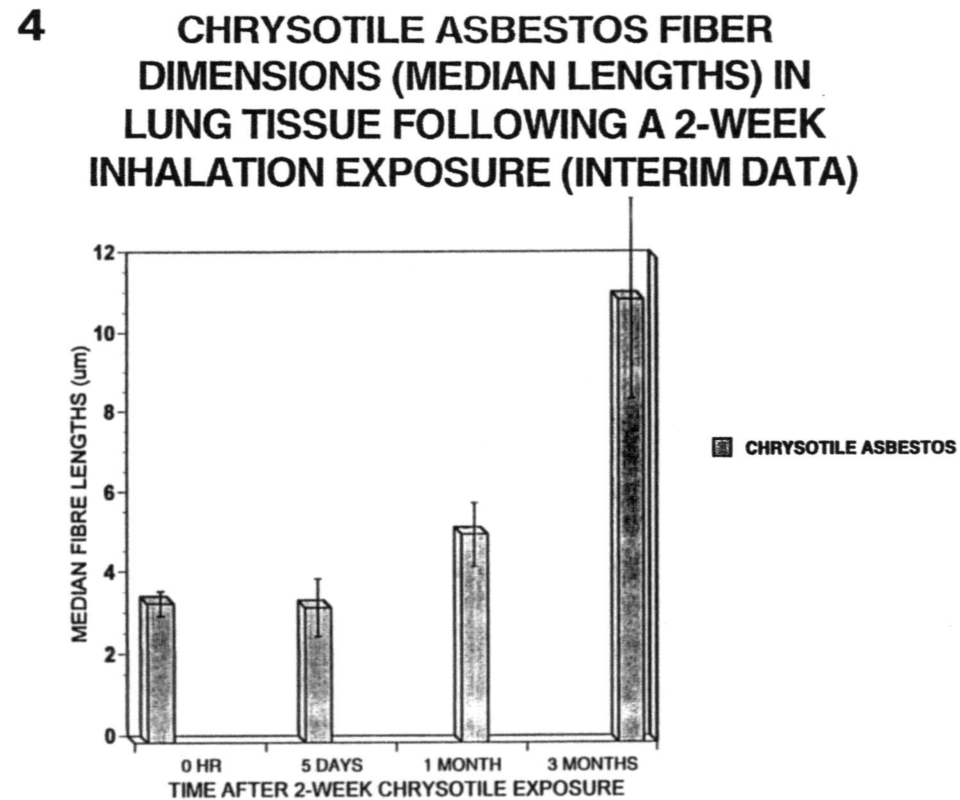

Fig. 4. Chrysotile asbestos fibre dimension (median lengths) in lung tissue following a 2-week inhalation exposure (interim data). Median lengths of chrysotile fibres were increased over time, suggesting that the short fiber were being preferentialled cleared.

Investigation of the Biodurability and Carcinogenicity of Different Man-Made Mineral Fibres

B. Bellmann, H. Muhle, and F. Pott[1]

Fraunhofer Institute of Toxicology and Aerosol Research
Nikolai-Fuchs-Straße 1
D-30625 Hannover
Germany

Introduction

The carcinogenicity of mineral fibres depends on length and diameter, however, the exact relationships are not yet known. Additionally the biopersistence is an important factor: the carcinogenic effect in the intraperitoneal injection test of fibre types with a short half-time in the lung was weak or not detectable in the case of gypsum, wollastonite, glass ATF 3101, HCl-treated E-glass, and HCl-leached chrysotile (Bellmann et al. 1987; Muhle et al. 1991; Pott et al. 1987, 1988, 1989, 1991, 1994). After elimination from the tissue either by translocation or by disintegration a fibre can no longer act as a carcinogen. The minimum period of persistence could be seen as a kind of threshold value. It will be difficult to detect this "threshold" before knowing the mechanism of fibre carcinogenesis. However, a comparison of data on biopersistence of fibre samples in the lung with carcinogenicity data received after i.p. injection may give another approach for recognizing the "threshold" which is important for the regulation of fibres. In the following, some new results on biopersistence are described and compared with carcinogenicity data.

Materials and Methods

The test materials of glass wool, stone wool, slag wool, and ceramic wool were samples which have been sized by Manville Technical Center (Table 1). For B-01/0,9 only milling in a

1 Medical Institute for Environmental Hygiene at the Heinrich Heine University of Düsseldorf, Auf'm Hennekamp 50, D-40225 Düsseldorf, Germany

NATO ASI Series, Vol. H 85
Cellular and Molecular Effects
of Mineral and Synthetic Dusts and Fibres
Edited by J. M. G. Davis and M.-C. Jaurand

ball mill but no sizing by sedimentation was necessary (Pott et al. 1994). Length and diameter were analysed for about 400 fibres (including fibres with L/D < 3/1) per test material sample by scanning electron microscopy (SEM), the number of critical fibres per mass and the dose used for instillation is shown in Table 2.

Table 1 Chemical composition of fibres

	Chemical composition [%]									
Material	CaO	MgO	Al_2O_3	TiO_2	FeOx	B_2O_3	SiO_2	Na_2O	K_2O	ZnO	BaO
B-01[a]	16.5	3.2				3.3	61.0	15.4			
MMVF11[b]	7.4	2.8	3.9	0.1	0.2	4.4	63.4	15.4	1.3		0.0
M-Stone[c]	24.7	11.4	9.9	2.0	8.0		38.4	1.6	0.8		0.1
M-Slag[c]	37.4	10.0	10.3	0.5	0.3		38.2	0.4	0.4		
RCF1[b]	0.1	0.1	48.0	2.0	1.0		47.7	0.5	0.2		

a chemical analysis by Bayer AG

b From TIMA (11.2.1992), sizing and chemical analysis by Manville Technical Center

c sizing and chemical analysis by Manville Technical Center

Table 2 Size distribution of fibres (L/D > 5), concentration of critical fibres (L>5 μm, D < 2μm, L/D > 5/1) and dose for intratracheal instillation

	Fibre length [μm]				Fibre diameter [μm]				Critical	
Material	10%<	50%<	90%<	99%<	10%<	50%<	90%<	99%<	fibres per ng	Dose [mg]
B-01/0,9	3.6	10.4	38.0	121.0	0.32	0.70	1.21	2.29	20	0.35
MMVF11	4.9	14.6	39.2	102.8	0.26	0.77	2.04	4.36	5.6	2.0
M-Stone	4.4	10.1	26.6	75.1	0.39	0.84	1.68	2.52	12	1.0
M-Slag	4.0	8.7	25.3	76.2	0.38	0.77	1.35	2.31	19	1.0
RCF 1	4.1	11.9	38.7	111.0	0.26	0.77	1.79	3.15	7.5	2.0
RCF[a]	3.4	13.1	51.1	140.8	0.25	0.84	2.23	4.65	1.7	

a Fibre sample for intraperitoneal injection test

Ten weeks old female Wistar rats were treated by intratracheal instillation of 0.35 - 2 mg test material in 0.3 ml NaCl per animal. 5 animals per group were sacrificed after 2 days, 1, 3, 6, and 12 months. An additional sacrifice date at 18 months after instillation is planned. The lungs were isolated, dried at 105 °C in an oven and subjected to low-temperature ashing. A fraction of the ashed lung was suspended in filtered water and filtered on a Nuclepore filter. These samples were prepared and analysed by SEM. For each lung sample the length and diameter of about 200 fibres was measured at different magnifications. For lower fibre count the analysis was terminated after scanning of 50 object screens per animal. The number of fibres per lung was calculated for each animal. The mass of fibres was estimated from length and diameter assuming cylindrical geometry. For calculation of clearance kinetics a regression analysis of logarithm of number or mass of fibres versus time after instillation was done using data of individual animals.

Results

The retention of fibres at different sacrifice dates up to 12 months is shown in Table 3.

Table 3 Analysis of number and mass of fibres in the lung ash. Means and standard deviations (S.D.) of 5 animals per sacrifice date

Sacrifice	Fibres analysed Mean	Fibres analysed S.D.	Fibres [10^6/lung] Mean	Fibres [10^6/lung] S.D.	Fibres (L>5µm, D<3µm) [10^6/lung] Mean	Fibres (L>5µm, D<3µm) [10^6/lung] S.D.	Calculated mass of fibres [mg] Mean	Calculated mass of fibres [mg] S.D.
B-01/0,9								
2 Days	210	6	9.100	1.279	4.803	0.817	0.088	0.033
1 Month	88	40	2.486	1.164	0.987	0.500	0.012	0.006
3 Months	16	5	0.458	0.147	0.178	0.086	0.003	0.001
6 Months	5	2	0.166	0.039	0.076	0.058	0.001	0.001
12 Months	1	1	0.035	0.015	0.018	0.015	0.000	0.000
MMVF11								
2 Days	208	7	29.170	3.375	22.908	2.012	2.05	0.51
1 Month	205	4	20.853	4.703	15.101	3.167	1.19	0.33
3 Months	209	4	16.794	2.287	12.162	1.316	0.73	0.15
6 Months	205	5	10.768	2.778	7.091	1.808	0.36	0.05
12 Months	200	4	8.076	1.600	5.559	0.967	0.23	0.03
M-Stone								
2 Days	205	4	9.895	1.485	8.108	0.872	1.04	0.18
1 Month	207	7	7.917	1.089	6.471	0.584	0.78	0.18
3 Months	207	4	6.365	0.499	5.253	0.410	0.48	0.13
6 Months	206	3	3.752	0.525	3.023	0.452	0.26	0.07
12 Months	116	42	0.993	0.331	0.759	0.250	0.06	0.01
M-Slag								
2 Days	210	11	27.034	2.561	17.160	1.343	0.918	0.114
1 Month	211	10	22.090	2.686	13.122	2.253	0.396	0.072
3 Months	212	5	15.186	2.342	8.274	1.205	0.171	0.018
6 Months	195	35	3.943	0.609	2.320	0.494	0.033	0.004
12 Months	25	9	0.443	0.171	0.329	0.121	0.004	0.003
RCF 1								
2 Days	210	11	19.617	1.622	15.095	1.351	1.70	0.24
1 Month	209	7	16.360	2.715	11.702	1.862	1.09	0.20
3 Months	205	5	15.311	1.076	11.986	0.893	1.11	0.15
6 Months	207	6	12.164	1.465	9.303	1.182	0.91	0.13
12 Months	206	7	9.118	0.615	7.347	0.638	0.56	0.10

The elimination of fibres from the lung is described as simplification by first order kinetics. In this case the elimination kinetics can be defined by only one parameter, the half-time as shown in Table 4.

Table 4 Half-time and 95% confidence limit (C.L.) for elimination of fibres in the lung ash (partly preliminary data, investigation after 18 months not completed)

	Half-time calculated from					
	Number of fibres		Number of fibres (L>5µm, D<3µm)		Mass of fibres	
Material	Mean	(95% C.L.)	Mean	(95% C.L.)	Mean	(95% C.L.)
B-01/0,9	46	(39 - 57)	49	(38 - 69)	40	(34 - 51)
MMVF11	206	(169 - 265)	188	(155 - 241)	121	(103 - 148)
M-Stone	109	(99 - 122)	105	(95 - 117)	89	(81 - 99)
M-Slag	59	(55 - 63)	62	(58 - 66)	45	(41 - 49)
RCF 1	350	(294 - 432)	389	(314 - 513)	266	(214 - 350)

Preliminary results of the intraperitoneal injection test are given in Table 5.

Table 5 Dose of fibre samples and tumour incidence after intraperitoneal injection in Wistar rats

Material	Dose [mg]	[10^9 Fibres][a]		Sex	Tumour rate [%]	Reference
B-01/0,9	20x25	10	2	♂	20	Pott et al. 1994[b]
	40x25	20	4	♂	59	
MMVF 11	2x35	0.4	0.13	♀	18[c]	Pott et al. 1994[b]
	6x30	1.0	0.34	♀	65[c]	
M-Stone	8.5	0.1	0.02	♀	3	Pott et al. 1994[b]
	8.5	0.1	0.02	♂	3	
	22.5	0.3	0.05	♀	31	
	22.5	0.3	0.05	♂	17	
	2x42.5	1.0	0.2	♂	63	
M-Slag	20	0.4	0.07	♀	0[c]	Pott et al. 1994[b]
	50	1.0	0.2	♀	5[c]	
	3x50	3.0	0.5	♀	13[c]	
RCF[d]	12	0.02	0.007	♀	47	Pott et al. 1991
	2x20	0.07	0.024	♀	81	
NaCl-solution	20x2ml	-	-	♀	2	Pott et al. 1994[b]
	20x2ml	-	-	♂	1	
	50x1ml	-	-	♀	0	Pott et al. 1991

a Fibre definition: D < 2 µm, L/D > 5/1, L> 5 µm (left column) or L> 20 µm (right column)

b Preliminary rough results according to macroscopic diagnosis, histology in progress

c Interim results 21 months after i.p. injection

d Fibre sample was not identical with that used in the biopersistence test

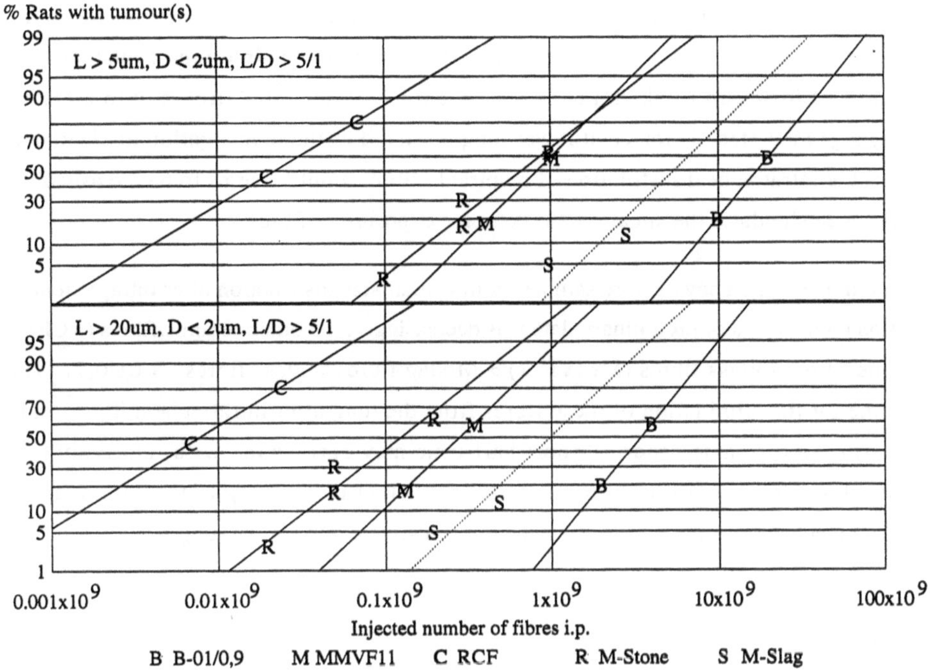

Figure 1 Dose response relationship for fibres after intraperitoneal injection (the tumour rates of the tested M-slag fibres may increase significantly up until end of study)

The tumour incidence versus injected dose of critical fibres with two different definitions (L > 5 µm, D < 2 µm, L/D > 5/1 and L > 20 µm, D < 2 µm, L/D > 5/1) is shown in a logarithmic probability plot in Figure 1. The slope of the dose response line was very similar for RCF, M-stone, MMVF11 and B-01/0,9. Because for M-slag wool only one value of the tumour incidence was significantly above control values a mean slope of the other fibre samples was assumed in Figure 1.

Discussion and conclusions

For the elimination of MMVF 11 fibres also data from chronic and from short-time inhalation are available (Hesterberg et al. 1993; Musselman et al. 1993). The half-time of about 250 days calculated from data with 12 months inhalation and 12 months recovery time is quite similar to the half-time of about 200 days in this study. In contrast the half-time calculated from lung retention data of fibres after 5 days inhalation was only about 100 days. This faster clearance could be due to a higher fraction of mucociliary clearance of long and thick fibres in the first weeks after inhalation and to a faster macrophage mediated clearance at lower lung burden. For the investigation of clearance kinetics the intratracheal instillation of respirable fibre samples seems to be a reasonable model which simulates the situation after chronic exposure of fibres.

The carcinogenic potency of fibre samples with similar size distribution after intraperitoneal injection on the basis of preliminary data was decreasing in the order ceramic fibres (RCF) > M-stone fibres ≈ glass fibres (MMVF11) > M-slag fibres > glass fibres (B-01/0,9). The half-time for the elimination of these fibres from the lung after intratracheal injection is decreasing in the same order. A short half-time may correspond with relatively low biodurability. Up to now it is an open question whether the ranking of the carcinogenic potency detected after intraperitoneal injection is also relevant for the lung.

References

Bellmann B, Muhle H, Pott F, König H, Klöppel H, Spurny K (1987) Persistence of man-made mineral fibres (MMMF) and asbestos in rat lungs. Ann Occup Hyg 31:693-709

Hesterberg TW, Miiller WC, McConnell EE, Chevalier J, Hadley JG, Berstein DM, Thevenaz P, Anderson R (1993) Chronic inhalation toxicity of size-separated glass fibers in Fischer 344 rats. Fund Appl Toxicol 20:464-476

Muhle H, Bellmann B, Pott F (1991) Durability of various mineral fibers in rat lungs. In: Brown RC, Hoskins JA, Johnson NF (eds) Mechanisms in fibre carcinogenesis. (NATO ASI series, Vol. 223) Plenum Press New York London, pp 181-187

Musselman RP, Miiller WC, Easters W, Hadley JG, Kamstrup O, Thevenaz P, Hesterberg TW (1993) Biopersistence of man-made vitreous fibers (MMVF) and crocidolite fibers in rat lungs following short-term exposures. Environ Health Perspect (in press)

Pott F, Ziem U, Reiffer F-J, Huth F, Ernst H, Mohr U (1987) Carcinogenicity studies on fibres, metal compounds, and some other dusts in rats. Exp Pathol (Jena) 32:129-152

Pott F, Matchek A, Ziem U, Muhle H, Huth F (1988) Animal experiments with chemically treated fibres. Ann Occup Hyg 32:353-359 (Suppl 1)

Pott F, Roller M, Ziem U, Reiffer F-J, Bellmann B, Rosenbruch M, Huth F (1989) Carcinogenicity studies on natural and man-made fibres with the intraperitoneal test in rats. In: Bignon J, Peto J, Saracci R (eds) Non-occupational exposure to mineral fibres. IARC Sci Publ No 90, Lyon, pp 173-179

Pott F, Roller M, Rippe RM, Germann P-G, Bellmann B (1991) Tumours by the intraperitoneal and intrapleural routes and their significance for the classification of mineral fibres. In: Brown RC, Hoskins JA, Johnson NF (eds) Mechanisms in fibre carcinogenesis. (NATO ASI series, Vol. 223). Plenum Press New York London, pp 547-565

Pott F, Roller M, Kamino K, Bellmann B, Althoff GH, Ulm K (1994) Beurteilung der Kanzerogenität inhalierbarer Fasern. VDI Berichte 1075 (in press)

Quantitative relationship between emphysema and lung minerals in man

James Leigh, Tim Driscoll, Bradley Cole, Rolf Beck
Epidemiology Unit
National Institute of Occupational Health and Safety (Worksafe Australia)
GPO Box 58
Sydney NSW 2001
Australia

Introduction

The relationship of emphysema to lung coal dust and silica dust content in coal workers exposed to mixed dust remains controversial, in spite of considerable recent research (Oxman et al, 1993). Well designed large studies using post-mortem pathological material have shown that emphysema, especially centriacinar emphysema, is qualitatively related to coal dust content of lungs (Ruckley et al, 1984; 1989) and quantitatively related to pneumoconiosis severity, years at the coal face and indices of obstructive lung disease obtained during life (Leigh et al, 1982; 1983). Recent studies of coalworkers have suggested that emphysema is relatively less likely with high lung silica content (Leigh, 1992). In contrast, emphysema has been found to be related to silica dust exposure in goldminers (Hnizdo et al, 1991).

Underground coalworkers in the New South Wales (Australia) coal industry are exposed to high rank bituminous coal (percentage carbon dry ash free 82-90%). Silica content of the New South Wales coal seams is on average less than two per cent. Some workers would have been exposed to higher silica levels when working in shaft development and in other mines, predominantly in the United Kingdom. Some of these workers would have been exposed to much higher dust levels than those obtaining today. The aim of this paper was to determine the relationship between quantified emphysema and quantified lung content of coal dust and silica dust in coalworkers.

Methods

The data analysed in this study came from 264 workers in the New South Wales coal mining industry who died during the period July 1966 to January 1983, representing about 20 percent of all coalworkers who died in this period. Measurements of FEV_1 (percentage predicted: Ferris et al, 1965) and data on lifetime average tobacco intake

NATO ASI Series, Vol. H 85
Cellular and Molecular Effects
of Mineral and Synthetic Dusts and Fibres
Edited by J. M. G. Davis and M.-C. Jaurand

were obtained at Joint Coal Board routine clinical examinations within five years of death. Both lungs were inflated by the Gough method (Gough and Wentworth, 1960) using ten percent acetate formalin. Emphysema was quantified by using a modification of the Heard method (Heard, 1969). This was done on macroscopic examination of the surfaces of at least three cuts in the sagittal plane in each lung. Each cut was divided into three segments of approximately equal area. In each segment the percentage of lung area involved as abnormal holes was measured and the mean for each of the three or more cuts calculated. The mean for all the cuts from both lungs became the final percentage emphysema score. This was then coded 1 - 7 (1: nil; 2: 1-10%; 3: 11-20%; 4: 21-30%; 5:31-40%; 6:41-50%; 7: >50%). Macules/nodules were classed as sparse (equivalent to ILO 1/0 or less), moderate (less than ILO 2/2) or profuse (ILO 2/2 or greater), and massive lesions documented. Distinction was made between coal only (macules), mixed coal/silica (nodules, palpable) and silica only changes.

Specimens were homogenized, dried on shallow trays at 105°C and ground in a mill. Powdered lung samples were hydrolysed using a modified version of the procedure of Guest (Guest, 1976). Total lung dust, total coal and total mineral matter (ash) were determined gravimetrically: the total lung dust after first drying the crucible contents at 110°C, the total mineral matter after ashing the crucible contents at 600°C overnight and the total coal by calculating the difference between total lung dust and total mineral matter. Finally, silica (quartz) was determined in the ash by Fourier Transform Infrared Spectroscopy (FTIR) using the potassium bromide disc technique (1 mg ash per 250 mg disc), essentially using the method of Dodgson (Dodgson and Whittaker, 1973). The main results consider the total amount of coal and silica in the two lungs of each subject, standardized by predicted vital capacity (using height and age) (Lazarus, 1982). For comparison purposes results were also expressed as proportions of coal or silica by dry weight of lung, and as absolute amounts of coal and silica.

Results

The frequency distribution for emphysema score is shown in table 1. Means and standard deviations for the main variables are given in table 2. The majority of subjects had minimal or mild pneumoconiotic changes and there were only four subjects with silica-only changes. Mean lung coal content and mean lung silica content were highest in those subjects with a mixed pneumoconiosis, next highest in those subjects with primarily coal pneumoconiosis, and lowest in subjects with minimal changes. All differences were statistically significant (2 sided).

Table 1
Emphysema score and percentage emphysema
(frequency and percent)

	Emphysema Score							
	1	2	3	4	5	6	7	Total
Emphysema (%)	Nil	1-10	11-20	21-30	31-40	41-50	>50	
Number	33	86	57	37	31	11	9	264
Percent	12	33	22	14	12	4	3	100

Table 2
Means and standard deviation of main variables

	Mean	SD	N
Emphysema score	3.1	1.5	264
Adjusted lung coal content (g/l*)	0.89	0.81	264
Absolute lung coal content (g=)	3.93	3.44	264
Adjusted lung silica content (g/l*)	0.04	0.04	261
Absolute lung silica content (g=)	0.18	0.17	261
Tobacco smoked (oz per week)	3.1	2.2	264
Age at death (years)	63.4	10.4	264
FEV_1%+	78.7	25.5	184

* Adjusted by predicted vital capacity
= Absolute amount of lung coal or silica per subject
\+ Measured FEV_1 as a percentage of predicted (using height and age) (Ferris et al, 1965)

Multiple regression analysis showed that the extent of emphysema (E score) had a strong positive quantitative relationship with lung coal content ($p<0.0003$), age ($p<0.0001$) and smoking ($p<0.0001$), with a significant negative interaction between lung coal content and smoking ($p<0.004$) (E score = -1.79 + 0.62 coal + 0.06 age + 0.21 smoking - 0.17 coalxsmoking, adjusted $R^2=0.25$). The relationship for these models was basically unchanged by adding a lung silica content term or any of the other potential interactions (coalxsilica; coalxage; silicaxage, silicaxsmoking; and agexsmoking). When the 40 life-long non-smokers were considered separately, the positive effect on emphysema score of lung coal content and age was striking, with these two variables accounting for about two thirds of the variation in emphysema score (coal: $p<0.001$; age: $p<0.0002$; E score = -1.56 + 0.78 coal + 0.06 age, adjusted $R^2=0.66$). Lung silica content and all interaction terms did not contribute significantly to the model. Similar results for all models were obtained when the coal and silica contents were expressed in an unadjusted form (as absolute weight per subject) or as a proportion of the dry weight of lung.

Emphysema score was highly negatively correlated with FEV_1 ($r=-0.44$, $p<0.0001$) and significantly higher in subjects with pneumoconiotic change compared with subjects with minimal or no pneumoconiotic change ($t=4.90$, $p<0.0001$). The degree of lung fibrosis was highly positively associated with lung silica content ($X^2_1=12.9$, $p<0.0003$), but not lung coal content, when examined in the same model.

Discussion

The results from this study show clearly that lung coal content is quantitatively related to the development of emphysema, taking into account age and smoking. This is true whether the effect of smoking is controlled using regression methods or by examining only non-smokers. The results indicate that a non-smoking miner with about the average lung coal content would have emphysema involving an extra seven to ten percent of the whole lung mass compared to a non-smoking person of the same age who was not exposed to coal. The influence of coal appears less marked in smokers, possibly due to a negative interaction between smoking and coal, but the very high R^2 in the regression equation for non-smokers strongly suggests that coal exposure and age are the main determinants of emphysema in non-smoking coalworkers. The negative interaction between coal and smoking does not have a clear explanation. It may be due to the fact that smokers have more bronchitis, and that the increased mucus production traps more coal in the bronchi before it reaches the lung parenchyma, where any pathological effects that might result in emphysema presumably occur. Alternatively, the negative interaction may be due to a form of competitive inhibition, with the individual effects of coal and smoking having a final common pathway, so that the presence of one interferes with the effect of the other.

Silica content in the lung was not found to be significantly associated with emphysema in this study. The lack of an emphysema-silica association may be due to relatively low silica exposures and correspondingly less silica-related fibrotic lung disease.

Emphysema score had a highly significant negative relationship with $FEV_1\%$ in this study. This result has been found previously in a larger group of coalworkers which included the subjects from this study (Leigh et al, 1983) and in other studies of coalworkers (Ruckley et al, 1984). The regression coefficient suggests that the emphysema found in coalworkers is of some functional relevance, with an increase of emphysema of ten percent leading to a decrease of predicted $FEV_1\%$ of about seven percent. This same negative relationship between $FEV_1\%$ and emphysema was suggested in the analysis which included only non-smokers, but the small number of subjects with available FEV1 measurements makes firm conclusions difficult.

As in previous studies of coalworkers most emphysema was of the centriacinar type, with relatively little panacinar emphysema. In advanced cases, centriacinar emphysema was seen to extend to involve whole lobules and to coalesce between lobules, ultimately involving the whole lung. In accordance with current thinking (McLean et al, 1992), we regarded the two types of emphysema as part of a spectrum and hence made no attempt to separately quantify panacinar emphysema. Dust analysis was performed using minor modifications of established techniques, with good agreement between duplicate samples, and analysis was done without knowledge of the emphysema measurements made on each specimen. Lung content of coal and silica can be regarded as the best quantitative estimate of exposure as, once clearance mechanisms are saturated, accumulation of dust in the lungs is proportional to further exposure. Lung coal and silica contents were adjusted using predicted vital capacity, which was used as an estimate of lung volume. This provides a measure of the concentration of coal and silica in the lung which is not subject to the criticism that emphysematous lungs contain less total lung tissue and hence, for a given amount of total lung dust, would appear to have a higher concentration of dust per dry weight of lung, thereby potentially introducing a spurious positive relationship between dust content and amount of emphysema.

In conclusion, the results from this study provide strong support for the hypothesis that severity of emphysema in coalworkers is causally related to coal content in the lung and thus to exposure to coal in life. In this regard it is of interest that emphysema has recently become a prescribed disease for underground coal miners in the United Kingdom (IIAC, 1992). The importance of age and smoking in emphysema severity is also confirmed. The results do not support a relationship between silica exposure and the development of emphysema. However, the lung content of silica of subjects may not have been high enough to demonstrate the true effect of silica on the development of emphysema.

Disclaimer: The views expressed in this article are those of the authors and do not necessarily reflect those of the National Occupational Health and Safety Commission.

References

Dodgson J, Whittaker W (1973). The determination of quartz in respirable dust samples by infrared spectrophotometry -I. The potassium bromide disc method. Ann Occ Hyg 16:373-87.

Ferris BG, Anderson DO, Zickmantel R (1965). Prediction values for screening tests of pulmonary function. Am Rev Respir Dis 91:252-61.

Gough J, Wentworth JD (1960). Thin sections of entire organs mounted on paper. In: Harrison CV , ed. Recent advances in pathology. 7th ed. London: Churchill Livingstone, 80.

Guest L (1976). The recovery of dust from formalin-fixed pneumoconiotic lungs: a comparison of the methods used at SMRE. Ann Occ Hyg 19:37-47.

Heard BE (1969). Pathology of chronic bronchitis and emphysema. London: Churchill Livingstone, 12.

Hnizdo E, Sluis-Cremer GK, Abramowitz JA (1991). Emphysema type in relation to silica dust exposure in South African gold miners. Am Rev Respir Dis 143:1241-7.

Industrial Injuries Advisory Council (1992). Chronic bronchitis and emphysema. Social Security Administration Act 1992. Department of Social Security; London.

Lazarus R (1982). Lung-function reference values from Victorian power-industry workmen. Med J Aust 2:121-4.

Leigh J, Outhred KG, McKenzie HI, Wiles AN (1982). Multiple regression analysis of quantified aetiological, clinical and post-mortem pathological variables related to respiratory disease in coal workers. Ann Occ Hyg 26:383-400.

Leigh J, Outhred KG, McKenzie HI, Glick M, Wiles AN (1983) Quantified pathology of emphysema, pneumoconiosis, and chronic bronchitis in coal workers. Br J Ind Med 40:258-63.

Leigh J (1992). Relationship between emphysema severity and histologic type of pneumoconiosis in coal workers. Eur Respir J 5:519s (Abst).

McLean A, Warren PM, Gillooly M, MacNee W, Lamb D. Microscopic and macroscopic measurements of emphysema: relation to carbon monoxide gas transfer. Thorax 47:144-9.

Oxman AD, Muir DCF, Shannon HS, Stock SR, Hnizdo E, Lange HJ (1993). Occupational dust exposure and chronic obstructive pulmonary disease. A systematic overview of the evidence. Am Rev Respir Dis 148:38-48.

Ruckley VA, Gauld SJ, Chapman JS, Davis JMG, Douglas AN, Fernie JM, Jacobsen M, Lamb D (1984). Emphysema and dust exposure in a group of coal workers. Am Rev Respir Dis 129:528-32.

Ruckley VA, Fernie JM, Campbell SJ, Cowie HA (1989). Causes of disability in coalminers; a clinico-pathological study of emphysema, airways obstruction and massive fibrosis. Edinburgh: Institute of Occupational Medicine (Report No TM/89/05).

CARCINOGENIC ASBESTOS FIBERS IN THE PARIETAL PLEURA

F. Rey[1], C. Boutin[1], P. Dumortier[2] , J-R. Viallat[3] , P. De Vuyst[2]

Françoise REY
Service de Pneumologie
Hôpital de la Conception
147, boulevard Baille
13385 Marseille Cedex 5 - France

INTRODUCTION

The role of asbestos fibers in the development of lesions of the parietal pleura and especially mesothelioma is unclear. Based on the experimental findings of Stanton (1981) with intrapleural asbestos injections in animals, it is generally accepted that fibers more than 8µ in length and less than 0.25 µ in diameter are the most carcinogenic. More recently, the same toxicity was attributed to fibers 5µ or more long (Jaurand 1991). The presence of such fibers is controversial. In man (Boutin 1981, Sebastien 1980, Dodson 1990, Kohyama 1991) as well as in experimental animals (Rey 1993), most of the fibers observed have been short (<3µ). Studies have also demonstrated a lack of correlation between the concentration of asbestos fibers in lung tissue and the results of biopsy of parietal pleura samples. This discrepancy has been explained by the heterogeneous topographical distribution of fibers in the parietal pleura (Sebastien 1980). Thoracoscopic observation of "anthracotic

1 Service de Pneumologie, Hôpital de la Conception, 147, boulevard Baille, 13385 Marseille Cedex 5, France
2 Hôpital Erasme, Route de Lennik 808, 1010 Brussels, Belgium
3 Institut Paoli-Calmettes, 232 Bd. de Ste. Marguerite, 13009 Marseille, France

Work supported by INSERM n° 921002

NATO ASI Series, Vol. H 85
Cellular and Molecular Effects
of Mineral and Synthetic Dusts and Fibres
Edited by J. M. G. Davis and M.-C. Jaurand

spots" near lymphatic vessels suggest that exogenous particles may collect there. The purpose of this study was to detect the presence of potentially carcinogenic asbestos fibers (i.e. 5μ or longer) in the parietal pleura of patients with a history of exposure to asbestos and to determine if fibers collect preferentially in anthracotic areas.

MATERIAL AND METHODS

Patient Population: This study included 10 patients (8 men, 2 women, mean age 61.7 ± 3.4 years) in whom anthracotic spots were observed during thoracoscopic examination. Six patients had a history of exposure to asbestos. Three had confirmed mesothelioma, two had pleural metastases, and five had non-malignant diseases.

Samples: Specimens were obtained by thoracoscopy from anthracotic areas of the parietal pleura (APP) and non-anthracotic areas of the parietal pleura (Non-APP) in all patients. Additional specimens (LP) were collected from the lung parenchyma of 8 patients. Mean sample weight was 3.65 ± 0.48 mg dry tissue for LP, 5.35 ± 1.3 for APP and 3.3 ± 0.96 for Non-APP.

Fiber Detection: Observations were made by transmission electron microscopy (TEM). LP samples were prepared by Na hypochlorite digestion. Parietal pleura samples were dehydrated and defatted in alcohol and xylene, then dried and ashed in a LTA. Particles were collected on 0.45 μm porosity cellulose ester filters (Millipore®). Pieces of the filter were carbon coated and transferred to grids. Observations were performed at two magnifications, i.e. 22000x (fibers longer than 1 μm) and 4600x (fibers longer than 5 μm). Under these conditions, the limits of detection are 200000 fibers/gm dry tissue for fibers longer than 1 μm and 100000 fibers/gm dry tissue for fibers longer than 5 μm.

RESULTS

1. Fibers with a length of 1 μm or longer: Asbestos fibers of 1 μm or longer were detected in 9/10 APP samples, 4/10 non-APP samples and 5/8 LP samples. The concentration of fibers in non-APP specimens was low (mean $0.26 \pm 0.15 \times 10^6$ fibers/gm). In contrast, the concentration in APP specimens was sometimes as high as 8.28 million fibers/g (mean $1.62 \pm 0.87 \times 10^6$ fibers/gm). In 5 patients, the concentration in APP was equal or greater than the concentration in LP. The mean number of fibers was similar in LP and APP ($1.96 \pm 0.87 \times 10^6$ fibers/gm). Fiber concentration was correlated with the history of asbestos exposure: all patients with a known history of exposure had more than 1 million fibers/gm in APP and LP specimens. Amphiboles accounted for almost 100% of fibers in LP, 90% in APP and 66% in non-APP. Generally the same type of fibers were found in LP, APP and non-APP specimens. We did not note a higher concentration of chrysotile in pleural specimens.

2. Fibers with a length of 5 μm or longer: Asbestos fibers of 5 μm or longer were detected in 5/8 LP samples, 4/10 APP samples, and 1/10 non-APP samples. In the one non-APP case, the fiber concentration was at the lower limit of detection. The mean concentration in the APP specimens was $0.29 \pm 0.21 \times 10^6$ fibers/gm, which was not significantly lower than in the LP specimens ($0.73 \pm 0.29 \times 10^6$ fibers/gm). In the parietal pleura long fibers were found only in APP where they accounted for 21.6% of fibers identified. "Stanton" fibers accounted for 8.6% of fibers. The maximum fiber length observed in APP was 20μ. There was also a good agreement between the concentration of fibers longer than 5μ in APP or LP and asbestos exposure. Long fibers were always amphiboles.

DISCUSSION

These findings not only document the presence of potentially carcinogenic asbestos fibers (≥5μ) in the parietal pleura but also indicate that these fibers concentrate in "hot spots" characterized by anthracosis. In previous studies such fibers have been reported to account for between 3 and 24% of fibers found in the parietal pleura (Dodson 1990, Sebastien 1980). The concentration of fibers is higher in samples taken near costal or diaphragmatic pleural plaques than in samples taken from normal areas of the pleura, but "hot spots" were non identified (Sebastien 1980).

What anatomic structures of the parietal pleura do these anthracotic "hot spots" where fibers collect correspond to? Are they Wang pores, Kampmeier foci or milky spots? These structures that have been identified in man and animal at the lower edge of the ribs, in the pleural layers behind the heart and on the peritoneal and pleural surface of the diaphragm are known to implicated in the clearance of particles (Wang 1975, Kanasawa 1985). Experiments involving injections into the pleural, pericardial or peritoneal compartments (Pereira 1992, Takada 1991, Moally 1987) show that particles are rapidly trapped in lymphatic pores or milky spots, either free or in macrophages. After subcutaneous or intravenous injection of asbestos fibers in mice, Kanasawa (1985) found fibers only in milky spots of the parietal pleura. Interactions between the fibers and cells in these areas could trigger factors involved in a pathogenic mechanism leading to fibrosis and cancer (free oxygen radicals, chemotactic factors, growth factors...)(Sahn 1984).

CONCLUSION

Although our observations do not allow us to define whether anthracotic spots correspond to Wang pores or milky spots, they do document the existence of pleural "hot spots" where carcinogenic asbestos fibers accumulate. Without anthracosis, such spots are practically impossible to identify. The interaction between asbestos fibers and immunocompetent cells in these areas could lead to the development of pleural plaques and mesothelioma.

REFERENCES

Boutin C, Sebastien P, Janson X, Viallat JR (1981) Métrologie des fibres minérales dans les biopsies thoracoscopiques pulmonaires et pleurales. Résultats préliminaires. Poumon-Coeur 37:253-257.

Dodson RF, Williams MG, Corn CJ, Brollo A, Bianchi C (1990) Asbestos content of lung tissue, lymph nodes, and pleural plaques from former shipyard workers. Am Rev Respir Dis 142:843-847.

Jaurand MC (1991) Observations on the carcinogenicity of asbestos fibers. Ann N Y Acad Sci 643:258-270.

Kanasawa K (1985) Exchanges through the pleura. Cells and particles. In : The pleura in health and disease. J Chretien, J Bignon, A Hirsch (Eds) Marcel Dekker Inc. New York 30:195-231.

Kohyama N, Susuki Y (1991) Analysis of asbestos fibers in lung parenchyma, pleural plaques, and mesothelioma tissues of north american insulation workers. Ann N Y Acad Sci 643:27-51.

Moally PA, Macdonald JL, Goodglick LA, Kane AB (1987) Acute injury and regeneration of the mesothelium in reponse to asbestos fibers. Am J Pathol 128:426-445.

Pereira AS, Grande NR (1992) Particle clearance from the canina pleural space into thoracic lymph nodes : an experimental study. Lymphology 25:120-130.

Rey F, Viallat JR, Farisse P, Boutin C (1993) Pleural migration of asbestos fibres after tracheal injection in rats. Eur Respir Rev 3:145-147.

Sahn SA, Antony VB (1984) Pathogenesis of pleural plaques. Relationship of early cellular response and pathology. Am Rev Respir Dis 130:884-887.

Sebastien P, Janson X, Gaudichet A, Hirsch A, Bignon J (1980) Asbestos retention in human respiratory tissues : comparative measurements in lung parenchyma in parietal pleura. In : Biological effects of mineral fibres. JC Wagner (Ed.) IARC Scientific Publication Lyon, France 30:237-246.

Stanton MF, Layard M, Tegeris A, Miller E, May M, Morgan E, Smith A (1981) Relation of particle dimension to carcinogenicity in amphibole asbestoses and other fibrous minerals. J Natl Cancer Inst 67:965-975.

Takada K, Otsuki Y, Magari S (1991) Lymphatics and pre-lymphatics of the rabbit pericardium and epicardium with special emphasis on particulate absorption and milky spot-like structures. Lymphology 24:116-124.

Wang NS (1975) The preformed stomas connecting the pleural cavity and the lymphatics in the parietal pleura. Am Rev Respir Dis 111:12-20.

Pleural Lesions induced by Mineral Dusts, Fibers and Chemicals

Anand Prakash Sahu
Preventive Toxicology Division
Industrial Toxicology Research Centre
Post Box No. 80, LUCKNOW-226 001, INDIA

Pleural fibrosis (PF), Pulmonary Interstitial Fibrosis (PIF), lung cancer and malignant mesothelioma are well documented changes due to asbestos-related exposure. PF is increasingly more common than PIF in many asbestos-exposed populations. This was found especially true in populations with less intense exposure (Miller and Miller, 1993). India is a developing country, rapidly industrializing and demand for asbestos and its production has, therefore, greatly increased mainly for construction activity such as cement-asbestos industries. The asbestos exposure, mechanism for the development of disease and prevention needs greater attention to safeguard the exposed workers.

For malignant mesothelioma, the possibility of a synergistic effect between exposure to asbestos and smoking has been rejected by Elmes (1980). Lack of interaction between exposure to asbestos and smoking both in mesothelioma and pleural thickening suggest that separate mechanism act on pleural changes (Yano et al.,1993). The incidence of malignant mesothelioma is strongly increasing in various industrialized countries (Bianchi et al.,1993). On the other hand, the rate of non-asbestos related mesothelioma incidence is not known or presently not measurable because definite etiological factor(s) are not known. Although two major hypotheses Oxygen radicals and Growth factors are presently in favour to explain the fibrogenic and carcinogenic response induced by asbestos fibers. The hydroxyl and superoxide radical generation which are cytotoxic can injure pulmonary cells and then the induction of fibrogenic response and lastly the resultant effect can be carcinogenic (Mossman and Marsh, 1988). The Platelet-Derived Growth Factor (PDGF) is most potent inducer of mesenchymal cell proliferation and Transforming Growth Factor β (TGF - β) in mediating the synthesis of extracellular matrix proteins (Brody, 1993). Moreover, how the cell injury leads to fibrogenic or neoplastic disease remain undefined. In this study the observed pleural reactions by non-asbestos exposure and along with asbestos are described which were observed during the course of different experiments.

NATO ASI Series, Vol. H 85
Cellular and Molecular Effects
of Mineral and Synthetic Dusts and Fibres
Edited by J. M. G. Davis and M.-C. Jaurand

METHODS

The animals used in this study were obtained from the Animal Breeding Facility of Industrial Toxicology Research Centre, Lucknow. The animals were kept on stock laboratory diet, leafy vegetables and drinking water ad libitum. The animals were inoculated with the dusts, chemicals and exposed to biomass-smoke in different experiments as given in Table 1. As per experimental details given in Table 1 the animals in each group were killed and lungs were fixed by 10 % formol saline. Paraffin embedded blocks were prepared and multiple sections (5 µm thickness) were stained with haematoxylin and eosin, silver impregnation for reticulin and Perl's stain for asbestos bodies.

RESULTS

The pleural reactions observed in each group of different experiments are summarized in Table 2. Moreover, the details of the observed changes are also described. In group 1 (choline alone) at 90 days, the collections of macrophages with pigments as inclusion were found at pleural surface. By 330 days, slight thickening of the pleura with the collection of cells were observed. Only few fibrotic reactions at the pleura and near the pleural surface were observed in group 2 (mica dust). However, the combined effect of choline and mica dust (group 3) the induced changes were collection of mesothelial cells, fibroblasts and small cells at 330 days (Figure 1). In group 5 (choline alone) in guinea pig at 570 days induced a marked cellular lining of the pleural surface. The bronchiole and lymphatic vessel were seen at the outer limiting of the lung lobe and an extension connected by a band of cellular mass with hyperchromatic cells (Figure 2). In group 6 (choline and chrysotile asbestos) the thickening of pleural surface which were more prominent at some places and where it appears neoplastic. In group 8, the cow-dung cake smoke exposure induced thickened mass at the pleural surface containing small cells, fibroblasts and necrosed cells (Figure 3). In group 12 (SpA and BLM combined group) at 30 days induced a connecting cellular mass of hyperchromatic cells attached to the lung lobe were observed. The mass contained cells similar to mesothelial cells and necrosed cells (Figure 4). These changes observed on the pleural surfaces were found in small number of animals.

Table 1. Experimental Details

Group	Animals	TREATMENT	Period of study
Experiment No. I			
1	Rat	Choline Chloride (CC), 25 mg in 1.0 ml Distilled water (DW), 5 days a week, 40 injections by intraperitoneal route (IP)	90, 180 and 330 days
2		Mica dust, 25 mg in 1.0 ml Normal Saline (NS) by Intratracheal route (IT)	
3		CC + Mica dust, same as in group 1 and 2	
4		Control	
Experiment No. II			
5	Guinea pig male	CC, 50 mg in 1.0 ml DW 5 days a week, 40 injections by IP	570 and 680 days
6		Chrysotile asbestos, 15 mg in 1.0 ml NS by IP after 210 days of CC treatment	
7		Control	
Experiment No. III			
8	Rat female	Cow-dung cake smoke, 1 gm dried fuel was burned for 15 minutes and exposure was done, 5 days a week in dynamic inhalation chamber for 30 days	30 and 60 days
9		Control	
Experiment No. IV			
10	Mice male	*Staphylococcus* Protein-A (SpA) 6 µg in 0.2 ml NS, twice a week for 4 weeks by IP	30 days
11		Bleomycin (BLM), 0.5 mg in 0.2 ml NS, twice a week for 4 weeks by IP	
12		SpA + BLM same as in group 10 and 11	
13		Control	

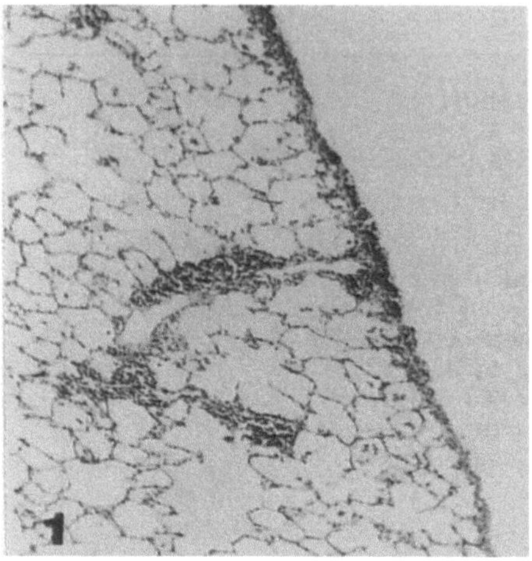

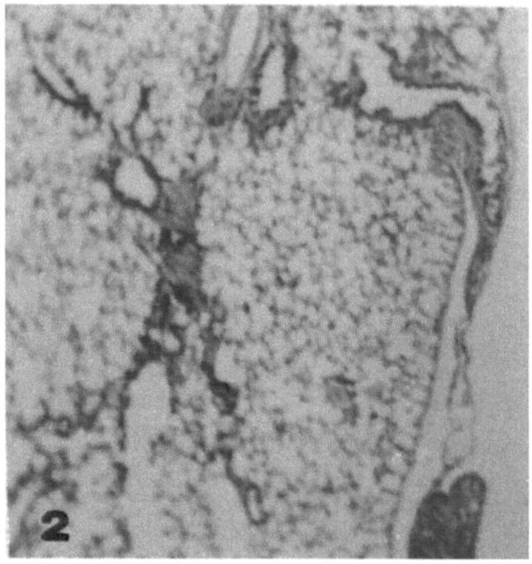

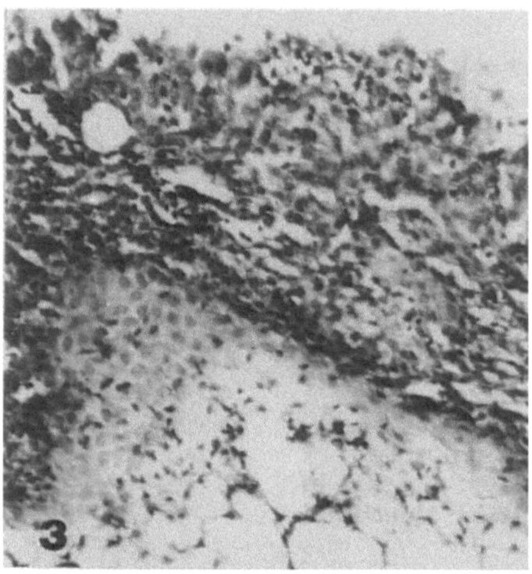

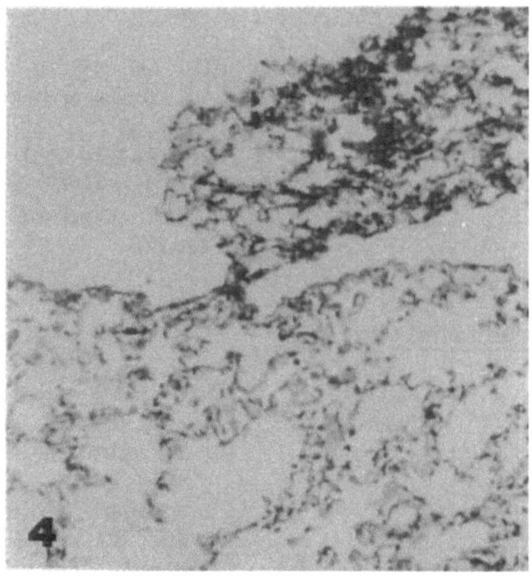

Figure 1. Lung of rat at 330 days following simultaneous administration of choline and mica dust (group 3). Hypercellular pleural reaction. Haematoxylin and Eosin (H & E), X 80

Figure 2. Lung of guinea pig at 570 days after parenteral administration of choline (group 5). At the outside of lung lobe, cellular mass containing hyperchromatic cells connected with the cellular band. H & E, X 60

Figure 3. Lung of rat at 30 days exposed to Biomass smoke (Cow-dung cake) in group 8. Collection of the mesothelial cells, fibroblasts and necrosed cells at the pleural surface. H & E, X 310

Figure 4. Lung of mice at 30 days , the simultaneous administration of Staphylococcus Protein-A and Bleomycin (group 12). The cellular mass with hyperchromatic cells and necrosed cells connected with the lung lobe. H & E, X 195

Table 2. The Histopathological Grading of Pleural lesions observed in the groups of the different Experiments

Group	PERIOD OF OBSERVATION		
Experiment No. I	90 days	180 days	330 days
1	+	+	++
2	+	+	+
3	+	++	+++
4	-	-	-
Experiment No. II	570 days	680 days	
5	++	++	
6	+++	+++	
7	-	-	
Experiment No. III	30 days	60 days	
8	++	++	
9	-	-	
Experiment No. IV		30 days	
10		-	
11		-	
12		+	
13		-	

- = Absent ; + = Slight ; ++ = Moderate ; +++ = Marked

DISCUSSION

In the present study the observed changes were mostly reactive lesions on the pleural surface induced by different agents. The pleural changes were also observed in female rats due to biomass smoke inhalation. Significantly, the prevalance of male/female ratio shows marked variation in the incidence of mesothelioma. Some of the World's highest rates of lung cancer in female population were found in Northern China mainly due to smoke exposure. Exposure to environmental or occupational air contaminants do not occur singly but to a multiplicity of agents and disease processs are frequently markedly affected by the interaction of a variety of factors. This fact suggests the role of factors other than asbestos in the development of asbestos-related mesothelioma. In the pathogenesis of mesothelioma the co-factors remains to be identified. The excess availability of choline produced lung lesions and pleural lesions in rats and guinea pigs and combined effects of fibrogenic and carcinogenic substances produced inhanced lesions (Sahu and Shanker, 1984). Experimental data indicate that asbestos and nitrosamines may co-operate in inducing pleural mesothelioma (Katada et al.,1983). Methylamines formed after choline ingestion and which possibly act as substrate for the formation of nitrosamine, showed marked carcinogenic activity (Zeisel, 1983). However, most choline is consumed in the form of lecithin in food such as liver, eggs, soyabeans and peanuts but largely absent from most fruits and vegetables (Sahu et al., 1986). Many studies have indicated that consumption of vegetable and fruits or some vegetable-related constituents have protective effects on the risk of developing mesothelioma.

The pleural and peritoneal lining mesothelial cells utilized choline to synthesized a phosphatidylcholine-rich material that resembles the surfactant of the human lung (Hills et al., 1982; Maher, 1992). The rat treated with intraperitoneal phosphatidylcholine had significantly more adhesions after standardized abrasions instead of preventing adhesions, the instilled lubricants appeared to promote them (Rozga et al.,1990). In support, the effects on the lymphatics of lung and lymph nodes due to excess avalability of choline were also reported by us. The causative agent, with the help of factors which are presently undefined can produce pleural changes as evident in the present study that the biomass smoke and bleomycin are known peumotoxic agents, had produced pleural lesions. Although the toxic substances were available at the site of action (pleural and peritoneal cavity) but the secretory products and phosphatidylcholine-rich material produced by mesothelial cells might have modified the effects. Thus the continuous availability of toxic stimuli were not possible. While fibers have physico-chemical characteristics, the release of soluble portion of the fibers in the cavity and other factors in inducing the selective toxicity, which can produce the normal mesothelial cells to mesothelioma.

REFERENCES

Bianchi C, Brollo A, Ramani L, Zuch C (1993) Asbestos-related mesothelioma in Monfalcone, Italy. Am J Ind Med 24:149-160

Brody AR (1993) Asbestos-induced lung disease. Environ Health Perspect 100 : 525-535

Elmes PC (1990) Fibrous mineral and health. J Geol Soc 137: 525-535

Hills BA, Butler BD, Barrow RE (1982) Boundary lubrication imparted by pleural surfactants and their identification. J Appl Physiol 52 : 463-469

Katada H, Higashiguchi R, Maruyama H, Emi Y, Yokose Y, Takashi S, Mikami R, Konishi Y (1983) Effects of chrysotile asbestos on lung and pleural carcinogenesis of N-bis (2-hydorxypropyl) nitrosamine in rats. Cancer Lett 17: 313-320

Maher JF (1992) Lubrication of the peritoneum.Perit Dial Int 12:346-349

Miller A, Miller JA (1993) Diffuse thickening superimposed on circumscribed pleural thickening related to asbestos exposure. Am J Ind Med 23: 859-871

Mossman BT,Marsh JP (1991) Role of active oxygen species in asbestos-induced cytotoxicity cell proliferation and carcinogenesis. In: Harris CC, Lechner JG and Brinkley, eds, Cellular and Molecular aspects of Fiber Carcinogensis, pp. 159-168. Cold Spring Harbor Laboratory, Cold Spring Harbor, New York

Rozga J, Andersson R, Srinivas V, Ahren B, Bengmark S (1990) Influence of phosphatidylcholine on intra-abdominal adhesion formation and peritoneal macrophages. Nephron 54: 134-138

Sahu AP, Shanker R (1984) Experimental evidence of possible relationship to choline in pulmonary carcinogensis. In : Symposium on Tumor Promotion and Enhancement of Human and Experimental Respiratory Tract carcinogenesis. Williamburg, V A, p. 52, U S Environmental Protection Agency

Sahu AP, Saxena AK, Singh KP, Shanker R (1986) Effect of chronic choline administration in rat. Ind J Exp Biol 24: 91-96

Zeisel SH, Wishnok JS, Blusztajn JK (1983) Formation of methylamines from ingested choline and lecithin. J Pharmacol Exp ther 225:320-324

COMPARISONS OF *IN VITRO* CYTOTOXICITY RESULTS WITH *IN VIVO* PULMONARY EFFECTS FOLLOWING INHALATION EXPOSURES TO A VARIETY OF MINERAL DUSTS: HOW WELL DO THEY COMPARE?

D.B. Warheit, S.H. Gavett, I.S. Yuen, and M.A. Hartsky
DuPont Haskell Laboratory
P.O. Box 50, Elkton Rd
Newark, DE 19714 USA

INTRODUCTION

In vitro toxicity assays have been proposed as substitutes for acute *in vivo* studies during the early phases of inhalation toxicology testing. Short-term toxicity tests are needed to screen materials. Test systems generally are comprised of cultured lung cells and dusts. *In vitro* tests have strengths as well as limitations. These assays could be utilized in toxicology screening strategies and have the advantage of being simpler, faster and less expensive than their *in vivo* counterparts. In contrast, many determinants of dust toxicity are difficult or impossible to evaluate with cell cultures alone. The studies presented here were undertaken to compare *in vitro* results using lung epithelial L2 cells with a variety of particle-types which had been previously tested in an inhalation bioassay. Lung epithelial (L2) cells were incubated at various concentrations with either carbonyl iron (CI), zinc oxide (ZnO), crystalline silica (Si) or titanium dioxide (TiO_2) particles. Cytotoxicity was assessed using the MTT assay, as well as measuring for lactate dehydrogenase activity in the cell culture fluid. The results obtained in the *in vitro* assays were then compared to bronchoalveolar lavage fluid biomarkers of toxicity measured following 3-day inhalation exposures to each of the particulate materials described above. The results of *in vitro* studies indicated that titanium dioxide particles were the least toxic, carbonyl iron and silica were intermediate, and zinc

NATO ASI Series, Vol. H 85
Cellular and Molecular Effects
of Mineral and Synthetic Dusts and Fibres
Edited by J. M. G. Davis and M.-C. Jaurand

oxide particles produced the greatest toxicity. In contrast, the results of inhalation tests indicated that carbonyl iron and titanium dioxide produced little or no pulmonary effects. Silica produced a progressive pulmonary lesion characterized by inflammation, Type II epithelial cell cytotoxicity and consequent granulomatous formation. Inhalation of zinc oxide caused a potent but transient pulmonary inflammatory response, which resolved within a few days postexposure, due, in part, to its high solubility in the lung. The results with zinc oxide and carbonyl iron particles point out the limitations of the *in vitro* tests. The *in vitro* toxicity test does not provide for pulmonary clearance of exposed dusts, either by solubility (in the case of zinc oxide) or by macrophage phagocytosis (as was evident in the lungs of carbonyl iron-exposed rats). Perhaps a sophisticated co-culture system (e.g., epithelial cells and macrophages) would be more useful in predicting the fibrogenic potential of inhaled dusts.

METHODS

In Vitro Studies

Rat lung epithelial cell lines (Rat L2 cells) were purchased from American Type Culture Collection (Rockville, MD) and grown to log phase using Kaighn's Nutrient Mixture F-12 (Ham's F-12) (10% FBS + concentrated glutamine + antibiotic-antimycotic solution) at 37°C, 5% CO_2. Following passage from culture flasks, L2 cells were added to designated wells (1×10^5/well in 24 well plates (Corning Glass Works, Corning, NY; diameter = 16 mm) for LDH determination or 1×10^4/well in a 96 well plate (diameter 4.5 mm) for MTT determination plated at 1250 cells/mm^2 for the MTT assay and plated at 1000 cells/mm^2 for LDH analysis. Cells were allowed to adhere overnight in wells at 37°C and 5% CO_2. During the next day, autoclaved silica, zinc oxide, titanium dioxide, and carbonyl iron particles were added to designated wells at concentrations of 0, 50, 200 and 800 μg/ml. The cells and particles were incubated overnight (i.e., for 20 hrs, which had previously been established as the optimal time period). For the LDH assay, the culture supernatants were collected, centrifuged and refrigerated until analyzed for LDH activity. For the MTT assay, the plate was inverted, flicked, blotted, and 50 μl of a 1% MTT solution dissolved in culture media without serum was added and incubated for 3 hrs. Subsequently, the formazan was solubilized with propanol and the absorbances were read on a microplate reader.

The purpose of the MTT cytotoxicity assay was to measure the inhibition of oxidative metabolism of L2 lung epithelial cells in culture following addition of particles to the cell culture medium. The inhibition of oxidative metabolism may be a more sensitive indicator of cytotoxicity than release of cytoplasmic enzymes such as lactate dehydrogenase (LDH). MTT (3-(4,5-dimethylthiazol-2-yl)-2,5-diphenyl tetrazolium bromide) is converted by mitochondrial succinate dehydrogenase to dark insoluble formazan. Propanol was used to solubilize the formazan, which is measured spectrophotometrically on a microtiter plate reader at 560 nm.

In additional experiments, the effects of in vitro silica or carbonyl iron particle exposure to L2 cells alone were compared with effects of particles on alveolar macrophages alone or with L2 cells co-cultured with alveolar macrophages. As described above, L2 cells passaged from culture flasks were added to designated wells. In addition, freshly lavaged rat alveolar macrophages were plated at similar concentrations in other wells, and also added to some wells containing L2 cells (total number of cells/well = 2×10^5). Silica or carbonyl iron particles were added to designated wells in varying concentrations (0, 50, 200 or 800 μg/ml). Cell/particle preparations were incubated for 20 hrs and the supernatants were assayed for LDH activity as described above.

General Experimental Design for In Vivo Studies

Groups of male Crl:CDBR rats (7-8 weeks old, Charles River Breeding Laboratories, Kingston, New York) were used to assess the pulmonary effects of short-term, high-dose aerosol exposures either to 100 mg/m^3 of zinc oxide, carbonyl iron, titanium dioxide, or Min-U-Sil silica particles. CD rats were exposed for 3 days at aerosol concentrations of 100 mg/m^3 to each of the particle-types listed above. Following the termination of exposures, animals were sequentially sacrificed and the lungs washed at several time periods postexposure.

Inhalation Exposure

For each of the particle exposures, rats were placed in cylindrical polycarbonate or stainless steel holders equipped with conical nose pieces. The restrainers were inserted into face plates on the exposure chambers such that the nose of each animal protruded into the chamber. Dust atmospheres were generated with a K-tron bin feeder equipped with twin feed screws. The dust was metered into a polycarbonate transfer tube wherein high pressures of air swept the test material into the exposure chamber. Chamber concentrations were maintained by controlling the dust-feed rate into the generation apparatus, or by varying the flow rate.

For gravimetric analysis, samples of atmospheric silica, CI, ZnO, or TiO_2 were taken from the animal breathing zone at approximately 30-minute intervals by drawing calibrated volumes of chamber atmosphere through pre-weighed glass-fiber filters. Filters were weighed on a Cahn 26 autmomatic electrobalance. The atmospheric concentrations of silica particles were determined from the filter weight differentials before and after sampling. Particle size measurements of airborne particles in the test chamber were determined with a Sierra cascade impactor and reported as the mass median aerodynamic diameter (MMAD) and the percentage of particles with less than a 10 μm aerodynamic diameter.

Pulmonary Lavage

Bronchoalveolar lavage procedures were conducted according to methods previously described (Warheit et al., 1984). Briefly CD rats were euthanized by intraperitoneal injections of sodium pentobarbital (Nembutal®). The trachea, airways, and lungs were gravity infused with a warmed, (37°C) Ca^{++}-Mg^{++} free, phosphate-buffered saline solution. This bronchoalveolar lavage (BAL) procedure was carried out 6 times or until 50 ml of fluid were collected from each animal.

Lavaged fluids recovered from sham and particle-exposed rats were centrifuged at 250 x g, and the supernatant was removed and resuspended in Eagles Minimal Essential Medium (Eagles MEM F-11; pH= 7.2, GIBCO, Grand Island, NY) supplemented with penicillin and streptomycin. Cell numbers and viabilities were quantified using a hemocytometer and trypan blue solution. Cell differential counts were carried out on Diff-Quik®-stained (American Scientific Co.) cytocentrifuge preparations. A minimum of 500 cells per slide was counted for differential analysis.

Biochemical Assays on Bronchoalveolar Lavage Fluid

Lavaged fluids from the first wash were centrifuged and assayed. All biochemical assays were performed on BAL fluids at 30°C, using a semi-automated clinical chemical analyzer (Encore II®, Baker Instruments, Allentown, PA). Lactate dehydrogenase (LDH) was measured using commercially available reagent kits (Baker Instruments, Allentown, PA). Lavage fluid protein was measured using a commercially available reagent kit based on Coomassie

Blue dye binding (QuantTest®, Quantimetrix, Hawthorne, CA). All refrigerated lavaged samples containing enzymes and proteins were stable for a minimum period of 24 hours.

RESULTS AND DISCUSSION

In Vitro Studies

The results of *in vitro* studies with L2 lung epithelial cells using both MTT and LDH as indicators of cytotoxicity demonstrated that titanium dioxide particles were the least toxic, carbonyl iron and silica were intermediate, and zinc oxide particles produced the greatest cytotoxic effects (Figs. 1 and 2). In an additional experiment, the cytotoxic effects of carbonyl iron or silica particles were tested with L2 epithelial cells, alveolar macrophages, as well as a co-culture of L2 cells and alveolar macrophages. Lactate dehydrogenase release was utilized as an indicator of cytotoxicity. As demonstrated in Figs. 1 and 2, carbonyl iron particles produced significant cytotoxicity to L2 epithelial cells. However, this effect was diminished when CI particles were incubated either with alveolar macrophages or when CI particles were exposed to a co-culture of macrophages and L2 epithelial cells. These results suggest that the alveolar macrophages "scavenged" or significantly reduced the toxicity of carbonyl iron particles to L2 epithelial cells (Fig. 3). In contrast to the observed effects with carbonyl iron particles, silica particles produced enhanced toxicity *in vitro* to L2 epithelial cells when compared with silica-exposed alveolar macrophages. However, in contrast to the observed effects with carbonyl iron particles, when L2 cells and macrophages were co-cultured, there appeared to be an additive effect, i.e., the LDH activity of the co-culture was equivalent to the two groups combined. These results strongly suggest that alveolar macrophages do not diminish the toxicity for L2 epithelial cells following *in vitro* exposure to silica particles.

In Vivo Tests

Using bronchoalveolar lavage fluid analysis as a biomarker, the results of inhalation tests indicated that 3 day exposures to carbonyl iron or titanium dioxide produced little or no pulmonary toxic effects when compared to sham-exposed controls (Warheit *et al.*, 1991; Kellar *et al.*, 1993). Silica produced a progressive pulmonary lesion characterized by sustained increases in BAL fluid levels of lactate dehydrogenase and protein. We have previously reported that a chronic pulmonary inflammatory response occurs after 3-day exposures and this progresses to Type II epithelial cell cytotoxicity and consequent granulomatous formation (Warheit *et al.*, 1991). In addition, inhalation of zinc oxide caused a potent but transient pulmonary inflammatory response, which resolved within a few days postexposure, due, in part, to its high solubility in the lung (Warheit and Hartsky, 1988).

In vitro tests have strengths as well as limitations. These studies were designed to compare *in vitro* results with dusts using L2 lung epithelial cells with identical materials which had been previously tested in an inhalation bioassay. The L2 epithelial cells were incubated at various concentrations with either carbonyl iron, zinc oxide, crystalline silica or titanium dioxide particles. Cytotoxicity was assessed by the MTT assay, as well as with LDH acitivity in the cell culture fluid. The *in vitro* studies demonstrated a range of toxicities with TiO_2 < Si = CI < ZnO. Preliminary studies indicated that in a co-culture system of macrophages and L2 epithelial cells, alveolar macrophages scavenged the CI particle toxicity to L2 epithelial cells. In contrast, inhalation tests indicated that CI or TiO_2 produced no significant pulmonary effects. Silica inhalation produced a sustained pulmonary inflammatory response, and ZnO caused a potent but transient pulmonary inflammatory response, which was short-lived due, in part, to its high solubility in the lung. The results with zinc oxide and carbonyl iron particles point out the limitations of the *in vitro* tests. The *in vitro* cytotoxicity test does not provide for pulmonary clearance of exposed dusts, either by solubility (as in the case of ZnO) or by macrophage phagocytosis/pulmonary clearance (as in the case of CI-exposed rats). It seems likely that a more complex co-culture system (e.g., epithelial cells and macrophages) could be useful in predicting the fibrogenic potential of inhaled dusts.

Manuscript References

Kellar KA, Rennix DK, Hartsky MA, and Warheit DB (1993) Acute inhalation of high particle burdens on pulmonary macrophages (PM) clearance responses. Toxicologist 13:45A.

Warheit DB, Hill LH, and Brody AR (1984) Surface morphology and correlated phagocytic capacity of pulmonary macrophages lavaged from the lungs of rats. Exp Lung Res 6:71-82.

Warheit DB, and Hartsky MA (1988) Acute inhalation of nuisance dusts produces variable lung responses in exposed rats. Amer Rev Respir Dis 137:352A.

Warheit DB, Carakostas MC, Hartsky, MA, and Hansen JF (1991) Development of a short-term inhalation bioassay to assess pulmonary toxicity of inhaled particles: Comparisons of pulmonary responses to carbonyl iron and silica. Toxicol Appl Pharmacol 107:350-368.

TABLE 1.
BAL FLUID LACTATE DEHYRODENASE (LDH) AND PROTEIN VALUES IN RATS EXPOSED TO ZINC OXIDE, SILICA, TITANIUM DIOXIDE OR CARBONYL IRON PARTICLES

BAL LDH VALUES
% of Sham Control Responses (i.e., 100%)
(Time after 3-Day exposure at 100 mg/m³)

	0H	24H	8D	1M	2M
TiO2	214 ± 24	117 ± 18	148 ± 38		
Silica	361 ± 53	391± 45	433 ± 23	1322 ± 129	1030 ±15
CI	126 ± 22	85 ± 17	130 ± 32	95 ± 19	97 ± 18
	0H	**24H**	**8D**	**1M**	**3M**
ZnO	1500 ± 171	1919 ± 520	335 ± 197	62 ± 4	71 ± 2

BAL PROTEIN VALUES

% of Sham Control Responses (i.e., 100%)
(Time after 3-Day exposure at 100 mg/m³)

	0H	24H	8D	1M	2M
TiO2	155 ± 59	109 ± 19	126 ± 16		
Silica	158 ± 23	187± 21	358 ± 11	622± 54	338 ± 21
CI	98 ± 21	90 ± 7	85 ± 9	105 ± 8	108 ± 12
	0H	**24H**	**8D**	**1M**	**3M**
ZnO	1567± 369	1039±248	280± 65	74 ± 10	69 ± 15

OH, 24H, 8D, 1M, 2M, & 3M = immediately after, 24 hrs, 8 days, 1, 2, or 3 months after 3-day exposure.

1

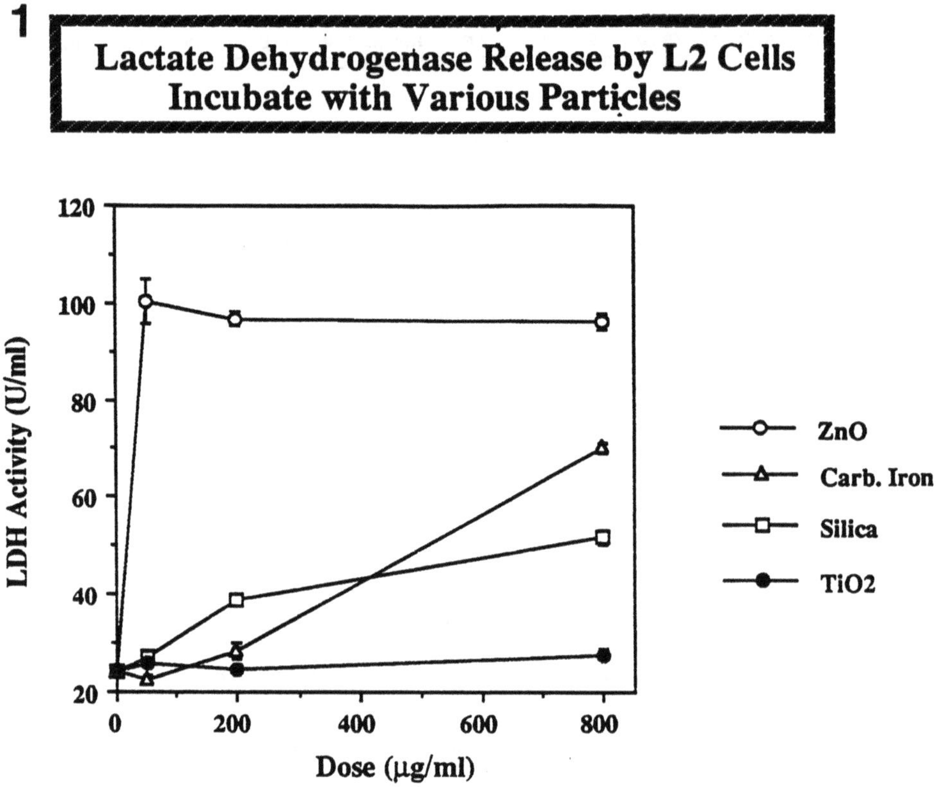

Fig. 1. Lactate dehydrogenase release by L2 cells incubated for 20 hrs either with zinc oxide, carbonyl iron, silica or titanium dioxide particles. In vitro exposures to zinc oxide particles caused the greatest cytotoxicity (i.e. increased LDH release). Carbonyl iron and Min-U-Sil silica were intermediate, and TiO2 particles were the least toxic.

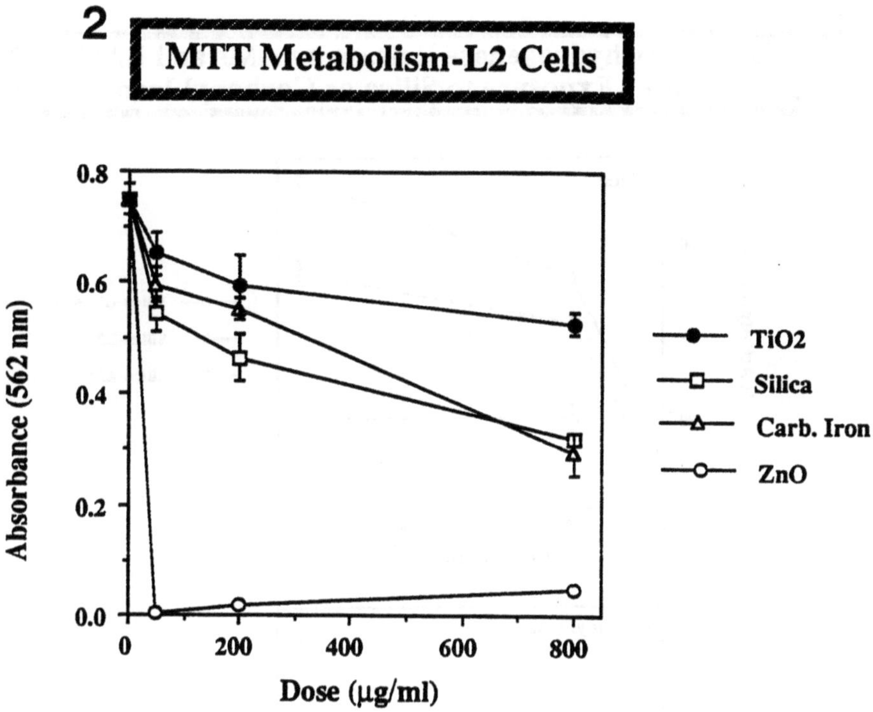

Fig. 2. MTT metabolism in L2 epithelial cells incubated for 20 hrs either with zinc oxide, carbonyl iron, silica or titanium dioxide particles. In vitro exposures to zinc oxide particles caused the greatest cytotoxicity (i.e. decreased absorbance or MTT). Carbonyl iron and Min-U-Sil silica were intermediate, and TiO2 particles were the least toxic.

3 **Lactate Dehydrogenase Release by AMs and L2 Cells after Exposure to Silica or Carbonyl Iron**

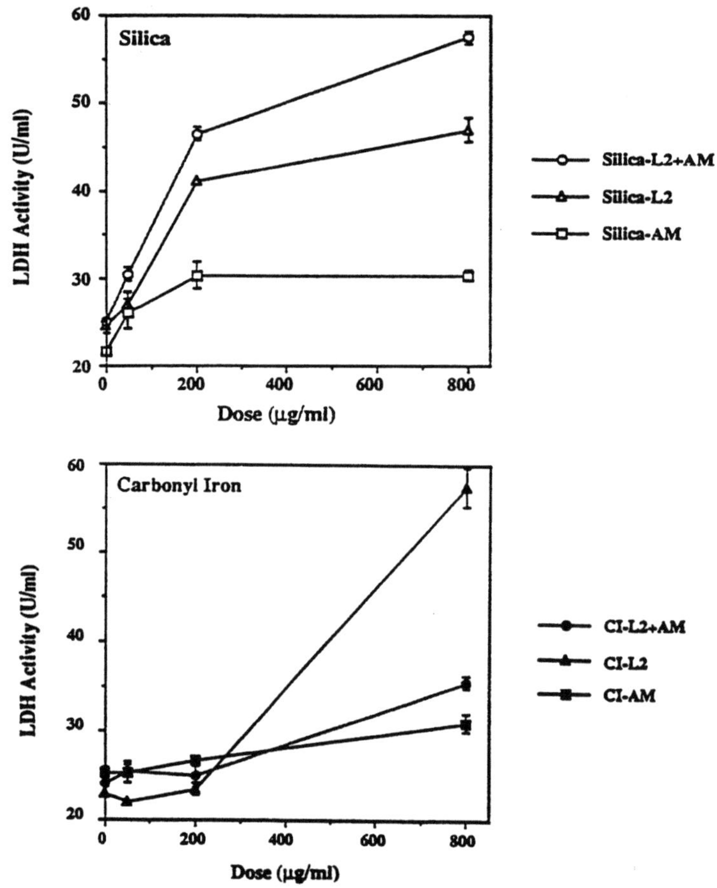

Fig. 3. Lactate dehydrogenase release by alveolar macrophages (AM) and L2 epithelial cells after in vitro exposures to Min-U-Sil silica or carbonyl iron (CI) particles. L2 cells and macrophages were individually or in combination (i.e., co-cultures AM + L2 cells) incubated with silica or carbonyl iron particles. In the case of carbonyl iron particles, the alveolar macrophages appeared to scavenge (or protect) the toxicity of CI particles to L2 cells (i.e., LDH values were decreased with AM or AM + L2 cells when compared to L2 cells alone). In contrast, the toxicity of silica particles for L2 cells or macrophages appeared to be additive.

Particle-Associated Hydrocarbons and Lung Cancer; the Correlation Between Cellular Dosimetry and Tumor Distribution

Per Gerde[1], Bruce A. Muggenburg, Rogene F. Henderson, and Alan R. Dahl
Inhalation Toxicology Research Institute
P.O. Box 5890, Albuquerque,
New Mexico 87185
USA

Abstract

The well-known correlation between patterns of particle deposition in the human bronchial tree and the apparent sites of origin of primary bronchial tumors is paradoxical because clearance of inhaled particles from the bronchi is comparatively rapid, leading to a short exposure time. However, there may be a dosimetric link between the deposition of highly lipophilic carcinogens such as polycyclic aromatic hydrocarbons (PAHs) reversibly adsorbed on inhaled particles, and the distinctive distribution of preneoplastic and neoplastic lesions in the lungs. We have demonstrated that once desorbed from the particles, highly lipophilic PAHs are diffusion-limited during clearance through the airway epithelium to the capillary blood. In the Beagle dog, clearance of benzo(a)pyrene (BaP) to the blood takes only minutes in the alveoli, while in the thicker epithelium of the bronchi, clearance may take hours. The slowed clearance results in a high concentration of the PAHs in the bronchial epithelium and an increased opportunity for metabolism to reactive forms. In contrast, less lipophilic substances are perfusion-limited during clearance and are likely to clear within minutes from all regions of the lungs to the circulating blood. The greater likelihood of highly lipophilic toxicants to induce first-pass toxicity during diffusion-limited clearance links the seemingly independent parameters of lipophilicity and particle-association and gives a plausible explanation for the relationship between particle deposition patterns and the distribution of tumors in the bronchial tree. This new information on microdosimetry of PAHs in the lungs should improve risk assessments of exposure to inhaled particle-associated hydrocarbons.

[1] Permanent address: National Institute of Occupational Health

S-171 84 Solna, Sweden

NATO ASI Series, Vol. H 85
Cellular and Molecular Effects
of Mineral and Synthetic Dusts and Fibres
Edited by J. M. G. Davis and M.-C. Jaurand

Introduction

The observation that primary lung cancers have a distinctive distribution within the lung was made decades ago, and speculation that the tumor pattern correlates with either patterns of particle deposition, clearance, or both, quickly followed (Auerbach et al. 1961; Hilding 1957; Kotin and Falk 1959; Macklin 1956). It was suspected that the agents which cause tobacco smoke-induced lung cancer involved polycyclic aromatic hydrocarbons (PAHs) generated during combustion. Based on the pattern of tumors and on the then-assumed behavior of PAHs in the tobacco tar particles, a dynamic picture of the dosimetry was hypothesized (Hilding 1957; Macklin 1956). Hilding mentioned the importance of the *time element* in the induction of bronchial carcinoma and that the chemical agents inducing cancer reached the airway epithelium during the period of rapid mucociliary clearance in the larger airways. These conjectures were based on observations of the behavior of inhaled particles deposited at reasonably low densities.

In order to investigate tumor patterns, a considerable experimental effort was devoted to finding a suitable animal model for induction of lung cancer using particle-associated PAHs (Henry et al. 1975; Saffiotti et al. 1968; Saffiotti et al. 1972). However, because of the comparatively short time span of the experiments and the low sensitivity of most species of laboratory animals to PAH-induced pulmonary carcinogenesis, large amounts of the PAHs were used. The dose received by the animals often exceeded realistic exposure levels in humans by 5-6 orders of magnitude, and the numbers of exposures were reduced from sometimes tens of thousands in humans to less than 20 in the animals. One conspicuous result of the high-dose instillation experiments was that coadministration with an inert carrier dust, such as iron oxide, increased the retention of the PAHs in the lungs and the tumor-inducing capacity. Based on these experiments, and in contrast with the earlier studies on particle deposition and clearance, a long retention of PAHs on the carrier particles was considered critical in the induction of lung tumors (Henry et al. 1975). This has been called the *slow release* hypothesis.

Because of the apparent contradictions in the dosimetry of particle-associated organic compounds between different experimental approaches to chemical-induced lung cancer, a new model was developed to describe the dosimetry of PAHs at low exposure levels (Gerde et al. 1991a). Their purpose was to develop and experimentally validate a mathematical model describing the dosimetry of PAHs at the low-level exposures typical in humans.

A mechanistic model of the respiratory tract dosimetry of particle-associated PAHs

This model incorporated the physicochemical properties of the toxicants and the physiological and biochemical properties of the tissues. Only dissolved PAHs were considered in the dose to lung. As a consequence, the potential for a given PAH molecule to induce toxicity was independent of how long it was adsorbed on its carrier particle. The prime determinant of dose according to the new model was the lipophilicity of the toxicant and thickness of the air/blood barrier at which the PAH is taken up in the lungs. The more lipophilic the PAH and the thicker the epithelium through which the substance is absorbed, the slower the clearance into blood. Therefore, the bronchial epithelium was predicted to be particularly vulnerable to exposure by highly lipophilic toxicants (Gerde et al. 1991a).

We validated the model for clearance of lipophilic organics from various regions the lungs in the Beagle dog. Alveolar clearance was measured by exposing dogs to a bolus of aerosolized PAHs in a single breath, then monitoring the appearance of the hydrocarbons in the circulating blood by repeatedly sampling blood as it entered and left the lung (Gerde et al. 1993a). Clearance of the highly lipophilic PAH, benzo(a)pyrene (BaP), and the moderately lipophilic PAH, phenanthrene (Phe), from the lung was rapid; the first 50% of BaP and Phe cleared within 2.4 and 1.0 min, respectively. However, whereas Phe was mostly perfusion-limited during clearance, the highly lipophilic BaP was clearly diffusion-limited during its clearance. Even so, at low doses, clearance of most organic solutes from the alveolar region, lipophilic or not, was rapid.

In the conducting airways, the model predicted that organic solutes may clear via two different routes: mucociliary clearance along the airways and penetration of the airway epithelium followed by the blood-borne removal of the toxicant and its metabolites. The relative importance of these routes were determined primarily by the lipid/aqueous partition coefficient of the toxicant. Highly lipophilic solutes dissolved readily in the thin layer of surfactant present on top of the mucous lining layer at the air interface. Highly lipophilic solutes may therefore penetrate only slowly through the mucous lining layer to the epithelium below and may move along the airways with the mucociliary escalator. In the Beagle dog, Phe penetrated the mucous lining layer almost instantaneously, whereas BaP was retained in the mucus with a half-time of approximately 10 min (Gerde et al. 1993b) The mucous lining layer is thus an important modifier of the dose of organic toxicants to the airway epithelium. With a normal secretion of mucus and surfactant, the mucous lining layer is essentially instantaneously permeable for substances with a $PC_{L/A}$ of less than approximately 10,000. Substances with a $PC_{L/A}$ exceeding approximately 100,000 will be increasingly retained in the mucous lining layer with a half-time that is proportional to the $PC_{L/A}$, the thickness of the mucous lining layer, and the lipid content of the mucous lining layer (Gerde and Scholander 1987a).

The fraction of the highly lipophilic toxicants that penetrates the mucous lining layer has a prolonged retention in the epithelium below. BaP has a half-time in the bronchial epithelium of the Beagle dog of approximately 1.4 hr, a period during which considerable metabolism of the hydrocarbon can occur (Gerde et al. 1993b). Figure 1 shows the calculated tissue concentration of BaP in three important tissue compartments following exposure of the lungs to a single dose of BaP at an even density of deposition over the entire surface area of the lungs.

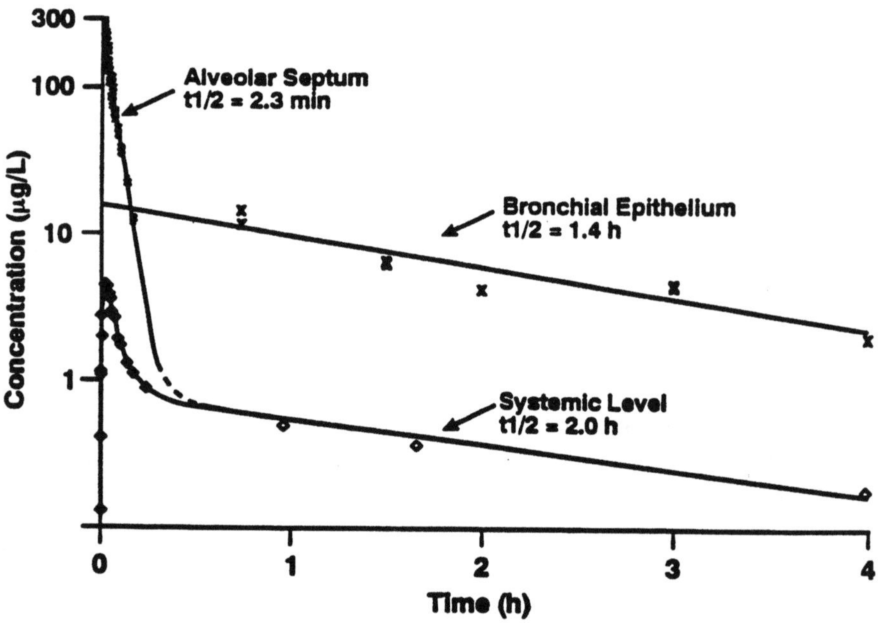

Figure 1. Calculated tissue concentrations of BaP-eq in the Beagle dog following an even deposition of 0.5 µg/m² of BaP over the entire surface of lungs and airways. The thicknesses of the alveolar air/blood barrier and the bronchial epithelium have been assumed to be 1.6 and 30 µm, respectively. Note the prolonged elevation of the concentration in the bronchial epithelium caused by the diffusion-limited absorption of the highly lipophilic carcinogen BaP. Based on data from Gerde et al. (1993a,b).

The dosimetry of particle-associated PAHs at high level exposures.

We will now examine how high- and low-level exposures influence dosimetry in light of the new dosimetric model for PAHs in the lungs. Because the *slow release* hypothesis has had such an influence on the perception of dosimetry of PAHs, it is important to evaluate its validity. The hypothesis had a dual impact on the thinking applied to PAH dosimetry. First, it was thought that the carrier particle influenced the dosimetry of PAHs for up to a week after deposition in the lungs. This shifted the perception of the critical time element in the dosimetry from minutes and hours to weeks, and particle retention was considered important in the dosimetry of PAHs. Second, the *slow release* hypothesis led to the perception that a slowly released fraction of PAHs on particles is more carcinogenic than fractions that elute rapidly from the particles, following deposition in the lungs (Creasia and Nettesheim 1974). Experimental evidence for both assumptions is likely to be an artifact of the high doses to which the experimental animals were exposed. Nonlinear behavior in the dosimetry of PAHs with increasing exposure levels may be caused by the limited solubility of most larger-sized PAHs in tissues. BaP, for example, has a very low solubility in water, so that despite a very high lipid/aqueous partition coefficient, its solubility in the lipid membranes of the cells is surprisingly low. The solubility can be estimated to be in the range of 0.02-0.03 g/g lipid (Patton et al. 1984; Shu and Nichols 1979). This solubility limit is not likely to be reached during realistic exposures in humans (Gerde and Scholander 1987b), but was probably greatly exceeded during the high-level animal exposures (Gerde et al. 1991b).

Typically 50/50 mixtures of BaP and a solid carrier dust such as ferric oxide were used in instillation experiments. If a typical specific surface area of the ferric oxide dust of 2 m^2/g is assumed and a monomolecular layer of BaP, at 100 $Å^2$/molecule, weighs approximately 0.5 mg/m^2, then the dust can adsorb at the most 1 mg BaP/g particles as a monolayer. BaP in excess of this amount will be crystalline. The important consequence of these estimates is that up to 99.9% of the BaP used in the old instillation studies was crystalline, so the observed tumors were most likely induced by crystalline BaP, despite the fact that BaP was retained by the particles for weeks. Because we now know that crystalline BaP inhaled at low exposure levels is dissolved almost instantaneously in the lung liquid lining (Gerde et al. 1993a), it must be concluded that the long retention of crystalline BaP observed in the high dose instillation experiments was most likely caused by the large amounts of BaP to which the animals were exposed; large aggregates of carrier dust were formed from which BaP dissolved slowly (Gerde et al. 1991b). When interpreted in the light of realistic human exposures, the slowly dissolved fraction of PAHs on particles was considered functionally identical with a slowly desorbed fraction of PAHs on the particle surfaces. However, one fundamental difference, brought on by the different processes, is that epithelial cells facing an aggregate containing slowly dissolving PAHs is likely to be immersed in a saturating plume of the toxicant, whereas epithelial cells facing a single particle with slowly desorbing PAHs should experience low exposures. Thus, the high dose experiments carried out with crystallinic PAH cannot necessarily be interpreted to mean that a certain particle-retained fraction of PAHs should be more carcinogenic than the rapidly released fractions also at low environmental exposures.

The dosimetry of particle-associated PAHs at low level exposures

The finding that organic chemicals are cleared from the lungs according to two different mechanisms profoundly influences the understanding of the dosimetry of organic chemicals. Low to moderately lipophilic organics are perfusion-limited during clearance from the lungs. The perfusion-limited substance diffuses rapidly through the airway epithelium and into the associated capillary bed. The clearance of an ideally perfusion-limited toxicant depends only on the tissue/blood partition coefficient and on the rate of perfusion of the tissue. The typical clearance half-time that can be expected for such a substance in the bronchi should be in the range of 1.4 -7 min (Gerde et al. 1993a). Thus, exposure at the portal of entry epithelium to perfusion-limited toxicant should be brief and not very different from that of tissues exposed via the systemic circulation. However, for diffusion-limited substances the situation is very different. The highly lipophilic toxicants dissolve readily in the lipid membranes of the portal of entry epithelium and diffuse only slowly to the capillary bed below. Exposure to the portal of entry epithelium can be considerably higher than that of tissues exposed via the systemic circulation (Figure 2).

The shift between the two different mechanisms of toxicant uptake in the lungs gradually occurs over a rather wide range of lipophilicities. However, one important observation is that in a homologous series of PAHs, the shift in uptake mechanism coincides with an increased tendency of the PAHs to adsorb onto airborne particles; typically the highly lipophilic PAH molecules are so large (Doucette and Andren 1987) that their low vapor pressure makes them prone to adsorb onto particles (Yamasaki et al. 1982). This covariance of seemingly independent variables gives a link between particle-association of lipophilic toxicants and their tendency to induce first-pass toxicity in thicker entrance epithelia, such as the bronchial epithelium.

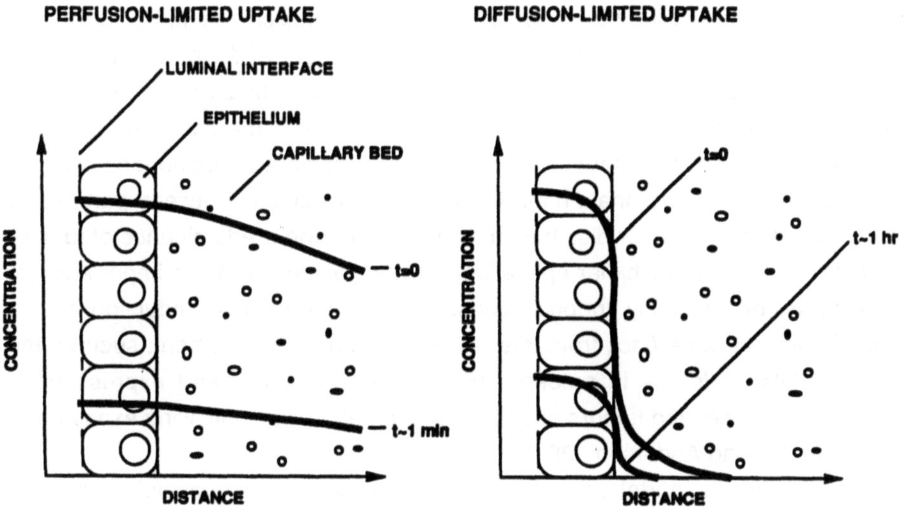

Figure 2. The two principal mechanisms through which organic toxicants may enter the organism via mucous membranes. A) Less to moderately lipophilic toxicants are likely to be perfusion-limited during uptake. Such substances are likely to diffuse rapidly through the epithelium into the capillary bed below and are taken up by the circulating blood within minutes. The exposure to the portal of entry epithelium will be brief and not very different from tissues exposed via the systemic circulation.
B). Highly lipophilic toxicants are likely to be diffusion-limited during uptake. Such substances diffuse slowly through the portal of entry epithelium along steep concentration gradients. Even after a brief exposure, the portal of entry epithelium can be selectively exposed for several hours at much higher toxicant concentrations than the rest of the organism (Gerde et al. 1993c).

Discussion and conclusion

With the advent of a mechanistically based, dosimetric model for highly lipophilic toxicants in the lungs, some of the apparent contradictions between the dosimetry of toxicants in high-level animal exposures and in low-level human exposures should be possible to resolve. The diffusion-limited clearance of highly lipophilic toxicants and the tendency of highly lipophilic hydrocarbons to be associated with airborne particles sets some fairly rigorous constraints to the possible dosimetry and, thus, to the expected toxic response following exposure. The very nature of lung cancer as a disease of the airway epithelium is an indication of a first-pass toxic response. This assumption is strengthened by the observation that lung tumors in humans appear mainly in the thicker bronchial and bronchiolar epithelium (Stinson and Saffiotti 1983), and by the observed correlation between patterns of deposition for particles carrying carcinogenic toxicants into the lungs and the distribution of tumors in the bronchial tree (Schlesinger and Lippmann 1978). If the toxicants inducing lung tumors require metabolic activation following their deposition, this would also correlate with diffusion-limited clearance. The diffusion-limited clearance of highly lipophilic and particle-associated carcinogens gives a mechanistically based explanation for the observed correlation between patterns of particle deposition and tumor distribution, as well as for the predominance of tumors in the bronchial and bronchiolar epithelium. Further evidence of the importance of PAHs (or possibly, similarly lipophilic compounds) in lung cancer induction comes from methods that *trace back* from later events in tumor development, such as the patterns of DNA adducts in smokers (Phillips et al. 1988), and expression of oncogenes in human lung tumors and in experimental animals following exposure to PAHs (Reynolds and Anderson 1991).

A mechanistic model describing the physicochemical, physiological, and biochemical processes involved in the uptake and distribution of organic carcinogens via the lungs is a necessary tool for extrapolating dose-response relationships, and eventually exposure-risk relationships, for lung cancer from laboratory experiments to humans. This is true despite the development of several alternative indicators of biological dose, such as the above-mentioned DNA adducts and oncogene expression. Particularly for highly lipophilic carcinogens, where strong nonlinear exposure-dose effects can be expected at low exposure levels, mechanistic models describing these nonlinearities can help to calibrate other dose indicators and make them more accurate.

Acknowledgement: Research sponsored by the Office of Health and Environmental Research, U.S. Department of Energy, Contract No. DE-AC04-76EV01013, the Swedish Work Environment Fund, Grant 91-0359, and by NIH, Grant RO1-5910 from NIEHS. Research conducted in facilities fully accredited by the American Association for the Accreditation of Laboratory Animal Care.

References

Auerbach O, Stout AP, Hammond EC, and Garfinkel L (1961) Changes in bronchial epithelium in relation to cigarette smoking and in relation to lung cancer. N Engl J Med 265: 253-267

Creasia DA, and Nettesheim P (1974) Respiratory cocarcinogenesis studies with ferric oxides: A test case of current experimental doses. In Experimental Lung Cancer, Karbe E, and Park JF (eds), pp 234-245. New York: Springer-Verlag

Doucette WJ, and Andren AW (1987) Correlation of octanol/water partition coefficients and total molecular surface area for highly hydrophobic aromatic compounds. Environ Sci Tech 21: 821-824

Gerde P, Medinsky MA, and Bond JA (1991a) The retention of polycyclic aromatic hydrocarbons in the bronchial airways and in the alveolar region - a theoretical comparision. Toxicol Appl Pharmacol 107: 239-252

Gerde P, Medinsky MA, and Bond JA (1991b) Particle-associated polycyclic aromatic hydrocarbons - a reappraisal of their possible role in pulmonary carcinogenesis. Toxicol Appl Pharmacol 108: 1-13

Gerde P, Muggenburg BA, Hoover MD, and Henderson RF (1993a) Disposition of polycyclic aromatic hydrocarbons in the respiratory tract of the Beagle dog: I. The alveolar region. Toxicol Appl Pharm 121: 313-318

Gerde P, Muggenburg BA, Sabourin PJ, Harkema JR, Hotchkiss JA, Hoover MD, and Henderson RF (1993b) Disposition of polycyclic aromatic hydrocarbons in the respiratory tract of the Beagle dog: II. The conducting airways. Toxicol Appl Pharm 121: 319-327

Gerde P, Muggenburg BA, and Henderson RF (1993c) Disposition of polycyclic aromatic hydrocarbons in the respiratory tract of the Beagle dog: III. Mechanisms of the dosimetry. Toxicol Appl Pharmacol 121: 328-334

Gerde P, and Scholander P (1987a) A hypothesis concerning asbestos carcinogenicity: The migration of lipophilic carcinogens in adsorbed lipid bilayers. Ann Occup Hyg 31: 395-401

Gerde P, and Scholander P (1987b) A mathematical model of the penetration of polycyclic aromatic hydrocarbons through the bronchial lining layer. Environ Res 44: 321-334

Henry MC, Port CD, and Kaufman DG (1975) Importance of physical properties of benzo(a)pyrene-ferric oxide mixtures in lung tumor induction. Cancer Res 35: 207-217

Hilding AC (1957) Ciliary streaming in the bronchial tree and the time element in carcinogenesis. N Engl J Med 256: 634-640

Kotin P, and Falk HL (1959) The role and action of environmental agents in the pathogenesis of lung cancer. Cancer 12: 147-163

Macklin CC (1956) Induction of bronchial cancer by local massing of carcinogen concentrate in outdrifting mucus. J Thoracic Surgery 31: 238-244

Patton JS, Stone B, Papa C, Abramowitz R, and Yalkowsky SH (1984) Solubility of fatty acids and other hydrophobic molecules in liquid trioleoylglycerol. J Lipid Res 25: 189-197

Phillips DH, Hewer A, Martin CN, Garner RC, and King MM (1988) Correlation of DNA adduct levels in human lung with smoking. Nature 336: 790-792

Reynolds SH, and Anderson MW (1991) Activation of proto-oncogenes in human and lung tumors. Environ Health Perspec 93: 145-148

Saffiotti U, Cefis F, and Kolb LH (1968) A method for the experimental induction of bronchogenic carcinoma. Cancer Res 28: 104-124

Saffiotti U, Montesano R, Sellakumar AR, Cefis F, and Kaufman DG (1972) Respiratory tract carcinogenesis in hamsters induced by different numbers of administrations of benzo(a)pyrene and ferric oxide. Cancer Res 32: 1073-1081

Schlesinger RB, and Lippmann M (1978) Selective particle deposition and bronchogenic carcinoma. Environ Res 15: 424-431

Shu HP, and Nichols AV (1979) Benzo(a)pyrene uptake by human plasma lipoproteins in vitro. Cancer Res 39: 1224-1230

Stinson SF, and Saffiotti U (1983) Experimental respiratory carcinogenesis with polycyclic aromatic hydrocarbons. In Comparative Respiratory Tract Carcinogenesis, Reznik-Schuller HM (ed) pp 75-93 Boca Raton, FL: CRC Press

Yamasaki H, Kuwata K and Miyamoto H (1982) Effects of ambient temperature on aspects of airborne polycylic aromatic hydrocarbons. Environ Sci Technol 16: 189-194

Which Surface Functionalities are implied in Dust Toxicity?

Bice Fubini
Dipartimento di Chimica Inorganica Chimica Fisica e Chimica dei Materiali
Università di Torino
Via Pietro Giuria 7
10125 Torino
Italy.

Introduction

Once inhaled a particle interacts at various stages, during its path in the respiratory tract, with cells, tissues and biological fluids. Each contact involves a specific functionality at the particle surface, so that the chemical features determining the overall pathogenicity of a given dust are probably more than one. Modulation of one surface property may thus end up in variation of the ultimate toxicity caused even by active surface sites which are unchanged. For this reason investigations on all surface properties of a toxic dust are required.

The surface of a particle bears several chemical functionalities such as Bronsted and Lewis acid or bases; H-bonding sites; dangling bonds; poorly coordinated metal ions; charges originated by lattice vacancies and defects or by electrolytic dissociation in water. Their characteristics and abundance does not merely depend upon the underlying crystal structure, but also upon the mechanical and thermal history of the dust, previous chemical treatments, aging, presence of impurities, form of the particle.

Data are reported on the following topics:

- effect of form on the surface reactivity of SiC and TiO_2 particles.
- surface properties of three cristobalite (SiO_2) dusts of various origin, relatable to the different biological responses elicited.
- mobilization of iron and release of free radicals from iron containing minerals.

Materials and Methods

Titania: anatase fine (250 nm particle size; Fisher Scientific, Springfield, NJ, USA) and ultrafine (20 nm particle size; Degussa AG, Frankfurt/Main, Germany) powders were kindly provided by Dr Oberdöster, (Oberdöster et al. in press)

NATO ASI Series, Vol. H 85
Cellular and Molecular Effects
of Mineral and Synthetic Dusts and Fibres
Edited by J. M. G. Davis and M.-C. Jaurand

Silicon Carbide: isometric ultrafine SiC powders was from Lonza. SiC whiskers were Tateho from Tateho Chemical.
Cristobalite: ground mineral (C&E Minerals, King of Prussia, PA), (Hemenway et al. 1990, 1993), CSR-m; pure quartz chips ground and heated (1300°C), CSR-q; heated diatomaceous earth (Dzwigaj et al., 1993), CSR-d;.
Iron minerals : haematite (Merk); magnetite (BDH); UICC chrysotile; UICC crocidolite;

The heat of adsorption of probe molecules were measured by means of a Calvet microcalorimeter connected to a volumetric device (Bolis et al. 1989, 1991; Fubini et al. 1992); the EPR spectra were recorded on a Varian E09 spectrometer (Fubini et al. 1989, 1990); free radical release was evaluated by means of DMPO as a spin trap (Giamello et al., 1990; Fubini et al., 1993) following the technique adopted by Pézerat and associates (Zalma et al. 1989). For more details on the experimental techniques employed see the references to previous papers.

Effect of form on surface reactivity

The interaction of a mineral particle with living matter much depends upon its form and its surface reactivity. The term "form" comprises crystal structure, size, shape and micromorphology of the crystallites while "surface reactivity" any surface arrangement liable to react or catalyse reactions at the liquid/solid or air/solid interface. The crucial role played by size and shape of a particle, particularly a fiber, in determining the pathogenicity of mineral particulate has been widely stressed in the past. In some cases we have found that the form may also modulate surface reactivity. This may be an additional reason for a close relationship between form and pathogenicity. We will discuss here two cases: TiO_2, a substantially inert dust, often used as non-toxic reference material, which - in overload conditions - is more toxic when in extremely fine form and SiC, which is pathogenic when in fibrous habit or as whiskers.

Titanium Dioxide

Two TiO_2 dusts, with different particle size, a fine one (ϕ = 250 nm) and an ultrafine one (ϕ = 20 nm), obtained by different chemical procedures (precipitation and ignition respectively), have been kindly provided by Dr Oberdöster. Under overload conditions the ultrafine one was more fibrogenic to rats than the fine one as toxicity turned out to parallel the surface area and not the weight of the inhaled dust (Oberdöster et al., in press). It has been previously found in our laboratory (Bolis et al 1989) that Lewis surface acidity of titania much depends on its origin and preparation route; we have therefore measured the characteristics and extent of acidic sites on the two samples by means of adsorption of carbon monoxide from the gas phase. The heat of adsorption and the relevant infrared spectra reported in figure 1 were obtained following the techniques previously described (Bolis et al. 1989). They clearly show that on the ultrafine sample two different types of Lewis acidic sites are present, which originate two separate IR bands, whereas just one band is visible on the fine one. Consequently uptakes and relevant heats are higher on the former by respect to the latter.

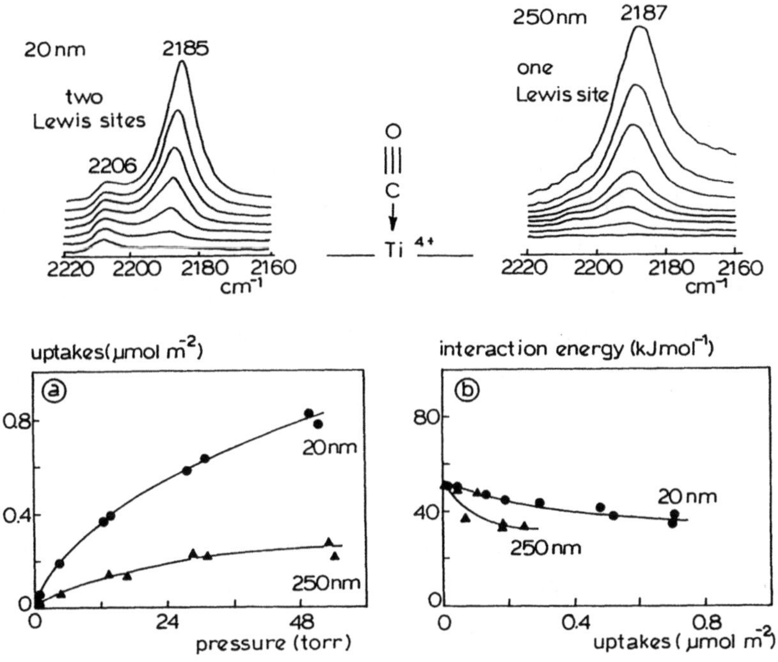

Figure 1. Adsorption of carbon monoxide on ultrafine (20 nm) and fine (250 nm) titania. Infrared spectra of CO stretching; adsorption isotherms (a) and related interaction energies (b).

The site typical of the ultrafine dust, characterized by the higher heat of adsorption and infrared frequency values, is likely originated either by a particular crystal plane exposed or by poorly coordinated Ti^{4+} ions at kinks, steps, edges of the crystallite. High resolution electron micrographs revealed that the exposed crystal face is the same on both samples but edges are much sharper on the ultrafine one. It is not possible to state whether the differences found in the number and kind of active acidic sites between the two samples might be also relatable to their different toxicity in inhalation studies. It has however to be borne in mind that strong Lewis acidic sites, such as those revealed by CO - a very weak base - are likely to be those involved in binding biomolecules at the surface through coordination bonds. This may end up in coating of the particle and/or denaturation of the molecule. In a systematic investigation on surface properties which may be involved in dust toxicity, these have certainly to be considered.

Silicon Carbide

Several papers have appeared pointing out a specific lung toxicity of silicon carbide (Funahashi et al. 1984; Birchall et al. 1988; Begin et al., 1989) which is mostly confined to fibrous by-product in the fine dust production and to SiC whiskers, largely used in manufacturing composite materials. Needless to say, the well known higher toxicity of fibrous vs isometric particles will be operating as with asbestos. As in the previous case (TiO_2), however, additional chemical effects may be originated by the peculiar form of whiskers. Whiskers are monocrystals grown along one defined crystal axis into a needle-like shape. Obviously the kind and extent of crystal faces exposed will vary from whiskers to isometric particles. Their capability of originating dangling bonds upon grinding and free radicals in solution has been previously reported by our group (Fubini et al., 1991), but the two properties have recently appeared unrelated as under some circumstances the dangling bonds persisted while the free radical production was eliminated (unpublished results).
For a thorough investigation on surface properties we have to consider that SiC exposed to air often appears at least partially covered by an SiO_2 layer (Dufresne et al., 1987), so that it is unclear whether its surface reactivity is related to silicon - carbon functionalities or to the thin oxidized layer. Adsorption of probe molecules on samples pretreated at various temperatures has shown that the surface behaviour of silicon carbide is quite different from the silica one (Fubini et al., 1992). In particular, opposite to silica, SiC develops a remarkable reactivity upon heating. This suggest that it is the unoxidized SiC surface to react. Alternatively it may be hypothesized that the oxidized patches, because of the underlying bulk phase, will exhibit a quite peculiar reactivity, typical of SiC and not of SiO_2. We have examined whether whiskers differed from isometric particles also in this reactivity.

Figure 2 compares the heat vs. pressure plots obtained with the two kind of dusts (heated at 800°C) upon adsorption of water and ammonia. The development upon heating of a high reactivity towards both adsorptives is much more pronounced on whiskers than on isometric dusts (more heat is released at the same equilibrium pressure, indicative of a higher uptake and interaction energy). The higher adsorption potential of whiskers towards NH_3 and H_2O is due to the development of particular surfaces, and to the presence of sharp edges where poorly coordinated atoms are located, which may be also more reactive in vivo.

As to the surface sites involved in toxicity, the difference in the pathogenic response elicited by SiO_2 and SiC points to an involvement of the silicon/carbon functionalities more than of the silica patches originated by oxidation, unless the latter, because of the underlying silicon carbide, exhibit a different behaviour from that of bulk silica.

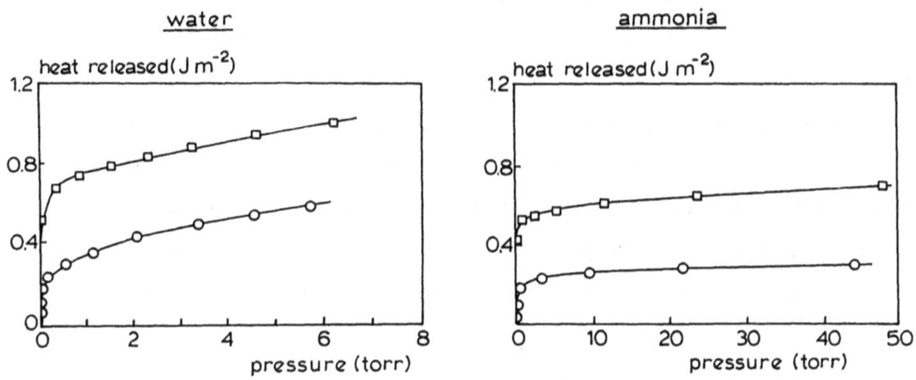

Figure 2. Heat released as a function of equilibrium pressure upon adsorption of water and ammonia on silicon carbide whiskers (□) and isometric particles (o).

Origin of the sample: the case of cristobalite

Cristobalite is the silica polymorph thermodynamically stable above 1470°C, which is considered the most fibrogenic form of "free silica". Once formed, its reconversion to quartz, which is the form stable at room temperature, is very slow and only occurs in geological times. For this reason large mineral deposits of it are present on the earth crust and silica converted into cristobalite e.g. by heating in furnaces, stays as such upon cooling down.

Cristobalite may also originate directly from biogenic silicas upon heating well below 1470°C, typically in the production of silica dusts from diatomaceous earths .

Three cristobalites of different origin have been examined in order to find a relationship between their surface properties and what is known on their toxicity:

- CRS-m a cristobalite mineral ground down to respirable size, found much more fibrogenic than Min-U-Sil quartz in inhalation studies on rats (Absher et al., 1989; Hemenway et al., 1990).
- CRS-d a diatomaceous earth found toxic to rats following intratracheal injection (Dzwigay et al., 1993)
- CRS-q a cristobalite powder prepared by us by thermal conversion above 1300°C from a fine quartz dust. It turned out to be biologically inert (no nodules following intratracheal injection, Honnons et al., submitted) partly because a fraction of the original particles did aggregate during the conversion but mostly because the thermal process inactivated the surface of the particles.

What happens to a silica surface upon heating? Surface radicals, if present, are progressively eliminated (Fubini et al. 1989) and silanols are converted into siloxane bridges with consequent loss of hydrophilicity (Bolis et al. 1991; Fubini et al.1992, Hemenway et al. 1993). Figure 3 reports the EPR spectra of CRS-m progressively heated: the major features of the spectrum are the lines arising from the sylilperoxide SiOO• and from the superoxide ions, $O_2^{-\bullet}$ (for more details see Fubini et al, 1990). The former is eliminated below 350°C, the latter progressively disappears, so that on samples heated at 800°C only a broad band of minor intensity, still to be assigned, is left at the surface. When the dust is reground the radicals are created again (Fig.3 f). These radicals, typical of ground dusts, are likely involved in acute damage to people exposed to freshly ground materials (Vallyathan et al. 1988; Fubini et al. 1989, 1990; Dalal et al. 1990). Long term silicosis, however, does not appear related to the actual presence in the inhaled dusts of these radicals. It has been recently found that some samples mostly deprived of any surface radical were fibrogenic (Hemenway et al., 1993, Wastiaux et al. 1993). Moreover, opposite to CRS-m, CRS-d and CRS-q were free of surface radicals visible in EPR but CRS-d was fibrogenic and CRS-q was not.

The extent of hydrophilicity is measured from the amounts of water adsorbed via H-bonding and its heat of adsorption: the higher these values the more hydrophilic the surface. The results of the adsorption of water on the three cristobalite samples of different origin, CSR-m,

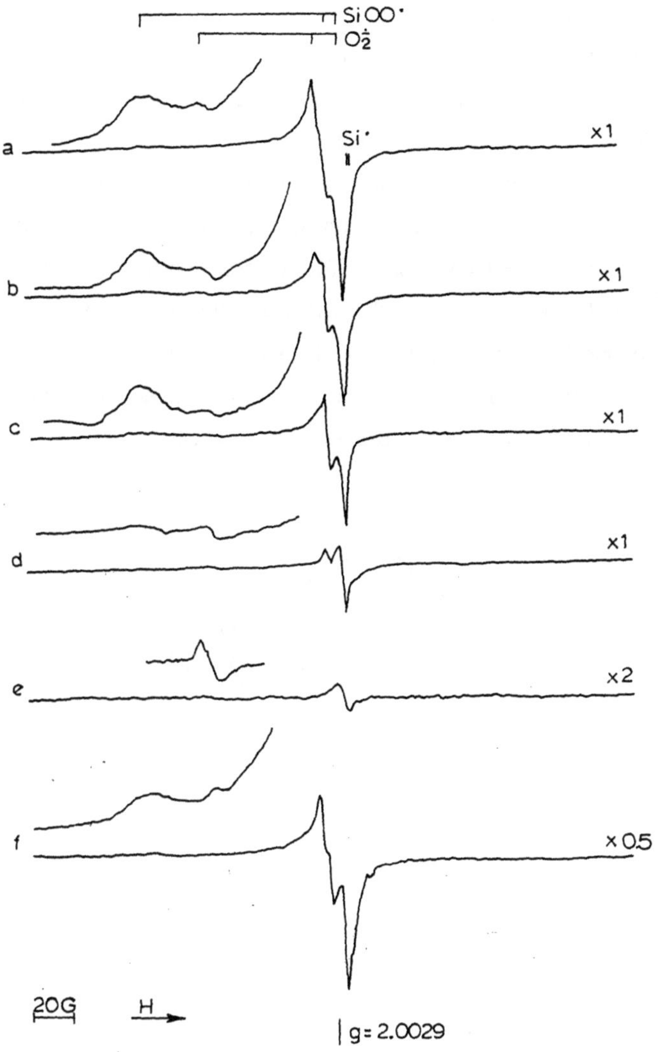

Figure 3. EPR spectra of CRS-m unheated (a) and heated two hours at 250°C (b), 350°C (c), 500°C (d) and 800°C (e). Spectrum f refers to a sample heated up to 800°C and subsequently reground.

CSR-q, CSR-d, are reported in figure 4.

Both adsorption isotherms and heat of interaction reveal remarkable differences between the three samples. CSR-m and CSR-d are very hydrophilic whereas CSR-q is fully hydrophobic (heat of interaction well below 44 kJ.mol^{-1}, heat of liquefaction of water). The higher adsorption capacity of CSR-d by respect to CRS-m may be simply due to metal oxide contaminants. It is here clearly demonstrated how the surface reactivity may differ, even within a set of samples of similar size and identical crystal structure. Hydrophilicity appears thus a characteristic required for the development of toxicity. It is noteworthy that a remarkable reduction in fibrogenicity upon heating quartz and cristobalite at high temperature has been also recently reported (Wiecek, 1993). A possible explanation is that unheated ground dusts or mildly heated ones still retain the active surface sites prerequisite to determine the biological responses leading to silicosis, whereas when heated at very high temperature, surface irregularities, radicals and charges are healed and silanols are wholly eliminated to form stable unreactive siloxanes, with consequent loss of the pathogenic potential.

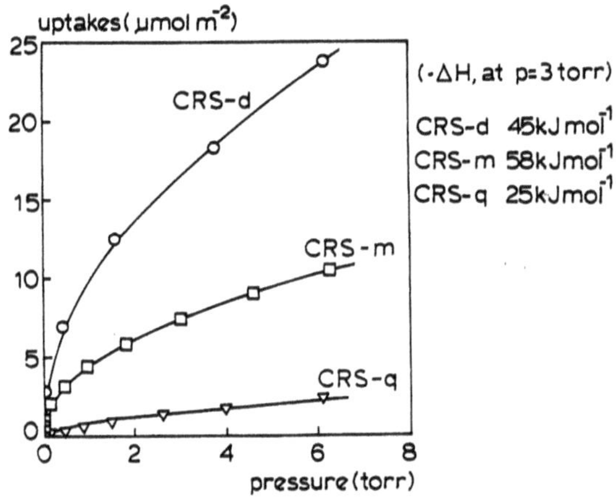

Figure 4. Adsorption of water vapour on the three cristobalite samples.

As fibrogenicity is the result of a complex chain of events in which several interactions with the silica surface take place, modulation of any of the surface properties involved, e.g. the capability of establishing strong H-bonds, may interfere deeply with the ultimate toxicity.

The role of iron in mineral dust toxicity

Iron deserves particular attention as far as surface reactive sites are concerned (Fubini, 1993). A crucial role played by iron derived AOS (active Oxygen Species) in fiber toxicity has been claimed in fact by various pathologists (Goodlick and Kane, 1986, Mossmann et al., 1986). Several in vitro tests (Lund and Aust, 1990,1991a; Zalma et al., 1988; Leanderson et al., 1988; Gulumian and Van Wyck, 1987) support this hypothesis. Cell free in vitro tests reveal a consistent release of free radicals in aqueous suspensions (Zalma et al., 1988) as well as mobilization of iron brought about by chelating agents, which readily damages DNA (Lund and Aust 1991b). Iron in the mineral appears thus involved in radical production. In the chain of events:

Fe/particle --> AOS --> target cell --> DNA damage --> disease

the initial reaction will be governed by the **surface chemistry** of the dust, no matter whether mobilized or surface bound iron is mostly involved. It is however still unclear which surface arrangement of Fe ions in the solid (redox state, coordination, clustering, etc) is active, as not <u>any</u> iron containing mineral is toxic. Moreover question arises on the possible catalytic role of the fibre surface in radical production. In order to clarify some of these aspects we have compared some iron oxides (haematite and magnetite) with iron containing asbestos (chrysotile and crocidolite) and with iron exchanged zeolites prepared in the laboratory as model solids for iron reactivity. We have measured the amount of iron removed following prolonged incubation in chelators and, by means of the spin trap technique applied to solid-liquid systems, we have investigated the radical release potential of the solids. The surface modifications brought about by prolonged suspension in aqueous solutions of various iron chelators (Desferrioxamine B, selective for Fe(III); Ferrozine, selective for Fe(II); ascorbate) are partly reported elsewhere in this book and will be the object of a future paper. One point only will be reported in the context of the present paper: the relationship between iron content, iron removed and free radical release, which is illustrated in figures 7 and 8.

The amount of iron present in the solid, by no means parallels that removed nor the capability of the solid surface to release free radicals in solution. It is clear that the state of iron in the solid (coordination, lattice energy) determines both the mobility by chelators and redox

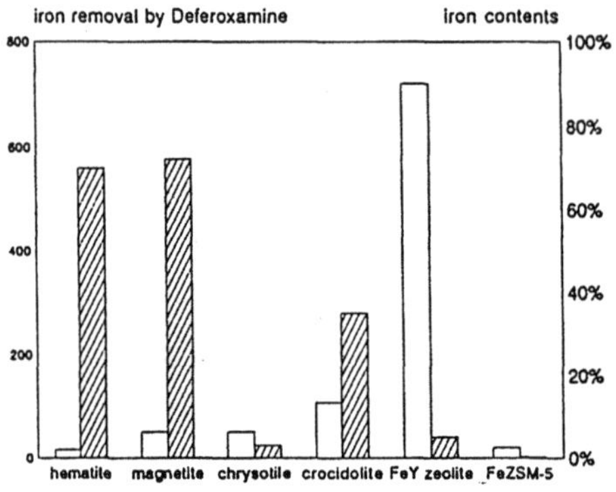

Figure 5. Iron contents of the solid (dashed) and iron removed by Desferrioxamine B (empty)

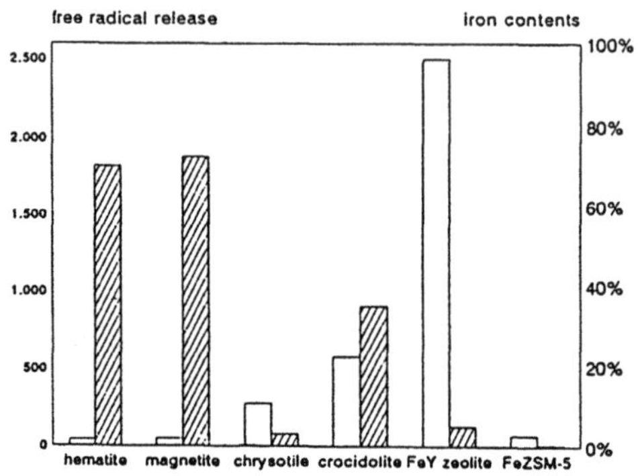

Figure 6. Iron contents of the solid (dashed) and free radical ib release (empty)

cycling, with consequent free radical release in solution. Iron in haematite, strongly held in the lattice, is unreactive opposite to the Fe exchanged Y zeolite, where not only, as expected, is easily removed, but is also redox active at a much larger extent even then on asbestos. Selective removal of Fe (II) by ferrozine or strong oxidation of the surface inhibits any radical release in aqueous solution. Further research is needed to identify the chemical structure of the actual active site. It certainly originates from aggregates of iron ions in more than one oxidation state and in a poorly coordinated location at the surface.

Acknowledgments

Financial support from CECA, convention No 7280/03/006, and from Italian Consiglio Nazionale delle Ricerche, (CNR) Progetto Finalizzato " Materiali Speciali per Tecnologie Avanzate" is here acknowledged. Thanks to Dr Hemenway for ground cristobalite, to Dr Oberdöster for the titania samples and to Dr Pézerat for the ex-diatomite cristobalite.
What reported in this paper is the result of a collective work of colleagues and students in the Chemistry department of the University of Torino: Prof E. Giamello, Prof V. Bolis, Dr M. Volante A. Cavenago, M. Giraudo, E. Merlo, L. Mollo, whose contribution is here deeply

References

Absher M.P., Trombley, L. Hemenway D.R. Mickey R.M., Leslie K.O. (1989) Biphasic cellular and tissue response of rat lungs after eight-day aerosol exposure to the silicon dioxide cristobalite. *Am. J. Pathol.* 134: 1243 - 1251.

Begin R, Dufresne A, Cantin A, Massé S, Sebastien P., Perrault G. (1989) Carborundum pneumoconiosis. Fibres in in the mineral activate macrophages to produce fibroblast growth factors and sustain the chronic inflammatory disease. *Chest* 95: 842-849.

Birchall J.D., Stanley D.R., Mockford M.J., Pigott G.H., Pinto P.J. (1988) Toxicity of silicon carbide whiskers, *J. Mat. Sci. Lett.* 7:350-352.

Bolis V., Fubini B., Garrone E., Morterra C. (1989) Thermodynamic and vibrational characterisation of CO adsorption on variously pretreated anatase. *J. Chem. Soc. Faraday Trans* I, 85:

Bolis V., Fubini B., Marchese L., Martra G., Costa D. (1991) Hydrophilic and hydrophobic sites on dehydrated crystalline and amorphous silicas. *J. Chem. Soc. Faraday Trans.*,87:497-505.

Dufresne A., Perrault G., Sebastien P., Adnot A., Baril M. (1987) Morphology and surface characteristics of particulate from silicon carbide industries *Am. Ind. Ass. J.*, 48: 718 - 729.

Dzwigaj S., Pézerat H., Wastiaux A., Sébastien P., and Honnons S. (1993) Silica and silicosis mechanism. The role of surface properties of particles and of the silica-supported iron. *Proceedings of the Eight International Conference on Occupational Lung Diseases*, Hurich et al. ed., ILO, Geneva, CH, p.1186-1191.

Fubini B., Giamello E., Pugliese L., Volante M. (1989). Mechanically induced defects in quartz and their impact on pathogenicity. *Solid State Ionics* 32-33: 334-343.

Fubini B.,Giamello E., Volante M., and Bolis V. (1990). Chemical functionalities at the silica surface determining its reactivity when inhaled. Formation and reactivity of surface radicals. *Toxicology and Industrial Health,* 6: 571-594.

Fubini B., Bolis V., Giamello E. and Volante M. (1991) Chemical functionalities at the broken fibre surface relatable to free radicals production in *"Mechanisms in Fibre Carcinogenesis"* (R.C. Brown, J.A. Hoskins, N.F. Johnson, eds.) NATO ASI series, Plenum press, New York, p.415-432.

Fubini B. Bolis V., Cavenago A. and Ugliengo P. (1992) Ammonia and water as probes for the surface re reactivity of covalent solids: the case of cristobalite and silicon carbide. *J.Chem.Soc. Faraday Trans.*, 88: 277-290.

Fubini B., Volante M., Giamello E., Bolis V.(1993) The role of surface chemistry in the pathogenicity of mineral dusts and fibres: some open questions"; in *"Proceedings of the Eighth International Conference on Occupational Lung Diseases"*,J.Hurich,M.Leasage, and A.David eds., ILO, Geneva, Switzerland, Vol. III, p.1180-1185.

Fubini B. (1993) The possible role of surface chemistry in the toxicity of inhaled fibers in *"Fiber Toxicology"*, D.B.Warheit ed., Academic Press, Chapter 11, p.229-257.85)

Funahashi A., Schlneter O.P., Pintar K., Siegesmund K.A., Mandel G.S., Mandel N.S. (1984) Pneumoconiosis in workers exposed to Silicon Carbide., *Am. rev. Respir. Dis.* 129(4): 635-640.

Giamello E., Fubini B., Volante M., Costa D. Surface oxygen radicals originating via redox reactions during the mechanical activation of crystalline SiO_2 in hydrogengen peroxide *Colloids and Surfaces*, 45, 155-165, (1990).

Goodlick L.A., Kane A.B. (1986) Role of reactive oxygen metabolites in crocidolite asbestos toxicity to mouse macrophages. *Cancer Res.* 46: 5558-5566.

Gulumian M , Van Wyk J. A. (1987) Hydroxyl radical production in the presence of fibres by a Fenton-type reaction. *Chem. Biol. Interactions*, 62: 89-97......

Hemenway D.R., Absher M.P., Trombley L., Vacek P.M.(1990) Comparative clearance of quartz and cristobalite from the lung. *Am.Ind.Hyg.Assoc.* 51(7):363-369.

Hemenway D.R., Absher M.P, Fubini B and Bolis V. (1993a) What is the Relationship between Hemolytic Potential and Fibrogenicity of Mineral Dusts? *Arch.Environ.Health*

Hemenway D.R., Absher M.P., Fubini B., Trombley L., Vacek P. Volante M., Cavenago A. Surface functionalities are related to biological response and transport of crystalline silica. 7th Intern. Symposium on Inhaled Particles BOHS, Edinburgh, 1991, *Ann Occup. Hyg.*, in press.

Honnons S., Oberson D., Wastiaux A., Dz Dzwigaj S., Pézerat H., Volante M., Fubini B., Porcher J.M. Pulmonary response to silica dust:the influence of crystalline structures and physico-chemical properties. Second International Symposium on Silica Silicosis and Cancer, San Francisco, October 1993

Leanderson P, Söderkvist, Tagesson C , Axelson O (1988) Formation of 8-hydroxydeoxyguanosine by asbestos and man-made mineral fibers. *Br. J. Ind. Med.*, 45:309-311.

Lund L.G. Aust A.E. (1990) Iron mobilization from asbestos by chelators and ascorbic acid. *Arch Biochem.Biophys.*, 278: 60-64. (1991a) Mobilization of iron from crocidolite asbestos by certain chelators results in enhanced crocidolite-dependent oxygen consumption. *Arch Biochem.Biophys.*, 287: 91-96.(1991b) Iron catalyzed reactions may be responsible for the biochemical and biological effects of asbestos. *Biofactors* 3:83-89.

Mossman B.T., Marsh J.P., Shatos M.A.(1986) Alteration of superoxide dismutase activity in tracheal epithelial cells by asbestos and inhibition of cytotoxicity by antioxidants. *Lab Invest.*, 54: 204-212.

Oberdöster G., Ferin J., Soderholm S., Gelein R., Cox C., Baggs R. Morrow P.E. 7th Intern. Symposium on Inhaled Particles BOHS, Edinburgh, 1991, Ann Occup. Hyg., in press.

Vallyathan,V., Shi, X., Dalal, N.S., Irr, W. and Castranova, V. (1988) Generation of free radicals from freshly fractured silica dust: potential role in acute silica-induced lung injury. *Am. Rev. Respir. Dir.* 138, 1213-1219.

Wastiaux A., Erre R., Pézerat H., Sebastien P. Quartz surface and fibrogenicity. 7th Intern. Symposium on Inhaled Particles BOHS, Edinburgh, 1991, *Ann Occup. Hyg.*, in press.

Wiecek E. The influence of real structure of quartz and cristobalite on their fibrogenic effects. Second International Symposium on Silica Silicosis and Cancer, San Francisco, October 1993

Zalma R., Bonneau L., Jaurand M.C., Guignard J. and Pezerat H. (1987) Formation of oxy-radicals by oxygen reduction arising from the surface activity of asbestos. *Can. J. Chem.* 652, 338-2341.

Role of Transition Metal Compounds in the Capacity of Dusts to Generate Electrophilic Species

Henri Pezerat
Laboratoire de Réactivité de Surface
Université P. et M. Curie - CNRS (URA 1106)
4 place Jussieu
75252 - Paris Cédex 05
France

The mechanistic approach to the study of the toxicity of transition metal compounds in the various compartments of the lungs requires taking into consideration the metabolic processes leading to the formation of electrophilic species and the kinetics of their appearance. Only these electrophilic species are able to trigger easily a great variety of agressions at the level of nucleophilic targets present on biologic macromolecules.

Consequently the first problem is to define the nature of the electrophilic species appearing during the metabolic process of the xenobiotics. Most of these species belong to the group of activated oxygen species (AOS).

The most important process allowing the AOS formation implies a series of reactions between the xenobiotics and molecular oxygen present in stationary concentration in all the biological compartments. In some particular cases it is also necessary to consider the role of the couple $O_2^{\bullet -}$ - H_2O_2 in these reactions.

The oxygen activation occurs only when the xenobiotics present reducing (i.e. electron donating) properties after a first metabolic step.

Then the two essential steps can be schematized as following :

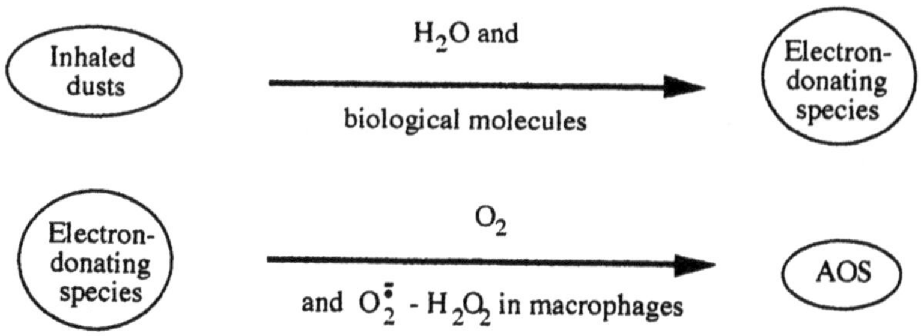

NATO ASI Series, Vol. H 85
Cellular and Molecular Effects
of Mineral and Synthetic Dusts and Fibres
Edited by J. M. G. Davis and M.-C. Jaurand

The first step of the metabolic process

All dusts, before their penetration into the lung medium, interact at various times with atmospheric molecular oxygen or with various other molecules (H_2O, CO_2, etc.). Then if reducing sites exist on the surface of the dusts after their mechanical formation or a condensation from vapours or fumes, these species react to give an oxidized coating on the particles.

The first step of the metabolic process implies an interaction in lung medium of this oxidized coating with water, favoured by the surfactant in the mucus, and with various biologic molecules which can contribute, by formation of hydrated species or soluble complexes, to destroying partially or completly this coating. Moreover the reducing properties of some biologic molecules can allow, in some cases, the appearance of some reducing species not present in the dusts.

On the other hand the interactions of the dusts with the components of the biological milieu can consolidate the oxidized coating and allow the formation of quasi inert particles, partially coated also with biologic inactivated molecules, from which the presence of millions of inorganic particles by gramme of dry pulmonar tissue in the majority of people, without a significant toxic activity.

After the first step of the metabolic process may appear three possible types of reducing species :

- Hydrated cations in solution after dissolution or leaching of the particles.
- Complexed cations, in solution, with reducing properties, a function of the nature of the ligands.
- Reducing species (essentially cations) bound to the surface of the particles, (producing a surface activity at the solid-liquid interface), also a function of the nature of the coordination of these species.

Metal compounds able to present electron-donating properties after the first step of the metabolic process

These precursors compounds can be subdivided in four groups All are potentially able to transfer electron to molecular oxygen after the different phenomena that may appear during the first step of the metabolic process.

First group :

Compounds with a metallic character, not only metals M° (such as Fe°, Cu°, Ni°, etc.), but also some compounds (arsenides, sulfides such as Ni_3S_2, NiAs, etc.) containing metallic bonds, e.g. Ni-Ni-Ni in both the above compounds.

Second group :

Compounds containing cations able to reduce molecular oxygen, such as Fe(II), Cu(I), Ti(III), Cr(II). Many iron-containing dusts belong to this group, sometimes with a low iron content (e.g. wollastonite, chrysotile, etc.).

Third group :

Compounds containing cations in a high oxidation degree, but easily reducible by biological reductants, with appearance of cations able to reduce molecular oxygen.Examples :

$$\mathrm{Cr(VI)} \xrightarrow{+\ \text{ascorbate}} \mathrm{Cr(V)} \xrightarrow{O_2} \mathrm{AOS}$$

(Lefebvre et al, 1992)

$$\mathrm{Cu(II)} \xrightarrow[\text{or } +\ O_2^{\bullet -}]{+\ \text{ascorbate}} \mathrm{Cu(I)} \xrightarrow{O_2} \mathrm{AOS}$$

$$\mathrm{Fe(III)} \xrightarrow[\text{or } +\ O_2^{\bullet -}]{+\ \text{ascorbate}} \mathrm{Fe(II)} \xrightarrow{O_2} \mathrm{AOS}$$

However in the last case some specific ligands bound to Fe(III) are necessary to improve the oxidising character of Fe(III) (Bielski, 1985, Perrin 1959).

Fourth group :

Compounds containing cations able to form complexes capable of reducing molecular oxygen, when hydrated cations cannot reduce it.
Example (J. Fournier, personal communication) :

$$Ni(II) \xrightarrow{\text{tetraglycin}} NiG4 \xrightarrow{O_2} AOS$$

Probably Co(II), and may be Cr(III), may lead to such complexes.

Activated oxygen species (AOS)

In the second step of the metabolic process, the reducing species react with dioxygen in several steps to give various species belonging to the AOS set, which can be subdivised in three subsets :

- Species containing either oxygen or oxygen and hydrogen, e.g. $O_2^{\bullet -}$, H_2O_2, $OH^{\bullet}$, O_3, $^1O_2$, $O_3^{\bullet -}$, etc.
- Species where oxygen is bound to an organic entity, e.g. ROOH, $ROO^{\bullet}$, $RO^{\bullet}$, etc.
- Species where oxygen is bound to a metal in an hypervalent state, e.g. ferryl, perferryl, Cu-oxo, Ni-oxo, etc.

In metal compound toxicity this last group is probably the most important, but less known. Among the AOS where oxygen is bound to iron, it is well established that ferryl ($Fe = O^{2+}$) and perferryl ($Fe = O^{3+}$) have deleterious effects on living celles. Bielski (1992) has shown the anionic nature of these species, for example $(H_2Fe^VO_4)^-$ and $(HFe^VO_4)^{2-}$for the perferryl, and its consequences on the sites of attack of the biologic molecules (Sharma and Bielski, 1991). Among various propositions concerning the process of formation of these species, it appears (Bielski, 1992) that there is some evidence that ferryl iron is formed via a dinuclear complex when ferrous iron reacts with dioxygen.
Ideally it would be desirable to characterize all the AOS appearing in the various metabolic processes, but it is only possible in particular cases. However the AOS can be classified with regard to their electrophilic properties, and therefore their toxicity. We have proposed three categories (Pezerat, 1991).

- The A* species, which are the most electrophilic, e.g. $OH^{\bullet}$, ferryl, perferryl, Cu-oxo, etc.

- The P* species, less electrophilic, only able to trigger a peroxidation process of polyunsaturated lipids, such as $^1O_2$, NiG4, etc.

- Some species with a weak electrophilic character, e.g. $O_2^{\bullet -}$, H_2O_2.

Various test reactions for quantifying the formation in aqueous medium of A* and P* species have been presented in preceding publications (Pezerat, 1991, D. Costa et al, 1990 ; Nejjari et al, 1993 ; Berger et al, 1993 ; J. Fournier et al, this volume). However it is probable that the various tests of lipid peroxidation are as yet insufficient to detect the P*"species appearing as surface sites on a particle, and not in solution.

Necessity of a dynamic speciation of dusts

Among the precursor compounds, that is compounds with a capacity, after a first step of the metabolic process, to present reducing properties, there is a great diversity of reactivity. Therefore there is a necessity of a dynamic speciation -case by case- of all the various metal compounds appearing among the pollutants.

This speciation is not only a physico-chemical characterization of there materials but, more precisely, a characterization of the evolution of their redox properties in the biologic compartment directly concerned by the pathology under study.

Moreover it is necessary to take into consideration various factors such as the evolution of the mineral compounds before inhalation, the nature and the strength of the different types of bonds and the density and the nature of the faults in the solid, the nature of the ligands around the cations which play a role on their redox potential, the biodurability of the particles, the eventual presence of reducing and/or complexing biologic molecules in the milieu used to mimic the biologic compartment, the pH of this compartment, the eventual presence of the couple $O_2^{\bullet -}$-H_2O_2, the nature and the concentration of the anti-oxidants etc.

Then it is theoretically possible to accede to a classification of the metal compounds in relation with their capacity to induce a pathology. In fact the number of factors is so great that it is now possible to propose only some probable relations between a pathology and various groups of pollutants able to generate electrophilic species in well defined conditions. The classification which is derived for this speciation, using various tests, takes into account the intensity of the oxidative activity and its kinetics as a function of the medium composition. If the process of dioxygen reduction is very slow, no oxidative stress is possible because, in all cases it is necessary to break down a balance between pro-oxidants and anti-oxidants.

Among the divalent iron compounds, several examples have been published showing that it is not the global content in Fe(II) in the particles that plays the principal role in the formation of very electrophilic species (A*), but its accessibility by dioxygen in solution, and probably also the nature of its ligands (Zalma et al, 1987, Pezerat et al, 1989 ; Costa et al, 1989 ; Costa et al, 1990 ; Fournier et al, 1991 ; Pezerat et al, 1992 ; Guignard et al, 1992).

An example which illustrates the interest of this speciation concerns the comparison between two samples of nickel oxide (black and green), with the same CAS number, almost the same chemical analysis and X-ray diffraction pattern. But the green NiO is obtained after heating at a higher temperature that the black NiO, and this last is reacher in faults. In an aqueous medium containing an oligopeptide such as tetraglycin, and linolenic acid, it is observed (J. Fournier, personal communication) that the black NiO is much more capable of triggering the lipid peroxidation than the green NiO. It is highly probable that there is not only a correlation but a causal relation with the fact that the black NiO is much more active than the green NiO to induce sarcomas after intra-muscular injection (Sunderman, 1990).

Relations between emphysema, fibrosis and redox properties of the dusts

Emphysema, in lung medium, implies a destructive activity of elastase, after $\alpha 1$ - Antitrypsin ($\alpha 1$ - AT) inactivation, in extracellular medium, in general by an oxidative process.

The inactivation of α-AT is necessarily limited in space and time, in account of its renewal by serum. Then this oxidative inactivation is only possible if the electrophilic species appear relatively quickly around the particles, just after their inhalation. Consequently this implies the presence of a soluble salt, able to generate rapidly electrophilic species in solution.

This is the case of some coal dusts that contain the compound $FeSO_4$, $7H_2O$, a product of pyrite oxidation, pyrite being a common coal contaminant. Only the acidic coal dusts can stabilize this Fe(II) sulfate which explains the great difference in emphysema incidence according to the coal basins (Huang et al, 1993 ; Huang et al, 1994).

In the case of fibrosis, the biologically concerned compartment subject to dust toxicity is intracellular : the macrophage phagolysosome. Moreover it is necessary to envisage a catalytic process for the AOS production which is probably at the origin of a permanent activation of macrophages. In this case the dynamic speciation of the dusts must take into account the presence of the couple $O_2^{\bullet -}$ - H_2O_2 in the phagosome.

The most appropriate catalytic cycle implies the Haber-Weiss reactions, where iron is grafted for example into the surface of a siliceous crystalline network which brings about the necessary coordination allowing the Fe(III) reduction by $O_2^{\bullet -}$ (Dzwigaj et al, 1992) :

$$LmFe(III) + O_2^{\bullet -} \longrightarrow LmFe(II) + O_2$$

$$LmFe(II) + H_2O_2 \text{ (or } O_2) \longrightarrow A^*$$

$$A^* + XH \longrightarrow X^{\bullet} + LmFe(III)$$

Lm symbolizes the appropriate ligands brought by the dust surface functionnalities, A* the ferryl and/or perferryl species, and XH a biologic target molecule, e.g. a polyunsatured lipid.
A similar scheme is probably to be envisaged, but not yet confirmed, for anthracosis where iron would be grafted on superficial functionnalities of hard coals. Two origins are possible for iron : exogeneous such as iron present in asbestos or on industrial quartz and cristobalite, or endogeneous, such as iron released inside the macrophages and able to react with dust surface after the phagocytosis.
In the field of these propositions for fibrosis mechanism, the dynamic speciation of dusts implies to find out whether the surface of the dusts possesses either the Fe(III) species or the precursor sites able to receive it, in a coordination permitting the Haber-Weiss reactions cycle.

Relations between lung cancer and redox properties of the divalent iron containing dusts

If the mechanism of lung cancer initiation in relation with asbestos, some iron ores, granit dusts, etc. is considered, the most probable processes involve (Elias et al, 1993, Guilianelli et al, 1993 and in this Volume) : (i) The internalization of the particles, particularly by an epithelial cell, (ii) The accessibility for O_2 of Fe(II) at the solid-liquid interface (absence of the couple $O_2^{\bullet -}$ - H_2O_2) ; (iii) The formation of electrophilic AOS during the step where the particles is near the genome (phase of cell division).
Other routes may be considered including complexes formation, with or without intervention of a biologic reductant (Lund and Aust 1990 ; Aust and Lund 1991 and Chouchane et al, this Volume). But it is necessary to consider that these last propositions imply a greater number of conditions : availability of some particular ligands to form the complexes, availability of

reducing species, and absence of interactions with transferrin and other iron trapping species. The greater is the number of necessary conditions, smaller is the probability of occurence of an event.

Références

Aust A.E., Lund L.G. (1991) Iron mobilization from crocidolite results in enhanced iron-catalyzed oxygen consumption and hydroxyl radical generation in the presence of cysteine. In Mechanisms in Fibre Carcinogenesis (Brown R.C. et al, eds), NATO 223, Plenum Press : 397-406

Berger M., Hazen de M., Nejjari A., Fournier J., Guignard J., Pezerat H., Cadet J. (1993). Radical oxidation reactions of the purine moiety of 2'- deoxyribonucleotides and DNA by iron-containing minerals. Carcinogenesis, 1 : 41-46

Bielski B.H.J. (1985). Fast kinetic studies of dioxygen-derived species and their metal complexes. Phil. Trans. R. Soc. Lond. B311 : 473-482

Bielski B.H.J. (1992). Reactivity of hypervalent iron with biological compounds. Ann. Neurol : S28-S32

Chouchane S., Guignard J., Pezerat H. (1994). Appearance of very electrophilic species generated by some iron oxides : effect of chelators and reducing agents (this volume)

Costa D., Guignard J., Zalma R., Pezerat H. (1989) Production of free radicals arising from the surface activity of minerals and oxygen. Part I. Iron mine ores. Toxicol. Ind. Health, 6 : 1061-1078

Costa D., Guignard J., Pezerat H. (1990). Oxidizing surface properties of divalent iron-rich phyllosilicates in relation to their toxicity by oxidative stress mechanism. In Health Related effects of Phyllosilicates (Bignon J., ed.) NATO G21, Springer-Verlag : 129-134

Dzwigaj S., Pezerat H., Wastiaux A., Sébastien P., Honnons S. (1992). Silica and silicosis mechanism. The role of surface properties of particles and of the silica-supported iron. In Proc. 8th Int. Conf on Occup. lung diseases (Hurych et al., eds.). Czech. Med. Soc., Prague : 1186-1191

Elias Z., Poirot O., Schneider O., Marande A.M., Danière M.C., Terzetti F., Pezerat H., Fournier J., Zalma R. (1993). Cytotoxic and transforming effects of some iron containing minerals in Syrian Hamster Embryo (SHE) cells. Cancer Detection and prevention. In press

Fournier J., Guignard J., Nejjari A., Zalma R., Pezerat H. (1991). The role of iron in the redox surface activity of fibers. Relation to carcinogenicity. In Mechanisms in fibre carcinogenesis (Brown R.C. ed.) Plenum Press, N.Y. : 407-414

Fournier J., Jadrijevic A., Nejjari A., Pezerat H. (1994). Oxidizing activity of some metallic dusts in relation to oxidative stress mechanisms (this volume)

Guignard J., Nejjari A., Costa D., Zalma R., Fournier J., Pezerat H. (1992). Divalent iron containing dust, oxidative stress and relation with etiopathogenesis of some occupational lung cancers. In Proc. 8th Int. Conf. on Occup. lung diseases (Hurych et al, eds.). Czech. Med. Soc., Prague : 1325-1330

Guilianelli C., Baeza-Squiban A., Boisvieux-Ulrich E., Houcine O., Zalma R., Guennon C., Pezerat H., Marano F; (1993). Effects of mineral particles containing iron on primary cultures of rabbit tracheal epithelial cells : possible implication of oxidative stress. Environ. Health Perspect, 5 : 436-442

Guilianelli C., Baeza-Squiban A., Pezerat H., Marano F. (1994). Involvement of the oxidative stress in the toxicity of iron-containing particles on tracheal epithelium in culture (This volume).

Huang X., Laurent P.A., Zalma R., Pezerat H. (1993). Inactivation of $\alpha 1^-$ Antitrypsin by aqueous coal solutions : Possible relation to the emphysema of coal workers. Chem. Res. Toxicol., 6 : 452-458

Lefebvre Y., Pezerat H. (1992). Production of activate species of oxygen during the chromate (VI)-ascorbate reaction : Implication in carcinogenesis. Chem. Res. Toxicol., 5 : 461-463

Lund L.G., Aust A.E. (1991). Mobilization of iron from crocidolite asbestos by certain chelators results in enhanced crocidolite-dependent oxygen consumption. Arch. Biochem. Biophys., 287 : 91-96

Nejjari A., Fournier J., Pezerat H., Leanderson P. (1993) Mineral fibres : correlation between oxidising surface activity and DNA base hydroxylation. Br. J. Ind. Med. 50 : 501-504

Perrin B.D. (1959). The oxidation-reduction potentials of metal complex ions. Rev. Pure Appl. Chem., 4 : 257-285

Pezerat H., Zalma R., Guignard J., Jaurand M.C. (1989). Production of oxygen radicals by the reduction of oxygen arising from the surface activity of mineral fibres. In Non-occupational exposure to mineral fibres (Bignon et al., eds.) IARC Scient. Public. n°90, Lyon : 100-111

Pezerat H. (1991). The surface activity of mineral dusts and the process of oxidative stress. In Mechanisms in Fibre Carcinogenesis (Brown R.C. ed.). Plenum Press : 387-399

Pezerat H., Guignard J., Cherrie J.W. (1992). Man-made mineral fibers and lung cancer : an hypothesis. Toxicol. Ind. Health 1/2 : 77-87

Sharma V.K. and Bielski B.H.J. (1991). Reactivity of ferrate (VI) and ferrate (V) with amino acids. Inorg. Chem. 30 : 4306-4310

Sunderman F.W.Jr, Hopfer S.M., Plowman M.C., Knight J.A. (1990). Carcinogenesis bioassays of nickel oxides and nickel-copper oxides by intramuscular administration to Fischer -344 rats.Res. Com. Chem. Pathol. Pharmacol., 1 : 103-113

Zalma R., Bonneau L., Guignard J., Pezerat H., Jaurand M.C. (1987). Formation of oxy radicals by oxygen reduction arising from the surface activity of asbestos. Can. J. Chem., 65 : 2338-2341

Surface Properties of Respirable Silicate and Aluminosilicate Dusts Affecting Bioavailability

William E. Wallace*†, Michael J. Keane*, Joel C. Harrison*, James W. Stephens*, Patricia S. Brower*, R. Larry Grayson†, V. Vallyathan*, Michael D. Attfield*
*National Institute for Occupational Safety and Health, Division of Respiratory Disease Studies, Morgantown, West Virginia, USA 26505
†West Virginia University, College of Mineral and Energy Resources, Morgantown, West Virginia, USA 26506.

Introduction:

Pure respirable quartz is a known pathogenic agent for lung disease. However, for mixed dusts, neither mineralogical nor *in vitro* cytotoxicity assays unequivocally predict the contribution of quartz to their pathogenic potential.

First, although Coal Workers' Pneumoconiosis (CWP) and Progressive Massive Fibrosis (PMF) prevalence have been found to correlate with total respirable coal mine dust exposure, and with coal rank (Attfield and Morring, 1992), epidemiological studies generally have not found the quartz component of coal mine dust as measured by X-ray diffraction or infrared spectroscopy (each of which essentially measures the mass-percent quartz in a bulk sample) to be clear predictors of CWP prevalence or incidence (Walton et al., 1971). Second, the properties which distinguish the extreme fibrogenic potential of quartz are not revealed by conventional *in vitro* bioassay comparison with other silicate mineral dusts. As an example, quartz and kaolin (aluminosilicate) clay *in vitro* cytotoxicities are comparable for lavaged primary pulmonary macrophages.

The studies discussed here address these two anomalies from a general hypothesis that the pathogenic activity of respired quartz particles is associated with properties of, and interactions with, the particle surfaces. One factor in the expression of toxic activity by a particle surface is the biological availability of that surface. Anomalous lack of correlation of disease prevalence with mass percent quartz in exposures to some mixed composition respirable-sized dusts suggests that individual particle surface features, invisible by bulk sample mass analysis, may alter the biological availability of a fraction of quartz particles in a dust. For pure-surfaced mineral particles, the comparably strong prompt membranolytic activities of quartz and kaolin *in vitro* suggest that upon deposition of such particles in the lung there must be other interactions with particle surfaces which distinguish quartz. This might be due to toxic surface-associated sites

NATO ASI Series, Vol. H 85
Cellular and Molecular Effects
of Mineral and Synthetic Dusts and Fibres
Edited by J. M. G. Davis and M.-C. Jaurand

specific to silica, or to interactions of other sites on other mineral particles which affect the biological availability of their toxic functional groups. For example, aluminol groups on clays affect the availability or activity of adjacent silanol groups. Surface activities which distinguish quartz pathogenicity might be manifest in one or more of the sequential interactions upon deposition in the lung; first with the pulmonary surfactant system, then with extracellular and phagolysosomal enzymatic digestive processes, or with subsequent cytotoxic processes.

Materials and Methods:

To investigate the mineralogical anomaly, research was designed to detect if significant surface contamination could occur on respirable quartz particles in mine dusts. Modified surfaces might affe-t the biological availability of the quartz component of the dust in exposures, and thus the epidemiological analyses of the role of quartz in CWP/PMF or other occupationally induced diseases. A technique was developed to detect substantial but sub-micrometer thick surface occlusions on respirable-sized mineral particles (Wallace et al., 1990). In this technique, scanning electron microscope - energy dispersive X-ray analysis (SEM-EDX) was modified by the use of multiple

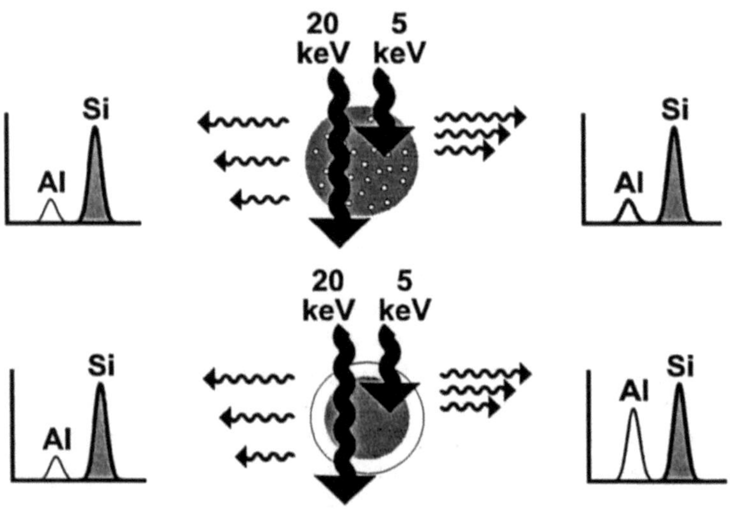

Figure 1. Depiction of multiple voltage scanning electron microscopy - energy dispersive x-ray analysis of respirable particle surfaces. Excitation of silicon (Si) (in gray) and aluminum (Al) (in white) X-ray lines occurs throughout a respirable-sized silicate particle by 20 keV electrons, and for a sub-micrometer surface layer by 5 keV electrons. This results in changes with voltage for the measured ratio of Si/Al for the heterogeneous (coated) structured particle.

electron beam accelerating voltages on each particle. At high accelerating voltages, the electron beam samples deeper into the particle than at low voltages.

Thus the ratios of the elemental x-ray signals are first measured at high electron beam voltage (20 keV). The electron beam energy is then reduced in one or several steps, covering the range between 20 and 5 keV. In the case of a homogeneous particle, the elemental ratios are expected to change slowly, whereas for a coated particle the ratios will exhibit a sharp change as the electron beam ceases to sample the core of the particle. An analytic model prediction for the silica fraction, F_{Si}, measured by SEM-EDX at an accelerating voltage V_e for a particle consisting of a silica core coated by aluminosilicate is

$$F_{Si} = I_{Si} / (I_{Si} + I_{Al})$$

where I_i is the intensity of the silicon (i = Si) or aluminum (i = Al) elemental X-ray given by

$$I_i = \frac{g_{Ci} w_i}{k_{xCi} + k_{eCi}} [1 - \exp[-(k_{xCi} + k_{eCi}) T]]$$

$$+ \frac{g_{Bi} w_i}{k_{xBi} + k_{eBi}} \exp[(-k_{eBi}(D-T) - k_{xCi} T)] [\exp[k_{eBi}(D-2T)] - \exp[-k_{xBi}(D-2T)]]$$

$$+ \frac{g_{Ci} w_i}{k_{xCi} + k_{eCi}} \exp[(-k_{eCi} D - k_{xBi}(D-2T) - k_{xCi} T)] [\exp[k_{eCi} T] - \exp[k_{xCi} T]]$$

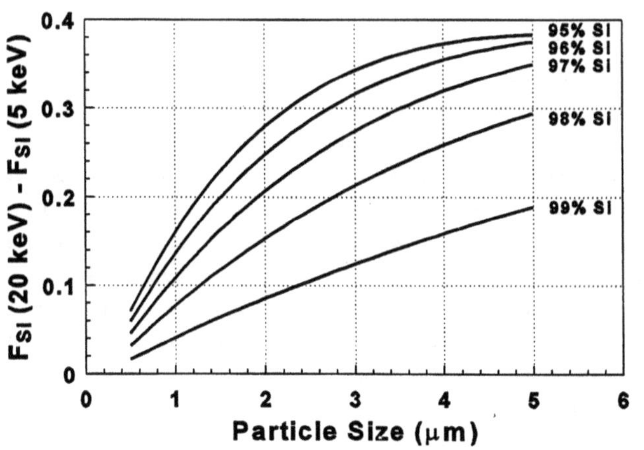

Figure 2. Theoretical model prediction for differences in Si/(Si+Al) measured at 20 and 5 keV versus particle thickness in micrometers, for five values of the true percent silica for a particle with a pure silica core and a coating of equiatomic Si and Al aluminosilicate.

where k_{Bi} and k_{Ci} are the extinction coefficients for electron beam intensity in the silica core (B) or clay coating (C), given by $3d_{B\ or\ C}[0.064(V_e^{1.68}-V_i^{1.68})]^{-1}$, where $d_{B\ or\ C}$ is the density of the silica core or clay coating, V_e is the incident electron beam energy, V_i is the energy of the silicon (i=Si) or aluminium (i=Al) X-ray line; k_{xBi} and k_{xCi} are the extinction coefficients for silicon i=Si) or aluminum (i=Al) X-rays in the core or coating; w_i is the probability of excitation of an X-ray rather than an Auger electron; and g_{Bi} and g_{Ci} are the excitation coefficients for electron ionization of silicon (i=Si) or aluminum (i=Al) inner shell electrons by the incident electrons; D is the particle thickness; and T is the thickness of the aluminosilicate coating on the particle surface, calculated from a model of equi-atomic silicon and aluminum in a top and bottom coating on a pure silica core giving the overall particle composition measured at high voltage.

To investigate the cytotoxicity anomaly and to seek factors and interactions responsible for the strong disease-inducing potential of quartz dusts, quartz and kaolin were compared and contrasted for their membranolytic potential under a series of treatments intended to model *in vitro* some early events occuring upon particle deposition in the lung. The respirable quartz dust used was Min-U-Sil (U.S. Silica Corporation, Berkley Springs, WV) ninety-eight percent <5 micrometers diameter, and a specific surface area of 3.97 m^2/g as determined by nitrogen adsorption isotherm measurement. The respirable kaolin dust used (Georgia Kaolin Mill) was ninety-nine percent <5 μm diameter with a specific surface area of 13.25 m^2/g. Research had shown comparable activities per unit surface area for the two dusts for prompt *in vitro* assays. Membranolytic activity against sheep erythrocytes was measured by hemolysis, and damage to lysosomal and plasma membranes was measured in lavaged rat alveolar macrophages by release of beta-glucuronidase, beta-N-acetyl glucosaminidase, and lactate dehydrogenase (Wallace et al., 1985). To model the first biochemical event upon deposition in the lung bronchio-alveolar region, dusts were incubated in a dispersion of L-alpha-dipalmitoyl phosphatidylcholine (DPPC) in physiological saline. This is a surrogate for the surfactant coating and dispersion on the hypophase of the alveolar surface. Silica or kaolin dusts were vortexed in ratios of 100 mg DPPC/g quartz and 200 mg DPPC/g kaolin, and incubated for 1 hr at 37°C. Treated quartz and kaolin dusts then were incubated in a cell-free system with phospholipase A_2 (PLA_2) enzyme. The particular PLA_2 used was a neutral pH optimum phospholipase. It was a commercial product, obtained from porcine pancreas. This was intended to provide a partial model for extracellular digestive processes at the periphery of some cells where such neutral pH optimum phospholipase is secreted, or for phagolysosomal digestion in interstitial cells with neutral pH lysosomal functions. However, it should be noted that lysosomal PLA_2 digestion in pulmonary macrophages occurs under acidic conditions. Controls were included for native dust, dust plus the maximum value of PLA_2, and dust plus DPPC. Membranolytic potentials of native quartz and kaolin dusts, DPPC coated dusts, and DPPC coated dusts subsequently incubated with PLA_2, were measured using erythrocyte hemolysis for 1 h incubation with dust at 37C.

Results:

Figure 3 demonstrates multiple voltage SEM-EDX analysis for a high silicon percentage mine dust particle. The predicted silicon fraction of signal, F_{Si}, versus electron beam accelerating voltage for a particle of 1.6 micrometer thickness and 96% silica fraction as measured at 20 keV, is shown as a solid line. The mean and standard deviation of ten measurements of F_{Si} each made at 20, 11, 9, 7, and 5 keV for a mine dust particle of that approximate thickness and composition are shown as the data points. This method was applied to respirable particles collected at a set of US coal mines. Prior epidemiological studies of miners in several U.S. coal mining regions have found that the prevalence of the various degrees of CWP, including progressive massive fibrosis, increases with cumulative dust exposure with coal rank as a parameter; the higher the rank, the greater the disease prevalence for a given level of cumulative exposure (Attfield and Morring, 1992).

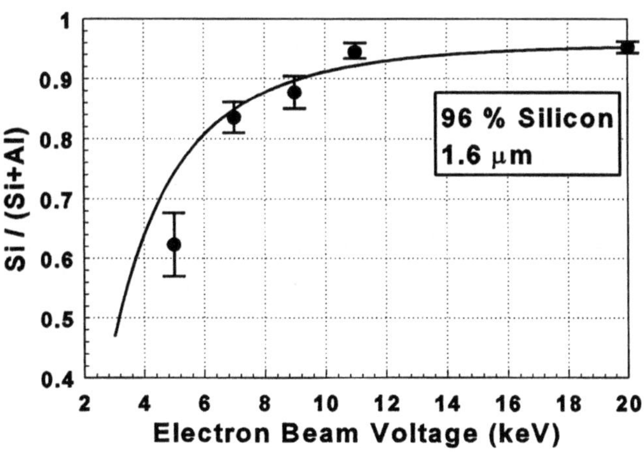

Figure 3. Data and theoretical prediction of measured silicon fraction of spectra, Si/(Si+Al), versus electron beam accelerating voltage. Data are for a mine dust particle of about 1.6 micrometer thickness and measured at 20, 11, 9, 7, and 5 keV.

Silicon and aluminum elemental ratios were determined with incident electron energies of 20 and 5 kiloelectron volts (keV) for some 20 respirable-sized, non-agglomerated particles with silicon fraction of signal of 75% or more (for elements above sodium) for each of 12 coal mine dust samples. Analyses were performed on two eastern Pennsylvania anthracite coal mine dust samples, three central Pennsylvania bituminous coal mine dust low-temperature-ash (LTA) samples and one non-ashed sample; two western Pennsylvania bituminous coal mine LTA samples and a non-ashed sample; two Illinois bituminous coal mine dust samples and a Colorado sub-bituminous coal mine dust. Five measurements were made at each voltage for each particle. Powdered aluminosilicate glass was used as the control to represent the behavior of particles with homogeneous aluminum contamination throughout the particle. Figure 4 shows the change in silicon fraction with reduction in voltage of the average repeated measurement for each particle in cumulative plots for the 12 samples grouped as three regions: eastern Pennsylvania anthracite samples; central Pennsylvania bituminous samples; and western Pennsylvania, Illionis and Colorado bituminous or sub-bituminous samples. The value of 0.029 for change in silicon fraction measured at 20 and 5 keV represents the 90th percentile of the change in silicon fraction for the homogeneous control group (glass). A majority of the particles analyzed in the anthracite samples, 80%, fall below this value, that is, manifest homogeneous rather than occcluded surface behavior. The remainder of the coal mine dust samples tend to have fewer particles below 0.029, especially for western Pennsylvania coal mine samples: 20% of the particles from western Pennsylvania dust samples and 46% of the

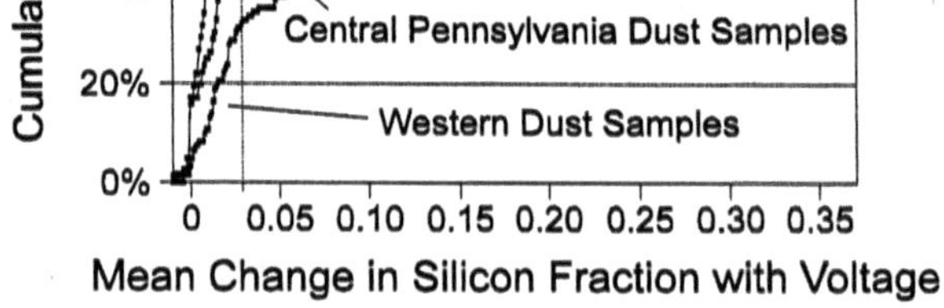

Figure 4. Cumulative frequency plot for percent particles with change in Si/(Si+Al) between measurements at 20 and 5 keV. The anthracite line contains measurements of 41 particles; the central Pennsylvania line 76 particles; the western line contains 61 western Pennsylvania, 41 Illinois, and 20 Colorado mine dust particles.

particles from central Pennsylvania dust samples were below the limit (p-value < 0.0001 in each case); 44% of the Illinois and Colorado particles were below the limit. Although there were significant numbers of particles which exhibited surface coating behavior, there were silica particles within each coal mine dust sample which did not exhibit surface coating behavior.

Upon incubation in DPPC dispersed in saline, both the quartz and kaolin dusts used in the investigation of membranolytic potentials promptly adsorb DPPC. An amount of adsorbed surfactant proportional to the dust's surface area resists saline elution (5.7 micromoles DPPC/ square meter of quartz surface and 8.5 micromoles DPPC/ square meter of kaolin surface) (Wallace et al., 1992). This incubation in DPPC neutralized activities of both dusts to control levels for erythrocyte hemolysis or macrophage membranolysis. Digestion of surfactant from both dusts proceeds at two distinct rates.

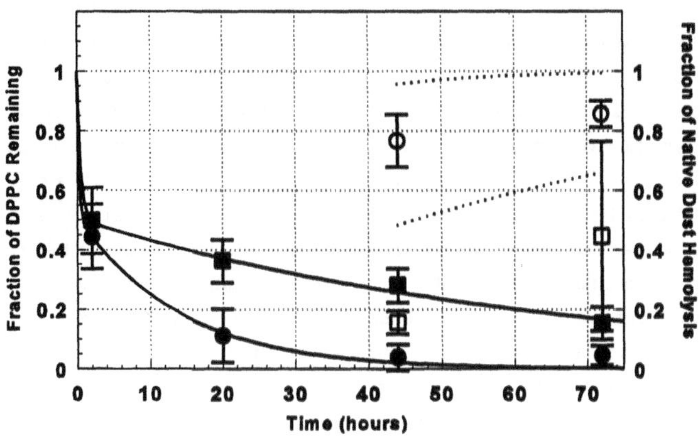

Figure 5. PLA_2 digestive removal of DPPC from a silica and from a kaolin sample; and hemolytic activity of treated dusts. Solid squares are fraction of initial DPPC remaining on kaolin; solid circles are for quartz. Solid lines are fits of a two exponential line to each data set. Open squares are the hemolytic activity of the kaolin dust incubated in DPPC and then in PLA2 for 44 and 72 h. Dotted lines are the hemolytic activities predicted by equating fractional activity with fractional available surface.

Approximately half the adsorbed surfactant is rapidly digested, as measured by subsequent organic solvent elution and thin-layer chromatography and phosphate quantitative analysis of the eluted material. The remaining half of the surfactant is removed much more slowly, and in a mineral-specific fashion, as shown in Figure 5. Results are normalized to the samples without enzymatic treatment. In the primary reference (Wallace, et al., 1992) data are provided for three applied activities of PLA_2. Here, data on the mid-activity value for both dusts are reviewed for incubation for 44, and 72 hr with applied PLA_2 enzyme activities of 2.72 units/ micromole DPPC adsorbed on quartz, and a comparable value of 2.90 units/ micromole DPPC adsorbed on kaolin. Each point in Figure 5 shows the mean and standard deviation for nine values. The smooth curves fit the data with a kinetics model of enzymatic digestion of an adsorbed DPPC bilayer assuming a two-exponential relaxation time fit. Rates of DPPC removal were greater for quartz-bound than for kaolin-bound DPPC. The first product of DPPC digestion by PLA2 is lysolecithin and a fatty acid residue. This production of lysolecithin must be accounted for in interpreting hemolytic potential measurements because lysolecithin itself is a membranolytic agent.

Figure 5 (open symbols) shows the hemolytic activities versus applied enzyme levels and incubation times, normalized to the hemolytic potentials of native dust controls used in each experiment. All DPPC and PLA_2 treated dusts showed an initial high membranolytic activity after 2 h PLA_2 incubation, which could be attributed to lysolecithin, the initial product of DPPC digestion by PLA_2. The dotted lines show for comparison the fraction of mineral surface bared of DPPC by PLA_2 digestion, calculated from the phosphate assay data. This is modelled as enzymatic digestion of an adsorbed bilayer, depicted in Figure 6, with $B(0) = S(0) = 0.5$ of the DPPC molecules. The model assumes that a surface adsorbed molecule is not accessible for enzyme digestion until the covering outer bilayer molecule is digested, and that when the intermediate enzyme-substrate complex, ES or EB, is formed then enzyme access to a number of adjacent substrate molecules, a[ES] or a[EB], is restricted due to the large size of the enzyme compared to substrate DPPC molecules, as shown in the model in Fig. 6. Simplification of the model for longer time batch digestion kinetics gives the expression for fraction of surface bare of DPPC at time t:

fractional loss of adsorbed DPPC = $1 - S(t)/S(0)$ =

$$1-\left[\frac{k_s'}{k_s'-k_B'}\exp[-k_B't]+\frac{k_B'}{k_B'-k_s'}\exp[-k_s't]\right]$$

where

$$k_B' = k_B(E-aE^2), \qquad k_S' = k_S(E-aE^2)$$

and k_B and k_S are determined from the phosphate assay data. At the applied specific activity level of approximately 3 units PLA_2/micromole adsorbed DPPC, the ratio of the quartz to kaolin digestion rate constants, $k_S'(\text{quartz})/k_S'(\text{kaolin})$, is 5.

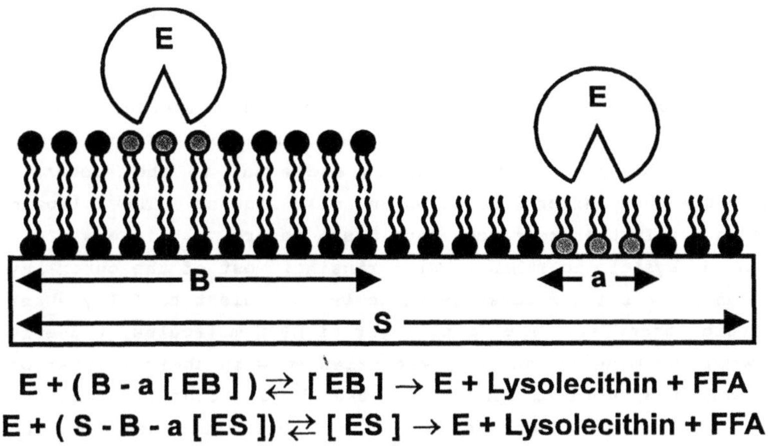

Figure 6. Depiction of phospholipase digestion of a bilayer of DPPC adsorbed on a mineral surface. "B" represents molecules in the outer layer, "S" molecules in the inner layer partially masked by B molecules, "a" represents the masking of DPPC molecules adjacent to a PLA_2-DPPC complex.

Discussion:

Analysis of respirable-sized silica particles by multiple voltage SEM-EDX indicated that a fraction of quartz particles can be surface occluded by aluminosilicate clay in dusts from some occupational settings. If the occlusion of the silica particle surface is persistent when the particle is retained in the lung, the biological availability might be modified relative to a pure silica surface. For dusts with a significant fraction of occluded particles the use of mass percent quartz content as a measure of dose might affect identification of a quartz component contribution to exposure and manifest disease. Additional research is needed to detail the nature of such occluded particle surfaces, the geological and processing factors which affect such occlusion, and the durability of such particles in tissue. Scanning Auger spectroscopy offers a complementary analytical method which can analyze for coatings only a few atomic layers thick on a surface. Limited tests have been performed on the durability of occluded silica particles upon a few hours incubation in surrogate pulmonary surfactant, and examples of both occluded and non-occluded particles have been observed in preliminary studies of tissue from the National Coal Workers Autopsy Study. Analyses of additional samples are needed to determine if surface occlusion is a factor in relationships of coal rank and quartz content of coal mine dust exposures and prevalence patterns for pneumoconioses and fibroses, and to determine mining situations or conditions which affect the fractional distribution of occluded silica particles. It should be carefully noted that surface occlusion does not indicate that these particles are incapable of inducing disease, but that some particles have more available silica surface than others.

The quartz and kaolin dusts used in this work are comparably hemolytic and cytotoxic on a surface area basis. The amounts of DPPC needed to fully suppress these activities of the dusts are equal to the amounts of DPPC retained on the dusts after incubation and mild rinsing. The following model is suggested for further research. Both dusts adsorb multilayers of DPPC from dispersion in physiologic saline. With rinsing, most of the outer layers are removed, leaving a bilayer coverage which is sufficient to fully suppress the membranolytic potential. The first layer of DPPC molecules, those in direct contact with the mineral surface, are oriented with their zwitterionic head groups toward the surface. These are backed by a second layer of DPPC, with acyl tails of the two layers associated and with ionic heads of the second layer pointed outward into the surrounding aqueous medium. This permits interaction of acidic or basic mineral surface sites with the cationic trimethylammonium or the acidic phosphate moieties of the DPPC zwitterionic phosphorylcholine head group, while minimizing contact of hydrophobic acyl groups with the surrounding aqueous media. In aqueous media the quartz forms acidic silanol surface sites and kaolin forms both acidic silanol and amphoteric or basic aluminol sites. Thus a silanol-choline ammonium acid-base association could form for both dusts with DPPC. But the kaolin alone could form an aluminol-phosphate acid-base bond. Phospholipase enzymes A_1, A_2, and C act, respectively, at the two fatty acid carbonyl ester linkages and on the phosphate of the phosphorylcholine ester linkages to the glycerol of DPPC. Thus, association of the DPPC phosphate with the kaolin surface could provide an additional hindrance to phospholipase enzymatic digestion of the adjacent palmitic acid carbonyl ester linkages. Comparison of the removal of DPPC at longer digestion times with measured restoration of membranoltic activity indicates that DPPC removal may equate to restoration of biological availability of active sites for the quartz surface. The situation is not as clear for kaolin. Analysis of possible retention of products of DPPC digestion, e.g., fatty acid association with kaolin aluminol basic surface groups is needed. Neutralization of dust membranolytic potential *in vivo* would involve adsorption of a variety of phospholipids and possibly of lipoproteins and other biochemical constituents of the pulmonary alveolar hypophase. Digestion of that surfactant coating could occur by the action of membrane-associated phospholipase and phagolysosomal phospholipases of alveolar macrophages or of interstitial cells.

REFERENCES

Attfield MD, Morring K (1992) An investigation into the relationship between coal workers' pneumocomiosis and dust exposure in U.S. coal mines. American Industrial Hygiene Association Journal 53(8):486-492.

Wallace WE, Vallyathan V, Keane MJ, Robinson V (1985) In vitro biologic toxicity of native and surface-modified silica and kaolin. Journal of Toxicology and Environmental Health 16:425-424.

Wallace WE, Harrison J, Keane MJ, Bolsaitis P, Eppelsheimer D, Poston J, Page SJ (1990) Clay occlusion of respirable quartz particles detected by low voltage scanning electron microscopy - X-Ray Analysis. Annals of Occupational Hygiene 34(2):195-204.

Wallace WE, Keane MJ, Mike PS, Hill CA, Vallyathan V, Regad ED (1992) Contrasting respirable quartz and kaolin retention of lecithin surfactant and expression of membranolytic activity following phospholipase A2 Digestion. Journal of Toxicology and Environmental Health 37:391-409.

Walton WH, Dodgson J, Hadden GG, Jacobsen M (1971) The effect of quartz and other non-coal dusts in coalworkers' pneumoconioses. In Inhaled Particles IV, Vol.2,edited by W.H. Walton. Oxford, United Kingdom: Pergamon Press. pp. 669-689.

Significance of Surface Properties of Quartz and Quartz-Containing Mineral Dusts to the Pulmonary Toxicity.

Zoltán Adamis, Erzsébet Tátrai, Miklós Timár and György Ungváry
National Institute of Occupational Health
Nagyvárad tér 2
H-1096 Budapest
Hungary

Introduction

The techniques of intratracheal instillation and bronchopulmonary lavage have proved useful for investigating the cytotoxicities of mineral dusts in vivo. Adamis and Krass (1991) found that the relative cytotoxicities of the different types of ceramic respirable dusts can be determined by this method.
The aim of our study was to investigate in vivo cytotoxicities of different quartz and respirable mine dusts. Furthermore significance of the surface properties of these samples were also examined.

Materials and Methods

There were five male Sprague-Dawley albino rats (CFY strain, body weight 350-400 g) in each group of animals tested. Dusts were suspended in sterile isotonic saline, and were sonicated for 5 min by Soniprep 150 (MSE). The rats were anaesthetized with sodium pentobarbital (40 mg per 100 g body weight) and 10 mg dust was injected directly into the trachea in 0.5 ml sterile saline.

NATO ASI Series, Vol. H 85
Cellular and Molecular Effects
of Mineral and Synthetic Dusts and Fibres
Edited by J. M. G. Davis and M.-C. Jaurand

The incision was then closed by a suture. Bronchopulmonary lavage was carried out 15, 60, 120 and 180 days after the intratracheal instillation. The rats were killed by ether anaesthesia and exsanguinated by excising both kidneys. The trachea was cannulated, and the lung was immediately washed out (in situ) with 3x5 ml sterile Tyrode's solution containing 4 units of heparin per ml. Lavage fluid was allowed to remain in the lungs for 2 min, and then withdrawn by gentle massage. This was repeated three times, the washes were combined (total recovery: approximately 12 ml) and then centrifuged at 250 g for 10 min. The cell-free supernatant was used for other investigations. Lactate dehydrogenase (LDH) was assayed by the method of Wroblewski and La Due (1955), proteins were determined by the method of Lowry et al.(1951),and phospholipids by the method of Raheja et al. (1973). Lung dry weight, wet weight and the cells recovered in the lavage fluid were also measured.

Table 1. Dust samples used in this study and their SiO_2 content

Dust	Code	Origin	SiO_2 %
Quartz DQ 12	DQ	Dörentrup (Germany)	87
Quartz FQ	FQ	Fehérvárcsurgó (Hungary)	98.6
Quartz GQ	GQ	Germany	98
Quartz HQ	HQ	Fehérvárcsurgó (Hungary)	99
Respirable colliery I	I	Pécs (Hungary)	29.4
Respirable colliery II	II	Márkushegy (Hungary)	1.1

Samples of respirable colliery dusts were collected, using a TBF 50 type sampler. All particles were below 5 µm (median 1-2 µm). Quartz DQ 12 is used as a standard dust in experimental silicosis research (Robock, 1973). Quartz FQ contained, in addition to quartz, 0.8 % Al_2O_3. Quartz HQ was ground in an agate vibratory mill for 32 h (Juhász et al., 1978). Quartz GQ is used in ceramic factories in Hungary. The designations and SiO_2 % of the

dusts are given in Table 1.
X-ray diffraction analysis was carried out by the Redler method, using a JEOL JXD-35 X-ray diffractometer. Surface characteristics of the particles were examined by ESCA XPS X-ray photoelectron spectroscopy (Adamis et al., 1991).

Results

The results obtained by the above biochemical methods are summarized in Fig 1 and Table 2 (the results shown represent the mean±SD derived from five separate experiments). The quartz samples induced LDH, protein and phospholipid increase in the supernatant of the lavage fluid at different time intervals after intratracheal instillation.After 15 days of exposure only quartz DQ was effective. Quartz GQ caused higher values after 2 months, quartz FQ after 6 months. Quartz HQ, the sample ground in the agate mill for 32 h was still ineffective after 6 months(Fig 1).

Table 2. Surface properties of the particles and pulmonary toxicity (intratracheal inj.: 10 mg, time after instillation: 6 months).

Sample	LDH	Protein	Phospholipid	Quartz %	Quartz %
	$mU\ cm^{-3}$	$\mu g\ cm^{-3}$	$\mu g\ cm^{-3}$	X-ray d.	ESCA
Quartz DQ	1200±361	2015±490	704±122	87	83
Quartz FQ	661±142	1293±365	847±160	98.6	68
Resp.col.I	135± 62	620±227	188± 27	29.4	15
Resp.col.II	46± 14	524± 90	198± 33	1.1	0
Control	55± 14	580± 69	134± 34	-	-

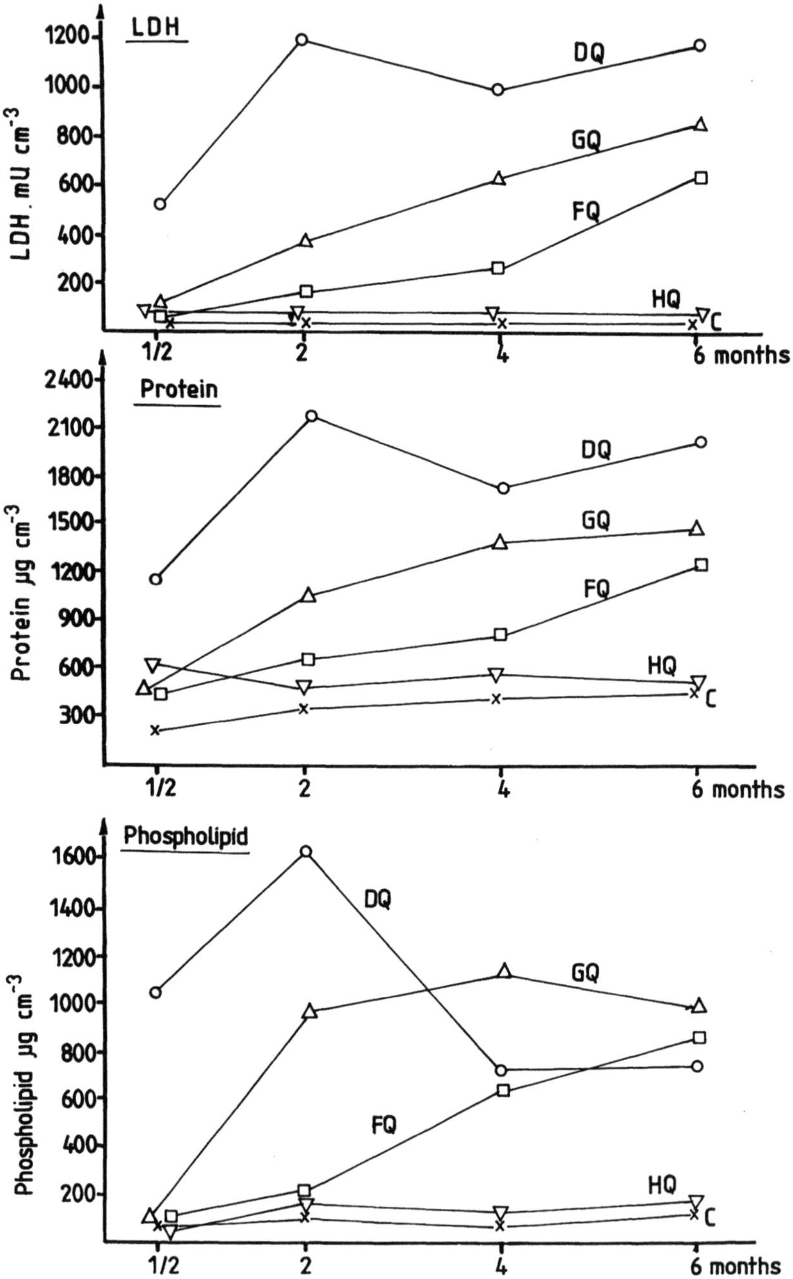

Fig. 1. LDH activity, protein and phospholipid in lavage fluid from rats 15, 60, 120 and 180 days after instillation of 10 mg quartz dust (abbreviations as in Table 1).

The amount of crystalline silica in the surface layer of particles is of crucial importance in relation to in vivo cytotoxicities as shown Table 2. LDH, protein and phospholipid increase in the supernatant of lavage fluid are well correlated with the SiO_2 content in the surface layer.

Discussion

The use of intratracheal instillation and bronchopulmonary lavage provided us with a sensitive and reliable experimental system to study the effect of different mineral dusts (Sykes et al. 1983). These in vivo investigations proved that quartz dusts induced cytotoxic reaction in the lung after different latency periods. The quartz dusts examined were divided into four groups according to their in vivo cytotoxic effect. The classification was based on the following parameters: the appearance of the inflammatory reaction, the degree of the inflammatory response characterized by the activity of LDH, the amount of protein and phospholipid in the lavage fluid. This phenomenon can be explained by the surface properties of the particles. As verified by our results, the early biological effect of these dusts is determined by the surface layer of the particles.

Conclusion

The bronchoalveolar lavage procedure is a very useful method for screening. The quartz samples can be divided into four groups according to their in vivo cytotoxicities.

Surface characteristics of the particles and the composition of solid surfaces, down to a depth of 2-4 nm, determine the biological effect of the dusts. Physical treatment decreased the cytotoxic effect of the samples.

References

Adamis Z, Krass BK (1991) Studies on the cytotoxicity of ceramic respirable dusts using in vitro and in vivo test systems. Ann. occup. Hyg. 35:469-483.
Adamis Z, Timár M, Bertóti I, Tassy M, Komlósy KJ, Vékény H, Kiss J (1991) Investigations of surface properties and in vitro biological effect of respirable dusts originating from Hungarian mines. Egészségtudomány. 35:146-151.
Juhász Z, Timár M, Adamis Z (1978) Wirkung mechanischen Aktivierens auf Lebendzellen-Reaktionen mineralischer Staube. Staub Reinhaltung der Luft. 38:131-134.
Lowry OH, Rosebrough NJ, Farr AL, Randall RJ (1951) Protein measurement with the Folin-phenol reagent. J. biol. Chem. 193:265-275.
Raheja RK, Kaur C, Singh A, Bhatia IS (1973) New colorimetric method for the quantitative estimation of phospholipids without acid digestion. J. Lipid Res. 14:695-697.
Redler L (1986) Phase analysis of the raw materials containing clay minerals by quantitative X-ray diffraction. Épitőanyag 38:235-240.
Robock K (1973) Standard quartz DQ 12 <5 µm for experimental pneumoconiosis research projects in the Federal Republic of Germany. Ann. occup. Hyg. 16:63-66.
Sykes SE, Morgan A, Moores SR, Holmes A, Davison W (1983) Dose-dependent effects in the subacute response of the rat lung to quartz. I. The cellular response and the activity of lactate dehydrogenase in the airways. Exp. Lung Res. 5:229-243.
Wroblewski F, La Due IS (1955) Lactic dehydrogenase activity in blood. Proc. Soc. exp. Biol. Med. 90:210-213.

The cytotoxicity of respirable coal dusts

Janick Massé[1], Pierre Larivée[1], Patrick Sébastien[2] and Raymond Bégin[1]
service de pneumologie et unité de recherche pulmonaire
CHU Sherbrooke, Québec, Canada J1H 5N4
and
Centre d'Études et Recherches de Charbonnages de France
Verneuil - en Halatte, France.

INTRODUCTION

The pathogenesis of coal-workers pneumoconiosis has been related to three factors: total lung dust, rank of coal and the effect of quartz (Morgan et al, 1976,Parkes, 1982, Seaton, 1983), but the relative contribution of each constituent of the mineral dust remains incompletely understood as the incidence of disease in many collieries cannot be entirely explained by these factors (Policard et al, 1967; Seaton et al, 1981; Schlipköter et al, 1971; Hilscher and Schlipköter,1972; Walton et al, 1976, Reisner et al, 1976, Heppleston et al, 1988). The risk of pneumoconiosis varies between collieries and these factors are not always providing satisfactory explanations. Nonetheless, it is accepted that the mining of high rank coal is generally associated with an increased risk of pneumoconiosis. This lead to the hypothesis that differences in the physico/chemical properties of coal mine dust could be the determinants of the toxicity of the coal dust. Thus, the concept of specific toxicity of coal mine dusts emerged (Sébastien, 1990).

Studies on factors influencing the toxicity of coal mine dusts have been done. LeBouffant (LeBouffant et al 1968; 1987) found that quartz is the most cytotoxic particle and that clay minerals (kaolin and mica) produce a protective effect on the quartz contained in the coal dust by coating the quartz particles with aluminum. This effect is most potent with mica. Schlipköter (Schlipköter et al 1971) documented the PVN-oxide inhibitory effect on some of the coal dust samples and others have added to these observations. However, the large variability of the risk of disease within mines of countries such as France and Germany remains incompletely understood. The present study was therefore undertaken as part of a larger effort to further clarify the understandings of factors contributing to the risk of pneumoconiosis in the French collieries.

NATO ASI Series, Vol. H 85
Cellular and Molecular Effects
of Mineral and Synthetic Dusts and Fibres
Edited by J. M. G. Davis and M.-C. Jaurand

The specific toxicity of coal mine dust of three different collieries (Nord et Pas-de-Calais HBNPC, Lorraine HBL, and Centre Midi Gardanne HBCM) were selected for these investigations in parallel with in vivo animal experiments, to address the question of a specific cytotoxicity for the coal matter as opposed to that of the quartz content of the coal dust.

The respirable dust collection procedure, physico-chemical characterization of the dust and the animal studies have been reported in a previous report (Sébastien et al, 1990). We summarize the collection procedure and physico-chemical characterization of the dust in the material section. The animal studies were short term (9 month) follow-up studies after an intra-tracheal injection in rats. These documented a toxicity as determined by lung weigth, lipid and collagen content, macrophage derived fibronectin production and weight of lymph nodes to be in the following order of toxicity: Lorraine > Nord et Pas-de-Calais > Centre Midi Gardanne. These findings were related to the quartz contents of each type of dust.

In the present study, we evaluated the relative cytotoxicity of these coal mine respirable dusts using simple in vitro alveolar macrophage culture and verified the inhibitory effect of aluminum lactate treatment on these samples, as to determined the relative contribution of quartz to the overall activity.

MATERIALS AND METHODS

1. Dust collection and characterization

The dust samples were collected using several CPM-3 dust collectors (MSA, France) placed at head level of miners in each of the 3 mines, Nord et Pas-de-Calais, Lorraine, and Centre Midi Gardanne. The samples were collected over several days until enough materials were available for these researches. The dusts were studied for composition of mineral matters as previously reported (Robveille and Mazza, 1990) and by transmission electron microscopy, for counting, sizing and identification by energy dispersive spectrometry of X-rays (EDS). Silica content was also determined by infra-red analyses. Other analyses in secondary ion mass spectroscopy (SIMS) to study surface chemistry and in electron spin resonance (ESR) were done and reported in a prior report (Sébastien et al, 1987). For a better appreciation of our cytotoxicity results, we present in RESULT section a summary of the selected analyses of the dust, as of interest to the present work.

The characterization of the quartz particles (Minusil-5®, Pensylvania Glass Co, Pittsburg, PA) has been previously detailed (Bégin et al, 1986).

2. Culture of rat macrophages

The macrophages were obtained from normal rats by bronchoalveolar lavage with physiologic saline. Cell viability was determined with the trypan blue exclusion technique. The cells were then suspended in Dulbelco's modified eagle medium (DMEM) (Gibco Laboratories, Grand Island, NY) with 2% serum for a total concentration of 10^7 cells per ml. Then 0.5 ml of the cell preparation was seeded in 24-well tissue culture plates (Falcon, Lincoln Park, NJ) and incubated for 2 hours at 37°C and 10 % CO_2 (time required for macrophage adherence). Then, the macrophages were rinsed three times with PBS solution and maintained in culture in DMEM.

3. Concentration of the mineral dust samples

The dust samples above identified and characterized are hereafter called as Nord, Lorraine and Midi for simplicity. Concentrations of 50, 100 and 250 µg / ml of the coal dust samples with or without aluminum treatement (AL), (Nord, Nord-Al, Lorraine, Lorraine-Al, Midi, Midi-Al), were prepared in DMEM without serum. The two controls of Minusil-5® (Si) (positive) and Minusil-5®-Al (Si-Al) (negative) were prepared in the same concentrations as the coal dusts. Each of these samples were incubated 24 hours at 37°C and 10 % CO_2 with the macrophages.

4. Aluminum lactate treatment (Al)

Preparation of a 1.1g % aluminum lactate (Chronos, Paris, France) was done at a pH of 7.0. Then 1g of sample (Minusil-5®, Nord, Lorraine, Midi) was mixed with 10 ml of the aluminum lactate solution for 3 hours at room temperature. The samples were then washed with water 4 times and dried at 100 °C (Bégin et al, 1986, 1987; Larivée et al, 1990).

5. Verification of phagocytosis

To ascertain the phagocytosis of the dusts by macrophages, separate cell cultures were done, and in addition to the usual procedures, coverslips were sterilized and inserted in the culture plates. After the supernatant was centrifuged the coverslips were retrieved, put on glass slides and stained with Wright-Giemsa. The slides were examined under polarized light at 400 x magnification.

6. Measurement of the cytotoxicity index (CI)

After the 24 hour incubation time with dust samples, the supernatant was centrifuged to remove the dust in suspension. The cytoplasmic enzyme lactate dehydrogenase (LDH) was determined by a spectrophotometric method (Walker, 1956) at 340 nm to read the LDH value in mIU / ml. The cytotoxicity index was determined with the following formula (Larivée et al, 1990):

$$\frac{\text{LDH sample} - \text{LDH negative control}}{\text{LDH positive control} - \text{LDH negative control}} \times 100$$

RESULTS

The mineralogical content of the different particles was previously analysed at CHERCHAR and summarized in Table 1. It shows that the particles obtained from Nord had the highest coal content followed by Midi and Lorraine. The highest quartz content was in Lorraine sample followed by Nord and Midi samples. The percentage of mineral matter was highest in Lorraine with Midi and Nord being second and third respectively. The kaolin content was highest in Lorraine followed by Midi and Nord. The size distribution analysis revealed that the mean diameter of the particles was similar in the three collieries.

The microscopical examination of the slides showed that most of the particles were inside the cells, thus apparently phagocyted by macrophages.

A dose-response study of the quartz with and without aluminum lactate at 50, 100 and 250 μg / ml were obtained with 7 to 9 samples of each different concentration of the particles. The results graphed in figure 1 showed that at concentration of 50 μg / ml, LDH was 11.15 ± 1.17 mIU / ml for Si and 2.86 ± 0.73 mIU / ml for Si-AL. For 100 μg Si / ml, the LDH was 9.19 ±1.41 mIU / ml with Si and 1.28 ±0.24 mIU / ml with Si-Al. For 250 μg Si / ml, the total content of LDH was 11.57±1.20 mIU / ml for Si versus 2.27 ± 0.22 mIU / ml for Si-Al. The average of these results of Si constituted the 100% index of cytotoxicity and the average results of Si-AL the 0%.

The coal dusts produced little effects at concentration of 50 µg / ml but the cytotoxicity appeared at the higher concentrations 100 and 250 µg / ml. In the latter 2 concentrations, the Lorraine was more cytotoxic than the Nord dust and the latter more than the Midi coal dust. The differences were significant at $p<0.05$ or higher.

The experiment was repeated at the 100 and 250 µg / ml doses to verify the effect of aluminum lactate treatment. The results are shown in figure 2. The effect of the treatment of the Si particles are seen in figure 1 and not repeated after. With the particles of the different collieries, the reduction of toxicity was most significant with Lorraine dust where the difference in the cytotoxicity index was significant at $p<0.0001$ at both concentrations of the particles. The dusts from other collieries showed some reduction in cytotoxicity with Al but the differences were not statistically significant.

For the Lorraine sample at 100 µg per ml the coal dust was 5 x more cytotoxic than control (SI-AL) and at 250 µg per ml, it was at 10 x. For the Nord samples, it was at 100 µg per ml 11x and at 250 µg per ml 24 x control. For the Midi sample, it was at 100 µg per ml 13x and at 250 µg per ml 16 x control.

DISCUSSION

The study of respirable coal dust from various mine sites was undertaken with the support of the CECA to further address the question of variability in the relationship of coal dust exposure intensity - risk of pneumoconiosis for miners. In the initial study, the coal dust was collected, characterized and in vivo animal studies were carried out over a one year period. The data has been reported to CECA (Sébastien et al, 1990). It was recognized that in the study of pneumoconioses, the animal models have been particularly reliable to inform us on the pathogenic mechanisms and this is a generally accepted point, for diseases such as asbestosis or silicosis (Bégin et al, 1989; Rom et al, 1990). However, in the study of coal dust pneumoconiosis, the animal models have been relatively unreliable to reproduce the fundamental pathologic changes of the pneumoconiosis (Policard et al, 1967; Lebouffant et al, 1967, 1982, 1987; Sébastien et al, 1990) and thus, of limited interest in the study of coal dust pneumoconiosis, and often incapable to provide satisfactory explanation for the relative risk of pneumoconiosis.

We therefore addressed the question of relative toxicity of the coal dust, from mines with established distinct risks of pneumoconiosis, using simple in vitro cytotoxicity tests, using alveolar macrophages in 24 hour culture, in the line of prior investigations oriented at the major target cell of the lung defense, the alveolar macrophage (Bingham et al, 1977; Bowden and Adamson, 1978; Castranova et al, 1985).

In our cell culture system, the data was clear. We found a dose-response cytotoxic effect for all 3 coal dusts (fig 1), which appeared at doses higher than 50 µg per ml and increased with higher doses. Within the dose range that had cytotoxic effects, we also found higher cytotoxicity for the Lorraine dust than for the Nord and the Midi dusts. Given the higher coal content in the Nord sample (table 1), it is clear that the cytotoxicity observed in this study cannot be entirely explained by the rank of the coal.

We then looked at the potential role of the quartz content of these coal samples, recognizing that the Lorraine sample had the highest level of quartz content (table 1) and the highest level of cytotoxicity (figure 1). To inhibit the potential role of the quartz content of the samples, we treated each sample with aluminum lactate, treatment that has been documented to inactivate almost completely the toxic action of quartz in vitro and in vivo (Lebouffant et al, 1968; Bégin et al, 1986, 87). The results of this treatement presented in figure 2 clearly documented the significant reduction of cytotoxicity by the aluminum treatment of the Lorraine dust, whereas no significant reduction were found for the other 2 dusts. This observation suggested that only the Lorraine sample had sufficient quartz to contribute significantly to the cytotoxicity of the dust.

In all cases however, the cytotoxicity of the coal dust samples remained clearly well above the level of inert dust as represented in the figure 2 by the Minusil-5® treated with aluminum (SI-AL). For the coal content, the rank of the coal dusts were proportional to the cytotoxicity after correction for the SI effect. These data however do not explain the relative low risk of pneumoconiosis in the Midi miners.

In conclusion, we have studied the cytotoxicity of coal dusts from 3 collieries with distinct composition. We found the cytotoxicity in part related to the quartz content for the

Lorraine sample only, and in all cases, residual cytotoxic effect was proportional to the coal content of the dusts. The data support the general concept of a relationship between the rank of coal and the risk of disease and document a significant contribution of quartz to the toxicity of respirable dusts in some of the collieries .

REFERENCES

Bégin R, Massé S, Rola-Pleszczynski M, Martel M, Desmarais Y, Geoffroy M, Le Bouffant L, Daniel H, Martin J: Aluminum lactate treatment alters the lungs biological activity of quartz. Exp Lung Research 10: 385-399, 1986.

Bégin R, Cantin A, Massé S: Recent advances in the clinical assessment and pathogenesis of the mineral dust pneumoconiosis. Eur Respir J 2: 988-1001,1989.

Bingham E, Barkley W, Murthy R, Vassalo C. Investigation of alveolar macrophages from rats exposed to coal dust. in Inhaled Particles III, vol II; Walton WH ed. Surrey, England, Unwin Bros, 1977, 543-550

Bowden DH, Adamson IYR (1978) Adaptive responses of the pulmonary macrophagic system to carbon. Laboratory Investigation, 38:422-429

Castranova V, Bowman L, Reasor MJ, Lewis T, Tucker J, Miles PR (1985) The response of rat alveolar macrophages to chronic inhalation of coal dust and/or exhaust. Environmental Research, 36:405-419

Dubois F, Bégin R, Cantin A, Massé S, Martel M, Bilodeau G, Dufresne A, Perreault G, Sébastien P: Aluminium inhalation reduces silicosis in a sheep model. Am Rev Respir Dis 137: 1172-1179, 1988.

Heppleston AG (1988) Prevalence and pathogenesis of pneumoconiosis in coal workers. Environmental Health Perspectives, 78:159-170

Hilscher W, Schlipköter H-W (1972) Antagonistic factors in the pathogenesis of coal workers pneumoconiosis. Ann NY Acad Sci, 200:166-181

Larivée P, Cantin A, Dufresne A, Bégin R (1990) Enzyme activities of lung lavage in silicosis. Lung, 168:151-158

LeBouffant L, Daniel H, Bruyet B (1987) Facteurs influençant la nocivité des poussières des mines de charbon. Colloque INSERM, 155:243-264

LeBouffant L, Daniel-Moussard H, Martin J-C, Charbonnier J, Letort M, Policard A (1968) Etude de l'action inhibitride des charbons vis-à-vis des effets fibrogènes du quartz. C R Acad Sc Paris, 267:1879-1882.

Martin JC, Daniel H, LeBouffant L (1977) Short- and long-term experimental study of the toxicity of coal-mine dust and of some of its constituents. Inhaled Particles, 1977:361-370

Morgan WKC, Lapp NL. Respiratory disease in coal miners. Amer Rev Respir Dis 113; 531-559, 1976.

Parkes WR, Occupational Lung Disorders (2nd edition), 1982.

Policard A, Letort M, Charbonnier J, Martin J, Daniel-Moussard H (1967) Recherches sur les interactions charbon-quartz dans le développement des pneumoconioses des houilleurs. Archives des Maladies Professionnelles de Médecine du Travail et de Sécurité Sociale (paris), 28:589-594

Reisner MTR, Robock K. Results of epidemiological, mineralogical and cytopathological studies on the pathogenecity of coal-mine dusts. in : Inhaled Particles IV, vol 2; Walton WH ed. Oxford: Perganon Press, 703-715.

Seaton A, Dodgson J, Dick JA, Jacobsen M (1981) Quartz and pneumoconiosis in coalminers. The Lancet, Dec:1272-1275

Seaton A (1983) Coal and the lung. Thorax, 38:241-243

Sébastien P, Wastiaux A, Courbon P, Kovacs MP, Robveille J, Erre R, Pézerat H, Lambré C, Specific toxicity of coal mine dust. Rapport final, CECA n° 7260-03/ 041/ 03, 1990..

Schlipköter, Hillllscher, Pott F, Beck EG. Investigations on the aetiology of coal workers' pneumoconiosis with the use of PVN-oxide. in Inhaled Particles III, vol 1; Walton WH ed. Surrey, England, Unwin Bros, 1971, 379-390.

Robville J, Mazza M. Mise au point de techniques d'analyses fines de minéraux argileux sur de faibles quantités de poussières. Rapport final, CECA n° 7260-03/ 042/ 03, 1990.

Rom Wn, Travis WD, Brody AR. Cellular and molecular basis of the asbestos-related diseases. Am Rev Respir Dis 1991, 143, 408-22.

Sébastien P, Bossé J, Dubois F, Bégin R: Effet du traitement au lactate d'aluminium sur l'épuration alvéolaire du quartz injecté dans le lobe trachéal du mouton. In "Silicose et pneumoconioses à poussières mixtes". L. LeBouffant, Ed., Colloque INSERM, vol. 155, pp. 523-528, 1987.

Walker W, Ulmer DD, Vallee BL. Metalloenzymes and myocardial infarction: Malic and lactic dehydrogenase and zinc concentration in serum. N Engl J Med 225; 449-456, 1956.

Walton WH, Dodson J, Hadden GG Jacobsen M. The effect of quartz and other non-coal dusts in coalworkers' pneumoconiosis. Part 1: Epidemiological studies. in : Inhaled Particles IV, vol 2; Walton WH ed. Oxford: Perganon Press, 1976, 669-690..

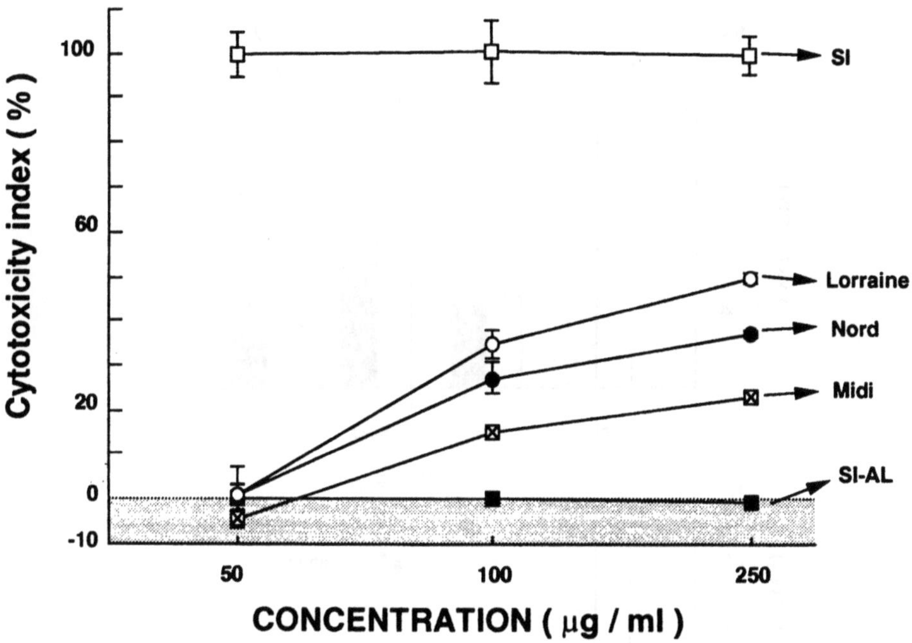

FIGURE 1

Dose-response cytotoxicity of quartz (SI) and coal dusts from the various collieries identified in the figure. The mineralogy of these mineral samples are in table 1.

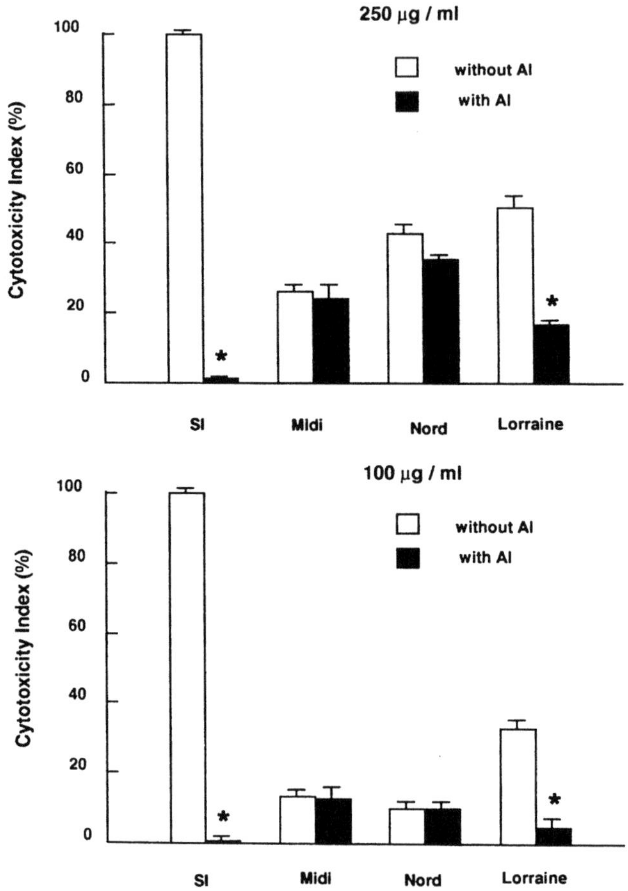

FIGURE 2

Effects of aluminum (AL) on cytotoxicity of the various coal dusts samples. Asterix indicates significant difference at $p < 0.05$.

Appearance of very electrophilic species generated by some iron oxides: effect of iron chelators and reducing agents.

Salem Chouchane, Joëlle Guignard and Henri Pezerat
Université P. et M. Curie, CNRS URA 1106
Laboratoire de Réactivité de Surface
4 place Jussieu, 75252 Paris Cédex 05
France

It is now well known that asbestos fibers are responsible of several lung diseases. This toxicity is suggested to be due to the presence of Fe(II) in the structure of the fibers, or in their contaminants (Zalma et al 1987). Moreover, an excess incidence of lung cancer has been found among workers in various metal mines (Mur et al 1987), in iron and steel foundries (IARC monographs 1984) and also among arc welding workers (IARC monographs 1990) pointing to possible carcinogenic properties of iron-containing mineral dusts. In the case of Fe(II) compounds, this toxicity is due to the activation of dioxygen in biological medium by divalent iron appearing at the interface solid-liquid or in solution, which may lead to the formation of various activated oxygen species (AOS). In the set of these AOS, the very electrophilic species are symbolized by A* (Pezerat 1991); they include ferryl, perferryl and $OH^{\bullet}$ (Yamazaki et al 1990 and Bielski et al 1992). The quantitation of the oxidizing power of the A* species is obtained through a reaction of H abstraction between A* and a target molecule, the formate anion (Zalma et al 1987). The production of the $CO_2^{-\bullet}$ is measured using a spin-trap agent, the DMPO (5-5'-dimethyl-1-pyrroline-N-oxide), giving an adduct $(DMPO, CO_2^{-})^{\bullet}$ whose life time allows its identification by ESR spectroscopy.

Minerals containing easily accessible Fe(II), such as crocidolite, amosite or nemalite in a medium model (aqueous phosphate buffer medium), are capable of oxygen activation (Fournier et al 1991 and Pezerat 1991) leading to the A* formation, but Fe(III)-compounds such as hematite (α-Fe_2O_3) and Fe(III)-Fe(II)-oxide such as magnetite (Fe_3O_4) are not capable of such activation in the same simple medium. Hematite and magnetite being pollutants very frequently present in the inhalable dusts in mines and industry, it appears necessary to take into account the possible role of biologic molecules, able to complex and/or to reduce iron in the first step of the metabolic process. Accordingly, the behaviour of these two iron oxides upon the A* formation in a medium containing some iron chelators (citrate, histidine, ADP, EDTA and NTA) and reducing agents (ascorbate and cysteine) was investigated, in comparison with the one of two Fe(II)-rich compounds such as crocidolite and vivianite (Fe(II) phosphate).

NATO ASI Series, Vol. H 85
Cellular and Molecular Effects
of Mineral and Synthetic Dusts and Fibres
Edited by J. M. G. Davis and M.-C. Jaurand

MATERIALS AND METHODS:

Minerals: Hematite (α-Fe_2O_3) from Merck; magnetite (Fe_3O_4) from BDH; crocidolite, an UICC sample from South-Africa, and vivianite, a naturel divalent iron phosphate $Fe_3(PO_4)_2,8H_2O$ from Cameroons.

Reagents: EDTA (ethylenediaminetetraacetic acid, disodium salt, dihydrate), NTA (nitrilotriacetic acid, disodium salt), ADP (adenosine-5'-diphosphate), L-cysteine, sodium ascorbate and DMPO (5-5'dimethyl-N-pyrroline-1-oxide) were from Sigma; L-histidine was from Fluka Biochemica; tri-sodium citrate dihydrate, sodium formate, K_2HPO_4, KH_2PO_4 and KOH were from Merck.

Carboxylate anion radical formation and detection: 2mL of reaction mixture contain: 45mg of mineral in a potassium buffered phosphate solution pH=7.4 (250mM) containing DMPO (50mM), sodium formate (1M) and eventualy an iron chelator (citrate, histidine, ADP, EDTA and NTA, 2.5mM) and/or reducing agent (ascorbate 1mM and cysteine 4.5mM). The mixture in an open reactor was protected from light exposure. The suspension was stirred at 37°C, aliquot of suspension was withdrawn after 30 minutes, filtered through 0.65 μm porosity filters and transferred into a flat quartz ESR cell, fitted into the cavity of the spectrometer (a Varian CSF spectrometer) and ESR spectrum was recorded (field set 3340 G, scan range 100G, microwave power level 10 mW and time constant 1 sec). The experiments were repeated twice or more to control the reproducibility of the tests, and also conducted in the absence of solids as the control tests. All the ESR measurements results are represented as the average of two experiments, and are expressed in arbitrary units (u.a). A signal of intensity of 1000 u.a corresponds aproximatively to 4.10^{18} spin L^{-1} (Zalma et al 1987). Crocidolite and vivianite were handground 3 minutes just before use to test the minerals with fresh surface.

RESULTS:

Effect of iron chelators on the reactivity of the minerals upon the A formation:* The results in figure 1 show that there is no significant oxidizing activity resulting in the A* formation with magnetite or hematite, either when the mineral is tested alone or in the presence of the different iron chelators used. On the other hand vivianite and crocidolite reveal a strong oxidizing activity, related to their easily accessible Fe(II). The A* formation decreases after addition of histidine, citrate, EDTA and NTA to crocidolite or vivianite, pointing to possible Fe(II) sites blocking-up by the chelators, preventing the accessibility of O_2 to these sites. The oxidizing activity of vivianite is about twice more importante than the one of crocidolite.

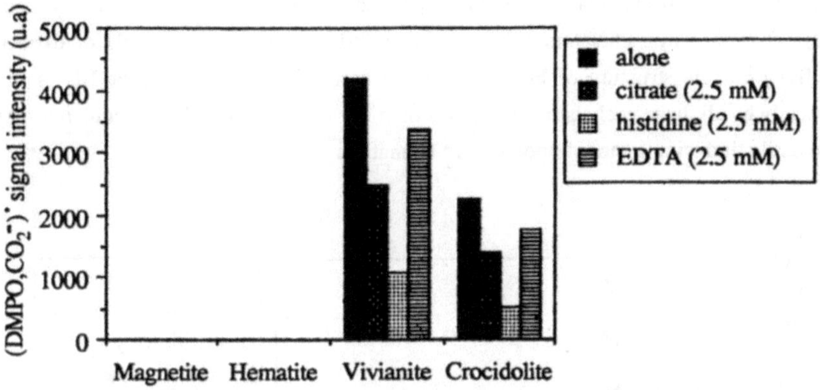

Figure 1: $(DMPO,CO_2^-)^{\bullet}$ signal intensity in buffered solution, with solid alone or in the presence of chelator.

Effect of reducing agents on the mineral reactivity upon the A* formation: The results in figure 2 show that the addition of ascorbate or cysteine to magnetite resulted in the appearance of an oxidizing activity, due to the formation of Fe(II) from the magnetite particle surface or in solution. There is a considerable decrease in the oxidizing activity of vivianite or crocidolite after addition of ascorbate or cysteine.

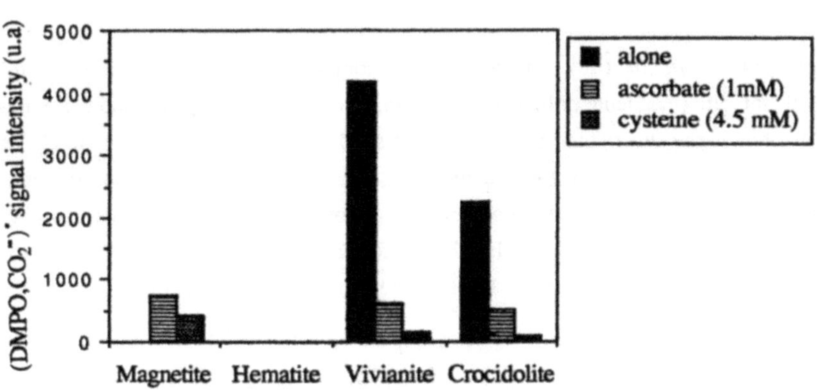

Figure 2: $(DMPO,CO_2^-)^{\bullet}$ signal intensity in a buffered solution, with solids alone or in the presence of reducing agent.

Effect of both iron chelator and reducing agent on the reactivity of the minerals upon the A* formation: The results presented in figure 3 show an increase of the oxidizing activity for magnetite and much more weakly for hematite in the presence of both iron chelator and reducing agent. Hematite purchases a significant oxidizing activity only after addition of EDTA and ascorbate. This suggest that EDTA which is a strong iron chelator can extract the Fe(III) from hematite, and this mobilized Fe(III) is afterward reduced by ascorbate to Fe(II). The other iron chelators NTA, histidine and citrate are not capable of extracting the iron(III) from the structure of hematite. In the case of magnetite, the Fe(III) is mobilized practically by all the iron chelators used, and then the iron(III) complexes reduction may occur more easily, inducing a more important A* formation.

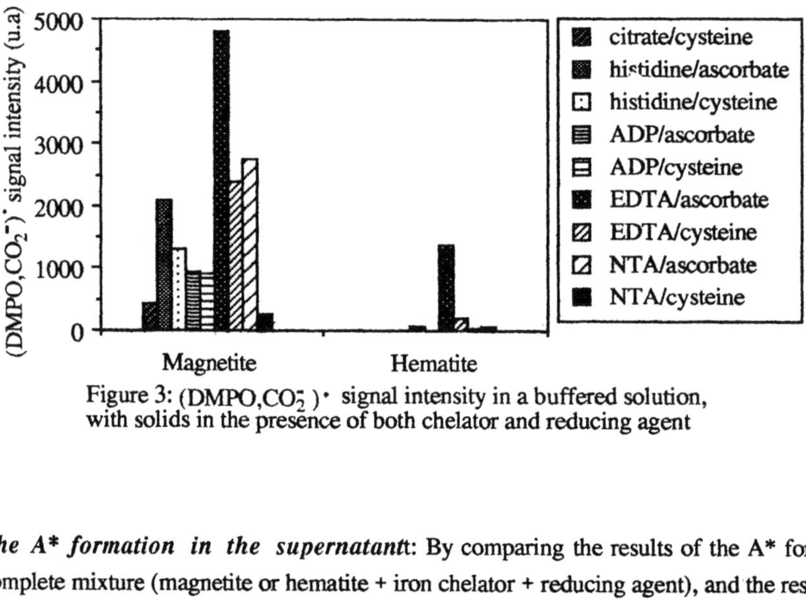

Figure 3: $(DMPO,CO_2^-)^\bullet$ signal intensity in a buffered solution, with solids in the presence of both chelator and reducing agent

***The A* formation in the supernatantt*:** By comparing the results of the A* formation in complete mixture (magnetite or hematite + iron chelator + reducing agent), and the results of the A* formation in the supernatant from which the solid -previously incubated with chelator- has been removed and at which we add reducing agent, DMPO and sodium formate, it was obseved an oxidizing activity of the supernatant almost the identical or higher than the one of the complete mixture; this suggests that the A* formation appears to be totally due to Fe(III) which was mobilized from the solid by iron chelators, then its transformation into Fe(II) by reducing agent.

DISCUSSION:

Hematite and magnetite are the two principal iron oxides playing a role as pollutants in many mines and industrial sites.Then, it is important to investigate their eventual role in oxidative stress processes after inhalation, in various conditions allowing to mimic the biological medium and to point out the appearance of very electrophilic species able to play a role of ultimate toxic agent.

It is known that the bonds in hematite structure are relatively strong, what explains its unsolubility and the difficulty to extract the Fe(III) cations, by complexing agents.The results quoted above show that essentially EDTA -a non biologic chelator- is able to release Fe(III) complexes in solution which is then reduced to Fe(II) complexe by ascorbate or cysteine. So, it is probable that hematite is not a potent agent able to participate in an oxidative stress process. But this does not exclude other roles in some mechanisms of toxicity, in relation with the capacity of hematite to adsorb selectively some biological molecules or organic pollutants, such as PAH (Saffiotti et al 1968).

Magnetite is not really a double oxide of Fe(III) and Fe(II). In its inverse spinel structure, the half of the Fe(III) are distributed among tetrahedral (A) sites, the second half of the Fe(III) and Fe(II) being distributed in octahedral (B) sites. It is known that there is a fast electron hopping between the Fe(III) and Fe(II) in next neighbours (B) sites (Jolivet et al 1990 and Allen et al 1982), making the iron cation in (B) sites with the valence 2.5+. Consequently there is no really Fe^{2+} in this oxide at room temperature, and correlatively in simple buffered solution, there is no possibility to obtain a redox potential for surface sites corresponding to the one of isolated Fe^{2+} cation. This is the principal reason for why this mineral is unable to allow the dioxygen reduction with the A* species formation.

Moreover it is also known that magnetite has not a stoechiometric composition, a part of its Fe(II) -particulary in the external sheets of the particles- being transformed into Fe(III). This evolution by oxidation involves a progressive structural transformation of the lattice toward the ones of maghemite, γ-Fe_2O_3, also with a spinel structure. Accordingly the magnetite particles are very rich in various faults, particulary in the sheets near of the surface of the particles. This high density of structural defects allows a partial accessibility of reducing agents to Fe(III) weakly bound, and the correlative formation of Fe(II) by reducing agents, necessary condition to the dioxygen activation and the A* species appearance.

In the absence of reducing molecules, vivianite and crocidolite -and not magnetite- are considerably more active to reduce dioxygen with the A* species formation, the reducing molecules being able to intervene to inhibite certain steps in the sequence of reaction beginning from the Fe(II) and ending to ferryl, perferryl or OH$^{\cdot}$ (A* species). This intervention is relevant to the antioxidant role of these reducing molecules (ascorbate and cysteine).

The last interesting situation is the one where magnetite is in the presence of complexing molecules (chelators) and reducing molecules. The high density of faults in the solid allows an easily release of Fe(III) in solution under the form of complexed Fe(III) which is then reduced to Fe(II) complexe able to react with dioxygen and to generate the A* species in solution. It is interesting to note that the four couples obtained with histidine or ADP and ascorbate or cysteine give significant results upon the A* formation, what proves the existence of possibilities for magnetite to generate an oxidative stress in biological medium. But it is also necessary to remark that in increasing the number of necessary conditions, it appears correlatively a weaker probability to acceed at the global oxidizing effect, more easily obtained with compounds able to present Fe(II) species at the interface solid-liquid.

REFERENCES:

Allen, G.C., Tucker, P.M. and Wild, P.K. (1982) Caracterization of iron/oxygen surface reactions by X-ray photoelectron spectroscopy. Philos. Mag. B. 46:411

Bielski, B.H.J. (1992) Reactivity of hypervalent iron with biological compounds. Ann. Neurol. 32:28-32

Fournier, J., Guignard, J., Nejjari, A., Zalma, R. and Pezerat, H. (1991) The role of iron in the redox surface activity of fibers. Relation to carcinogenicity. In "Mechanisms in Fiber Carcinogenesis" (Brown R.C. ed.) Plenum Press, New York.: 407-414

IARC Monographs (1984): On the evolution of carcinogenic risk to humans: Polynuclear aromatic compounds, part 3: Industrial exposures in aluminium production, coal gasification, coke production, and iron and steel founding. Lyon. France. 34:133-190

IARC Monographs (1990): On the evolution of carcinogenic risk to humans: Chromium, nickel and welding. Lyon. France. 49:447-526

Jolivet, J.P., Tronc, E., Barbe, C. and Livage, J. (1990) Interfacial electron transfer in colloidal spinel iron oxide: Silver ion reduction in aqueous medium. J. Colloid. Interface. Sc. 138:2:465-472

Mur, J.M., Meyer-Bisch, C., Pham, Q.T.,Massin, N., Moulin, J.J., Cavelier, C. and Sadoul, P. (1987) Risk of lung cancer among iron ore miners. J. Occupat. Med. 29:9762-768

Pezerat, H. (1991) The surface activity of mineral dusts and the process of oxidative stress. In " Mechanisms in Fiber Carcinogenesis". (Brown R.C. ed.) Plenum Press, New York. 387-395

Saffiotti, U., Cefis, F. and Kolb, L.H. (1968) A method for the experimental induction of bronchogenic carcinoma. Cancer Research. 28:104-124

Yamazaki, I. and Piette, L.H. (1990) ESR spin-trapping studies on the reaction of Fe^{2+} ions with H_2O_2-reactive species in oxygen toxicity in biology. J. Biol. Chem. 265:23:13589-13594

Zalma, R., Bonneau, L., Guignard, J., Pezerat, H. and Jaurand, M.C. (1987) Formation of oxy radicals by oxygen reduction arising from the surface activity of asbestos. Can. J. Chem. 65:2338-2341

Kinetics of Asbestos-Induced Haemolysis of Human Red Blood Cells

Robert Englman
Soreq N.R.C.
Yavne 70600
Israel

Encounters between chrysotile fibres and red blood cells (RBC) lead to the haemolysis of the latter and to the loss of capacity of the former to further damage the cells. Jaurand et al.'s *in vitro* experiments (1979) yield the evolution of concentrations of healthy RBC and of active chrysotile fibres. When it is assumed that once-reacted RBC are completely deactivated , the rate equations are solvable and show the observed concentrations. Moreover, they show precisely the experimentally observed linear relation beween the inverses of maximal haemolysis and of the ratios of fibre/RBC surfaxes. When the assumption of complete RBC deactivation is rescinded, systematic deviations from the data are found. Observation of the collision fingerprints on the fibres, e.g. by atomic force microscopes, is suggested.
Pre-incubation of fibres with RBC or liposomes leads to the inhibition of haemolysis. Here also, the proposed deactivation model accounts for the observed data.

1. Introduction

Biological damage casued in vitro by asbestos fibres has been studied for some decades (Mcnab and Harington 1967, Schnitzer and Pundsack 1970, Harington et al 1971, Beck et al. 1972, Light and Wei 1977, Iguchi and Shoshuka 1989, Benedek and Brody 1990). A review of the haemolytic activities of silicate minerals was given by Oscarson et al. (1986) and the molecular and environmental implications have been summarized at a meeting (Bignon et al. 1989, Jaurand 1989). It is known that asbestos is cytotoxic, carcinogenetic and haemolytic, but the relation betrween these manifestations is subject to debate. Correlations were stated to have been observed between the haemolytic activity of asbestos fibres and their cytotoxicities for macrophages in culture (Harington et al. 1971). Agents having symmetric

NATO ASI Series, Vol. H 85
Cellular and Molecular Effects
of Mineral and Synthetic Dusts and Fibres
Edited by J. M. G. Davis and M.-C. Jaurand

shapes like SiO_2, TiO_2 show fibrogenic activity that correlates positively with their haemolytic and cytotoxic effects. However, it is nor established whether and to what extent these properties are connected with the formation *in vivo* of malignancies by fibres.

It is nevertheless clear that, though these efects may be different, asbestos-induced haemolysis is a convenient and fast method to acquire data, through which to study the actions of mineral fibres and dust on biological membranes and to seek correlations between these effects (Beck et al. 1972). A source of kinetic data are the haemolysis studies of Jaurand et al. (1979), which will be interpreted in this paper with a theoretical model. Such data might also be useful to elucidate the early stages of haemolytic activities mediated by agents other than fibres (Malinski and Nelsestuen 1989, Burt et al. 1990). In fact, the special type of pores created in erythrocytes by the membrane attack complex of complement and by T-lymphocytes was the original motive for my undertaking the present investigation. [See Berke (1989).]

The phenomenology of asbestos-caused haemolysis relates to the sizes and times involved in the fibre-cell encounters. Lysis increases with fibre size to the extent that some researchers see no role for chemistry (Oscarson et al. 1986), a view disputed by others (Trosic et al. 1986). Most effective fibre dimensions are 4-8 microns in length and 0.1-1 microns across (Bignon et al. 1989). The absence of carcinogenesis by short fibres can be attributed to their engulfment by macrophages.

The back-effect on the fibres is no less remarkable. It is known that they become deactivated after contact with RBC, liposomes and a broad class of inhibitors (Jaurand et al. 1979). This effect resides in the adsorption at the fibres of glycproteins or glycolipids that stall the damage mechanism. Direct observation of glass or siliconized glass beads after flow in a suspension of washed human RBC revealed the presence of membrane proteins and of filamentous deposits (Borenstein and Brash 1986). Parsegian and Gingell (1980) proposed a tether-type erythrocyte-mineral imteraction by which part of the membrane (including integral membrane proteins) are extruded and attached to the glass surface. The area involved in the contacts is about 5 microns in diameters and appears to undergo, at the interaction stage, conformational and other changes at the molecular and elctronic levels . Interaction distances between cell are estimated, on the basis of free energy considerations, as ranging between 10 and 100 nm, depending on the ionic strength (Gingell 1990).

In this work we apply a rate equation method to the deactivation of chrysotile fibres upon interaction with human RBC. The nonlinear differential equations are known from Volterra's time (Chapman 1931), and are exactly solvable under the conditions extant in the experiments of Jaurand et al. (1979).

2. Reactant Concentrations

In the non-linear rate equations for the reaction kinetics the concentrations of the healthy red blood cells (RBC) and the active fibres (AF) are represented by B=[RBC], A=[AF] . With rate constants **a** and **b**, whose values will be determined from experimental results, the rate equations are:

(1) $dA/dt = -\mathbf{a}\, B.A$

(2) $dB/dt = -\mathbf{b}\, B.A$

The boundary values for $A=A(t)$ and $B=B(t)$ are: at t=0, $B=B_0$ and $A=A_0$.
At large times $B=B_\infty > 0$, $A \sim 0$.

The meaning of the rate equations is that the numbers of both the healthy RBC and the active fibres diminish due to randomly occurring conjunctions (collisions) between them , whereas meetings wih damaged cells (ghosts) or with inactivated fibres do not affect their numbers. We shall later modify this assumption, to test the sensitivity of the results to it.

From (1) and (2)

(3) $dA/dt - (\mathbf{a}/\mathbf{b})\, dB/dt=0$

whose solution is

(4) $A(t)= (\mathbf{a}/\mathbf{b})[B(t) - B_\infty]$

This can be substituted into (1) to give

(5) $d(B/B_\infty)/dt = -a\, (B/B_\infty)\, (B/B_\infty -1)$

(6) $a = \mathbf{a}B_\infty$

The solution is (Chapman 1931)

(7) $B(t) = B\ [\, 1- (\mathbf{b}/\mathbf{a})(A\ /B\ )]\, [\, 1- (\mathbf{b}/\mathbf{a})(A\ /B\ )e^{-at}]^{-1}$

(8) $A(t) = (A\ /B\ )\, B(t)\, exp(-at)$

The number of surviving cells is

(9) $B_\infty = B\ [\, 1-(\mathbf{b}/\mathbf{a})(A_0/B_0)]$

This relation shows the dependence of B_∞ on the kinetic parameters **a** and **b**.

The ultimate fractional decrease of the healthy RBC is

(10) $m_\infty = (1- B_\infty/B_O)$

One finds then the following relations.

(11) $B(t) = B_\infty (1- m_\infty e^{-at})^{-1}$

(12) $A_O = (\mathbf{a/b})(B_O - B_\infty)$

$A_O/B_\infty = (\mathbf{a/b}) [(B_O/B_\infty) - 1]$

(13) $= (\mathbf{a/b}) m_\infty /(1- m_\infty)$

The last relation is rewritten in terms of the ratio R of the chrysotile and RBC surfaces, defined by Jaurand et al. (1979) and given by

(14) $R = r_S (A_O/B_\infty)$

where r_S is the ratio of the surfaces of a typical chrysotile fibre and of a RBC. From (13) and (14) one obtains the remarkably simple relationship

(15) $m_\infty^{-1} = (\mathbf{a/b}) r_S R^{-1} + 1$

Such linear relationship was discovered experimentally by Jaurand et al. (1979) between the inverses of m_∞ (in their terminology: M, where M is $100m_\infty$) and of R.

3. Comparison with Experiment.

The curves shown in Figure 1 are the the haemolysis percentage

$100 m(t) = 100 [1 - B(t)/B_O]$

where $B(t)$ is the theoretical result given in (7), plotted against time for five values of the surface ratio R defined in (14), superimposed on experimental results (dots). The shape of the curves tallies also with the data of other investigators, e.g. of Burt et al. (1990), obtained from haemolysis following monosodium urate-RBC interaction.

To test the validity of Eq. (15) we have replotted in Figure 2 the M^{-1} vs R^{-1} data of Jaurand et al. (1979) (replacing their plot that contained M in excess of 100%), and fitted the linear regression of

(16) $m_\infty^{-1} = 0.35 R^{-1} + 0.88$

to the data. A non-linear fit of the form

(17) $m_\infty^{-1} = 0.28 R^{-1} + 0.15 R^{-2} + 0.93$

did not reduce the error (the sum of the squares of the residuals) in the fit. Therefore the existence of the linear relationship predicted by our model was

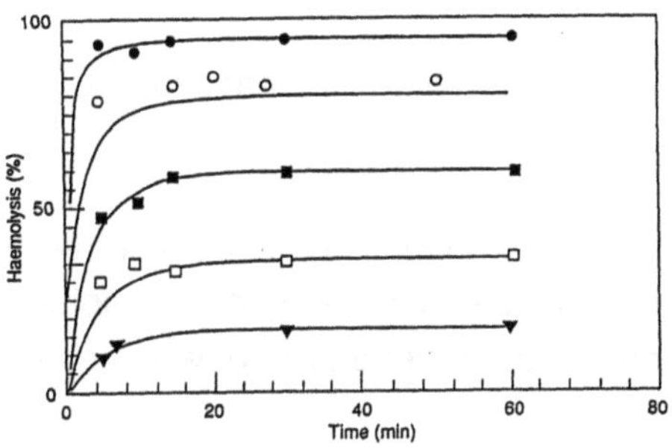

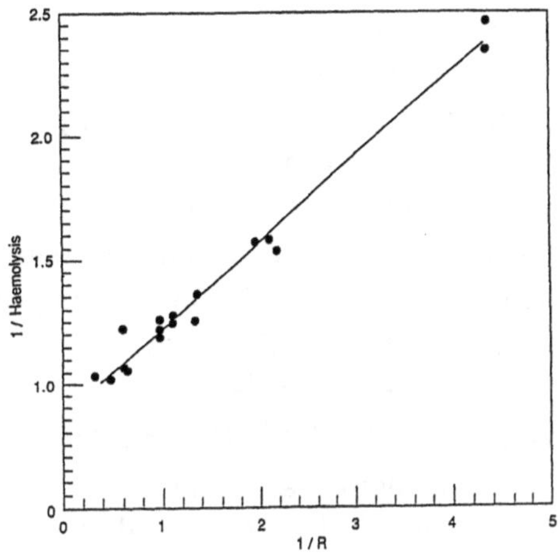

Fig. 1: Haemolysis curves from the model for five values of the ratio R, defined in (14). Dots are from experiments of Jaurand et al. (1979)

Fig. 2: Inverse maximal haemolysis m_∞^{-1} against R^{-1}. The line is the least square fit to the observational points, satisfying eq. (16)

taken as justified by experiments. [The relative standard error was calculated as .012. The discrepancy between the estimated (0.88) and predicted (1) intercepts is ascribed to some bias or to an additional mechanism outside the model.] The coefficient (0.35) of the term R^{-1} leads to an estimate

(18) $\quad a/b = 1.7$

for the ratio of the rate constants. [This was obtained from estimates for the surface density of the chysotile fibres (25 m^2/g), density of asbestos (2.5 g/cm^3), typical dimension of fibres (5x0.25x0.25 cube-microns) ,leading to surface area of 19 square-microns; the ratio of this to the RBC area (92 square microns) gives r_S =0.2.]

Membrane Inactivation

Perhaps the most remarkable conclusion from the data and their analysis is the realisation that a damaged RBC membrane fully loses its capacity to inactivate any further fibres after a previous collision. The molecular implicaions of this effect appear to be important, but have not been properly studied. The reverse effect of the fibres losing their haemolytic activity after contact with the RBC may well be as remarkable, but the plain fact of haemolysis saturation (namely, that no further RBC-damage occurs after inactivation of fibres) leaves no choice for this.

To investigate the consequences of relaxing the full membrane inactivation assumption, we replace eq. (2) by

(19) $dA/dt = -\mathbf{a}\, A\, [(1-q)B + qB_O]$

so that q represents the relative efficacy of damaged RBC, whose concentration is $B - B_O$, in deactivating asbestos fibres. (1) and (19) have analytic solutions for q=0 (treated in the previous section). In the other extreme when q=1, the form of the solution is as follows:

(20) $A = A_O\, e^{-\mathbf{a}B_O t}$

(21) $B = B_O \exp[-(\mathbf{b}/\mathbf{a})\, (A_O/B_O)\, (1 - e^{-\mathbf{a}B_O t})$

The final haemolysis ratio is

(22) $m_\infty = 1 - \exp(-\mathbf{b}A_O/\mathbf{a}A_O)$

(23) $= 1 - \exp[-\mathbf{b}R(1 - m_\infty)/\mathbf{a}r_S)$

which yields m_∞ as an implicit function of R. To simplify, we replace R by

(24) $\mathbf{R} = R\mathbf{b}/\mathbf{a}r_S$

and obtain by a perturbational expansion of (23), valid for small m_∞ and $\mathbf{R}$,

(25) $m_\infty^{-1} = \mathbf{R}^{-1} + 1.5 - (5/12)\, \mathbf{R} + O(\mathbf{R}^2)$

Comparing this with the theoretical relation in (15), obtained for q=0, we note that in the above the first term on the right is the same, the second (constant) term is larger and there are additional terms resulting in a negative (downward) curvature. We recall that the experimental results in Fig. 2 have shown a (constant) intercept term that is close to but *less* than 1, and possibly a very slight positive curvature. All these show that the model in this section agrees with experiments *worse* than the complete inactivation model of the previous section. Note, however that the experimental results have to be checked for accuracy before basing on them far-reaching conclusions.

One expects also that the intermediate cases of q, between 0 and 1, will also show worse fits than the complete deactivation case. Numerical integration of the rate equations confirms this, though the differences are slight. Figure 3 shows m_∞^{-1} as function of $\mathbf{R}^{-1}$ (for a rather wider range of values than in Fig.2), for a range of values of the relative effectiveness parameter q.

4. Haemolysis Inhibition

Crystalline and mineral cell killers are known to lose their cytolytic effect by pre-incubation with several types of inhibitors, such as Al, inorganic and organic salts, liposomes RBC, glutaraldehyde, neuraminidase. The procedure involves the bringing into contact of the cell killer with the inhibitor for some time (that may minutes or about an hour) and the the addition of RBC, whose lysis is then measured by the usual methods. The quantity of interest is the reduction of the final haemolysis as function of the concentration of the inhibitor in the pretreatment. Though it is not absolutely evident that the same mechanism operates between the fibre and the inhibitor at the first stage, as it does between the fibre and the RBC at the second stage, we shall assume this to be the case. We shall therefore use our model twice: first for describing the reduction of the cell killer (asbestos) concentration $A(t)$ due to its pretreatment with the inhibitor (concentration $I(t)$, initial concentration I_0) and then, using the value $A(t_p)$ for the cell concentration at the end ($t=t_p$) of the pretreatment as the initial value A_0 for the kinetics of of the haemolysis between $A(t)$ and $B(t)$ (these concentration having meanings as before). In consequence, the final value B_∞ or the haemolysis m_∞ will depend on the initial inhibitor concentration I_0 through the dependence of $A(t_p)=A_0$ on I_0.
Rather than giving the mathematics, we show in Figure 4 the relative reduction of the haemolysis as function of the inhibitor concentration I_0. The ratio of rate constants ***b/a*** was taken from fitting the haemolysis results in our earlier section. The agreement between the calculated curves and the data points of Jaurand et al. (1979) is only fair, but is sufficient to leave the theory credible.

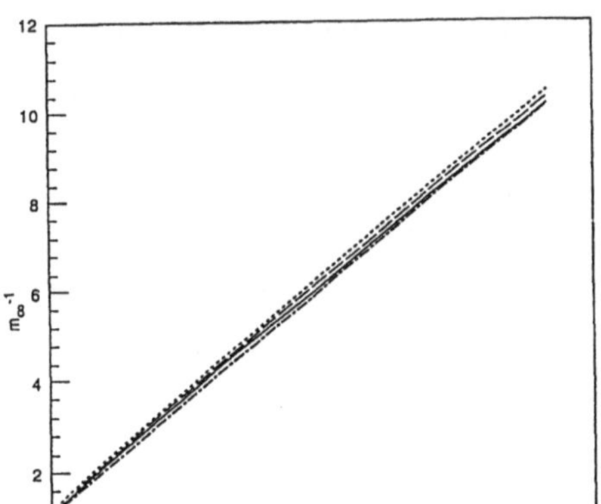

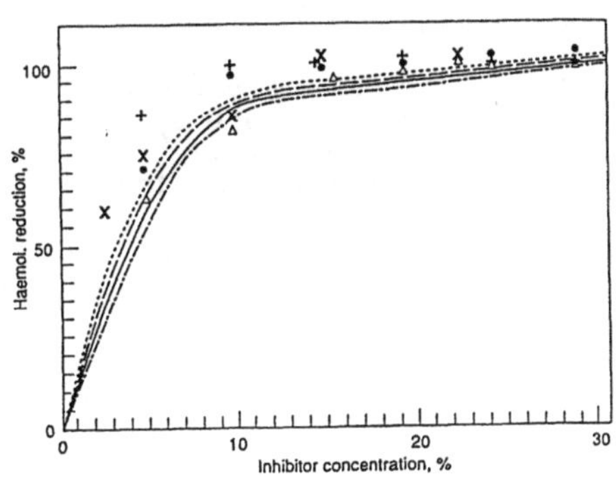

Fig.4: Haemolysis reduction *vs* I_0.
Data for R:+=.2, •= .3, x= .45, = .6

Fig.3: Maximal haemolysis vs the surface ratio for *q* (the damage effectiveness of once colllided RBC)= 1,.75,.5,.25 plotted in decreasing order of m_∞^{-1}

5. Conclusion

Our results have highlighted the link between RBC membrane lysis (involving pore formation through aggregation and lateral motion of integral membrane proteins) and asbestos-inactivation (arising from transfer of some proteins to the fibre) indicating that they are causalllly connected and simultaneous. Estimated interaction distances are about 0.1 micron (Gingell 1990). We suggest that the fingerprints of RBC-fibre collisions left on the latter (presumably in the form of protein islands) be viewed by atomic force microscope of fibres, in a manner that is easier than observing the cells.

Research supported by Israel Ministry of Science and Technology and by the Israel Cancer Research Fund.

References

Beck EG, Holt PF, Manojlovic N (1972) Comparison of effects of macrophage cultures of glass fibre, glass powder and chrysotile asbestos. British J Ind Med 29: 280-6
Benedek EG, Brody AR (1990) Changes in lipid ordering of phospholipid membranes treated with chrysotile and crocidolite fibres. Env Res 53: 152-67
Berke G (1989) in Fundamental Immunology. Raven Press New York: 753-64
Bignon J, Peto J, Saracci R (eds) (1989) Non-occupational exposure to mineral fibres. IARC Sci Publ Lyon
Borenstein N, Brash JL (1986) Red blood cells deposit membrane components on contacting surfaces. J Biomed Mat Res 20: 723-30.
Burt HM, Jackson JK, Kim KJ (1990) Role of membrane proteins in monosodium urate cystal-membrane interaction. J Rheumatology 17: 1353-8.
Chapman RN (1931) Animal Ecology. McGraw-Hill NewYork
Gingell D (1986) Cell contact with solid surfaces. in Biophysics of the cell surface (Glaser R, Gingell D, eds) Springer Berlin: 264-85.
Harington JS, Miller K, Mcnab G (1971) Haemolysis by asbestos. Environ Res 4: 95-117.
Iguchi H, Shoshuka K (1989)Possible generation of H_2O_2 and lipid peroxidation of erythrocyte membrane by asbestos. Biochem Inf 28: 981-90.
Jaurand MC (1989) Particulate-state carcinogenesis in non-occupational exposure to mineral fibres. In Bignon et al. (eds) ibid
Jaurand MC,Magne L, Bignon J (1979) Inhibition by phospholipids of haemolytic action of asbestos. British J Ind Med 36: 113-6
Light WG, Wei ET (1977) Surface charges and haemolytic activity of asbestos. Environ Res 13: 135-45
Macnab G, Harington JS (1967) Haemolytic activity of asbests and other mineral dusts. Nature 214: 522-3.
Malinski JA, Nelsestuen GL (1989) Membrane permeability to macromolecules mediated by the membrane attack complex. Biochem 28: 61-70.
Oscarson DW, van Scoyoc GE, Ahlrichs JC (1986) Lysis of erithrocytes by silicate minerals, clay and clay minerals. 34: 74-80
Parsegian NA, Gingell D (1980) Red blood celll adhesion J Cell Sci 41: 151-7
Schnitzer RJ, Pundsack FL (1970) Asbestos haemolysis. Environ Res 3: 1-13.
Trosic I, Horvat D, Racis J (1986) The effect of asbestos. Per Biol 88:269-76

Oxidising Activity of Some Metallic Dusts in Relation to Oxidative Stress Mechanisms

Jeanine Fournier, Anna Jadrijevic, Abdelkader Nejjari and Henri Pezerat
Université Pierre et Marie Curie
Laboratoire de Réactivité de Surface - CNRS URA 1106
4 place Jussieu
75252 Paris Cédex 05
France

The aim of this study is to research and to identify inorganic metallic compounds able to induce an oxidative stress, and then to contribute to a better understanding of epidemiological, in vivo and in vitro data.
Our toxicochemical approach detailed in Pezerat's article (this volume) is based on the study of the possible relations between the development of redox surface properties of inorganic compounds in aqueous medium and the toxic activity of these compounds during their metabolisation. This mechanistic approach requieres to research activated oxygen species (AOS) generated by these dusts after interaction with molecular oxygen dissolved in aqueous medium. Among oxidising AOS set, we have distinguished two subsets of electrophilic entities symbolized A* and P* respectively (Pezerat, 1991; Fournier et al., 1991). A* species are highly electrophilic entities, such as metal oxo, able to attack a great variety of biological macomolecules (proteins, ADN, etc). P* species are less oxidizing than A* and are only capable of triggering the polyunsaturated lipid peroxidation.
We have studied the capacity of some metals (Ni°, Cu°, Zn°, Fe° and Cr°) and also of two Ni compounds having a metallic character (Ni_3S_2 and NiAs) to generate A* and P* species. In the field of the peroxidation, we have used three complementary methods to reveal the presence of P* entities produced from reaction of metallic compounds with dioxygen.

INORGANIC COMPOUNDS AND METHODS

The inorganic compounds were commercial materials in powder: Ni° (Alpha; size 1 μm); Cu° (Alpha; size 10 μm); Cr° (Aldrich; size 50 μm); Fe° (Heraeus; size 10 μm); Zn° (Merck; size < 60 μm); Ni_3S_2 (INCO) and NiAs (K&K Laboratories) having an important granulometry (> 100 μm). Only the two last compounds have been handground (2 min.) and then conserved under argon atmosphere.

NATO ASI Series, Vol. H 85
Cellular and Molecular Effects
of Mineral and Synthetic Dusts and Fibres
Edited by J. M. G. Davis and M.-C. Jaurand

Formate Oxidation (Test A*): The experimental procedure of this test was previously described in detail (Zalma et al., 1987; Costa et al., 1989). The reaction mixture is a suspension of particles (45 mg) in 2 ml of 25 mM aerated phosphate buffer (pH 7.4) containing 1 M sodium formate and 50 mM of DMPO (5-5'-dimethyl-1-pyrroline-N-oxide) as a spin trap agent. After 25 and 60 minutes of incubation (37°C), aliquots of the suspension were withdrawn and filtered. Detection of radical adducts $(DMPO,CO_2^-)^\bullet$ was performed by EPR. The results are given as the mean of the $(DMPO,CO_2^-)^\bullet$ signal intensities after the two incubation times. The samples were considered as active when the signal intensity was higher than 300 a.u (arbitrary units).

Test of the Linolenic Acid Peroxidation according to Three Methods:

- **Dioxygen Consumption:** Inorganic solid was added to oxygenated phosphate buffer (25 mM), then linolenic acid (99%, Sigma) was introduced (1 mM) in the reactor thermostated at 37°C. The oxygen consumption was measured by a specific O_2 electrode. This measurement was monitored in function of the incubation time. Results are expressed as kinetics curves (consumed oxygen concentration in function of the time).

-**Determination of MDA (malonaldehyde):** The experimental method was close to the ones described by Ottolenghi (1959) and Gavino et al. (1981). After 1 and 18 hours of incubation at 37°C, the formation of MDA from the suspension (inorganic solid in aerated and buffered linolenic acid solution, 1 mM) is detected after reaction with thiobarbituric acid (TBA) to form a pink chromogen. This chromogen was extracted with butanol and was measured by Visible spectroscopy at 535 nm.

-**Determination of conjugated dienes and trienes:** Reaction mixtures were close to the one described above. After an incubation for different times (30, 60, 1020 and 2880 minutes) at 37°C, conjugated dienes and trienes were extracted from the suspension with butanol. The organic phase was analysed by UV spectroscopy. The absorbance measurements of conjugated dienes and trienes were performed at 235 and 268 nm respectively. The results are given as an average of four assays, with the corresponding confidence interval.

For all these tests, control tests were performed without solid particles.

RESULTS

Mineral-Induced Formate Oxidation:

Table I shows the results of the formate oxidation by the minerals, which are expressed as the intensity of the $(DMPO,CO_2^-)^\bullet$ signal with their confidence interval. The oxidising activity of these inorganic compounds to generate highly electrophilic species A* is related to the intensity of the radical adduct $(DMPO,CO_2^-)^\bullet$ EPR signal.

Sample	$(DMPO,CO_2^-)^\bullet$ Signal Intensity
Ni_3S_2	1070 (100)
NiAs	900 (50)
Fe°	480 (80)
Cu°	450 (90)
Ni°	310 (30)
Zn°	90 (10)
Cr°	0-10
Control	0-10

Table I: Mean Oxidising Activity of Metallic Compounds in Formate Oxidation

Mineral-Induced Linolenic Acid Peroxidation:

Kinetics curves (Figure I) reveals the consumption of dioxygen during linolenic acid peroxidation in the presence of some inorganic solids.

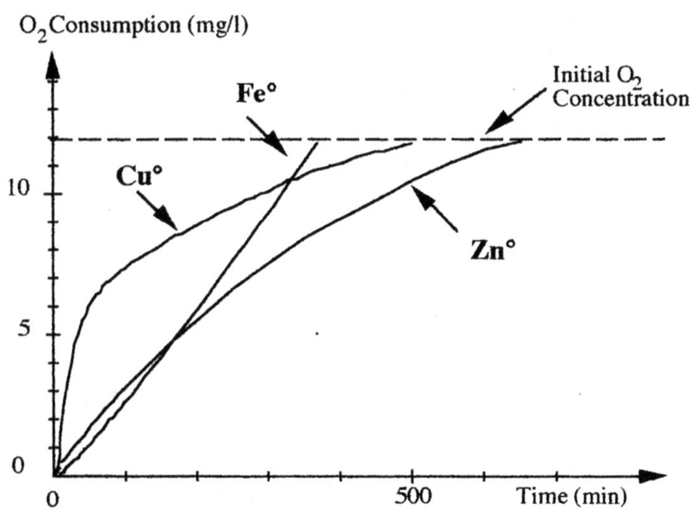

Figure I: O_2 Consumption in the presence of Metallic Compounds and Linolenic acid

Only the results obtained with Cu°, Fe° and Zn° are significant by comparison with the control data (linolenic acid alone, inorganic compound alone). The oxygen consumption kinetics with Cu° is faster than those with Fe° and with Zn°.

For the other metals and for the Ni compounds with metallic character, the dioxygen consumption in the presence of linolenic acid was either almost at the same level (Ni°, Ni_3S_2) or lower (Cr°, NiAs) than the consumption obtained with the controls.

In the detection of MDA, it is not possible to measure the pink chromogen with all metals: the organic phase presented different colors according to the nature of metallic compounds. However, with the Ni compounds with metallic character, the quantity of formed MDA is more important in the case of Ni_3S_2 (7.5 10^{-3}mM) than in the case of NiAs (3 10^{-3}mM) after 1hour of incubation. After 18 hours of incubation the oxidising activity from NiAs is no more high (3.5 10^{-3}mM) while for Ni_3S_2 the colour of the organic phase became yellow, characteristic colour of the predominant presence of monoaldehydes as degradation products from the peroxidation (Marcuse and Johansson, 1973).

In the measurement of conjugated dienes and trienes in the presence of metallic compounds, Figures II and III shows their absorbance evolution in function of the incubation time.

Globally Cu° appears as very active among all metals, with a fast kinetics of conjugated dienes and trienes formation. Except for Fe° which is inactive in the studied times, other metals are active but with a kinetics much slower by comparison to copper. If the formation of dienes and trienes for Zn°, Ni° and Cr° increases with the time, it appears for Cu°, from 60 minutes, a decreasing of dienes quantity while the quantity of trienes increases progressively. In the case of the two Ni compounds with metallic character, Ni_3S_2 (more active than NiAs) presents a kinetics of dienes and trienes formation close to the ones observed with Zn°, Cr° and Ni°.

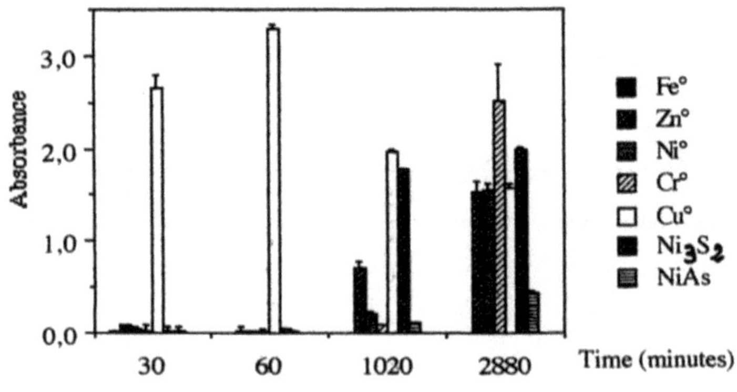

Figure II: Formation of Conjugated Dienes in function of Incubation Time

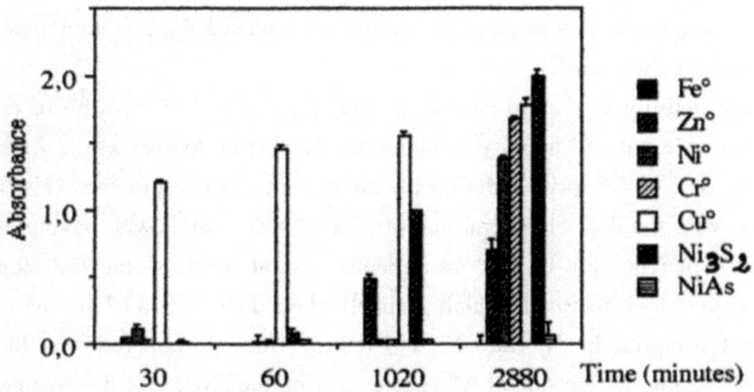

Figure III: Formation of Conjugated Trienes in function of Incubation Time

DISCUSSION

In the test of formate oxidation which allows to appreciate the formation of A* species, we can distinguish two groups of compounds: the inactive materials with Cr° and Zn°; the active materials with Fe°, Cu°, Ni°, NiAs and Ni_3S_2.

It appears much more difficult to appreciate the formation of P* species, less electrophilic than A* entities. Indeed the two types of species are able to trigger the lipid peroxidation and, in the presence of A* species it is difficult to know if P* species participate in the peroxidation process.

The most simple case is the one of Cr° and Zn° which are not capable of producing A* species in significant quantity. The formation of conjugated dienes and trienes from linolenic acid, with a kinetics faster with Zn° than with Cr° (this observation is confirmed by the oxymetry results), shows that these two metals can be implicated in an oxidative stress. It is possible to hypothesize that this reactivity, with a latence time inferior to one day, is in relation to appearance of metal fume fever after inhalation of fresh metal fumes (Vogelmeier et al., 1987). The latence time is probably related to slow dissolution or reorganization of the oxidised coating on the surface of particles.

Among the compounds considered as active to generate A* species, two appear as inactive to induce the formation of conjugated dienes and trienes: NiAs and Fe°. However NiAs is active to produce MDA from linolenic acid peroxidation. The surface oxidation of NiAs conducts probably to the production of arsenite and/or arsenate entities which may be interfere in the peroxidation process. With respect to Fe° it is probable that the formation of A* species brings about a peroxidation process, but for an unknown reason there is no a appearance of the conjugated dienes and trienes.

The considerable activity of Cu° in oxygen consumption and in formation of conjugated dienes and trienes, when its A* activity is moderate, allows to hypothesize the appearance of P* species by the side of A* species, during the dioxygen reduction process. This fact would be in agreement with the data of several authors (Eberhardt et al., 1988) having noted that, during the dioxygen reduction by Cu^{+} in aqueous neutral medium, the last step was the formation of a copper hydroperoxide which probably plays the role of a P* species.

For the two last compounds (Ni° and Ni_3S_2) it is difficult to conclude on the possible appearance of P* species by the side of A* species, during the process of dioxygen reduction. From our results it appears that it is very difficult, in absence of complementary experimental methods involving the detection of different degradation products (hydroperoxides, monoaldehydes, volatile products, ...), to appreciate the importance of the lipid peroxidation process.

Although data on the size distribution of the particles, our results bring about some evidence that all the above metallic compounds with a metallic character are able to react with dioxygen to produce AOS. So, after inhalation probably all these compounds are capable of participating to some oxidative stress processes.

REFERENCES

Costa D, Guignard J, Zalma R, Pezerat H (1989) Production of free radicals arising from the surface activity of minerals and oxygen. Part I. Iron mines ores. Toxicol Ind Health 5: 1061-1078

Eberhardt MK, Colina R, Soto K (1988) The reaction of Cu^{+}-O_2 and Cu^{2+} - ascorbic acid - O_2 with dimethyl sulfoxide. J Org Chem 53: 1074-1077

Fournier J, Guignard J, Nejjari A, Zalma R, Pezerat H (1991) The role of iron in the redox surface activity of fibers. Relation to carcinogenicity. In Mechanisms in fibre carcinogenisis. Brown RC et al. eds Plenum Press New York 23: 407-414

Gavino VC, Miller JS, Ikharebha SO, Milo EG, Cornwell DG (1981) Effect of polyunsaturated fatty acids and antioxidants on lipid peroxidation in tissue cultures. J Lipid Res 22: 763-769

Marcuse R, Johansson L (1973) Studies on the TBA test for rancidity grading: II. TBA reactivity of different aldehyde classes. J Am Oil Chemist's Soc 50: 387-391

Ottolenghi A (1959) Interaction of ascorbic acid and mitochondrial lipids. Arch Biochem Biophysics 79: 355-363

Pezerat H (1991) The surface activity of mineral dusts and the process of oxidative stress. In Mechanisms in fibre carcinogenisis. Brown RC et al. eds Plenum Press New York 23: 387-405

Vogelmeier C, König G, Bencze K, Fruhhmann G (1987) Pulmonary involvement in zinc fume fever Chest 92: 946-948

Zalma R, Bonneau L, Jaurand M-C, Guignard J, Pezerat H (1987) Formation of oxy-radicals by oxygen reduction arising from the surface activity of asbestos. Can J Chem 65: 2338-2341

IDENTIFICATION AND QUANTIFICATION OF FIBRE–INDUCED LIPID PEROXIDATION PRODUCTS BY GC–MSD

M. Gulumian[1,2], K. Ambaram[1], and G.H. Verdoon[3]

[1] National Centre for Occupational Health
P O Box 4788
Johannesburg 2000

[2] Department of Medical Biochemistry
University of the Witwatersrand
P O Wits 2050

[3] Department of Chemistry and Biochemistry
Rand Afrikaans University
P O Box 524
Aukland Park 2006
South Africa

INTRODUCTION

It is now well documented in the literature that mineral dusts and fibres induce the peroxidation of lipids present in many biological samples. The samples tested include among others, erythrocytes (Chvapil et al., 1976), subcellular fractions of silicotic lungs (Gupta and Kaw, 1982), lung homogenates of silica–exposed rats (Zsoldos et al., 1983), microsomal fractions of rat lung (Gulumian et al., 1983) and liver (Fontecave et al., 1987), liposomes (Weitzman and Weitberg, 1985) macrophages (Goodglick et al., 1989), and fibroblast cell cultures (Iguchi et al., 1993).

These reports have all employed the malonaldehyde–thiobarbituric acid (MDA–TBA) assay system to monitor lipid peroxidation. The latter is a complex process (Gardner, 1989) and MDA is only one of the several aldehydic end products obtained from the decomposition of lipid hydroperoxides. Moreover, oxidative injury of molecules other than lipids also promotes MDA formation (Gutteridge, 1979; Bucknall et al., 1978). For these and other reasons, Janero (1990) has concluded that the measurement of different classes of lipid peroxidation products is not only desirable but is virtually mandatory.

Methodologies which have so far been successfully employed by different authors to detect individual or complete spectrum of primary and secondary products of lipid peroxidation include HPLC (Esterbauer et al., 1982; Benedetti et al., 1984; Curzio et al., 1986; Lang et al., 1985; Yoshino et al., 1986; Esterbauer et al., 1984), TLC (Benedetti et al., 1982),

NATO ASI Series, Vol. H 85
Cellular and Molecular Effects
of Mineral and Synthetic Dusts and Fibres
Edited by J. M. G. Davis and M.-C. Jaurand

and GC–MSD (Van Kuijk et al., 1986; Norsten–Höög and Cronholm, 1990). Esterbauer and Zollner (1989) and Smith and Anderson (1987) have however agreed that GC–MSD might be the most promising procedure which can be employed with great confidence to detect the products especially the aldehydes from lipid peroxidation.

The present study, has therefore employed GC–MSD to identify and quantify the peroxidation products from arachidonic acid induced by three different asbestos fibres namely, crocidolite, amosite, and chrysotile.

MATERIALS AND METHODS

Hexanal, *trans*–2–hexenal, pentanal, and arachidonic acid were purchased from Sigma Chemical Co.. Butanol, pentanol, and sodium borohydride ($NaBH_4$) came from BDH. Dichloromethane and N,O–Bis(trimethylsilyl)tri–fluoroacetamide (BSTFA) were obtained from Merck and hexanol from Saarchem. Crocidolite, amosite, and chrysotile asbestos fibres tested were of UICC (Union Generale Contre le Cancer) reference samples.

Sample preparation for GC–MSD assay

Detection of aldehydic end products from the peroxidation of arachidonic acid in the presence and absence of different asbestos fibres were achieved by the GC–MSD after their reduction to stable alcohols as reported recently (Des Rosiers et al., 1993) with some modification. To detect aldehydes other than malonaldehyde, samples were incubated with 500 μl 1 M $NaBH_4$ for 30 min at room temperature. All aqueous samples were brought to pH < 2 and extracted two times with 10 ml dichloromethane. The combined dichloromethane extracts were evaporated under N_2 to a residual volume of about 500 μl and the residue was treated with 50 μl pyridine and 50 μl BSTFA. A 0.4 μl aliquot was injected into GC–MSD.

Sample incubation

Ten mg arachidonic acid in 2 ml of 0.25 mM Tris buffer pH 7.4 was incubated at 37° C for different intervals of time. To study the effect of fibres on the peroxidation of arachidonic acid, 1 mg/ml of UICC crocidolite, amosite or chrysotile was added to this mixture.

GC–MSD assay

All samples were analyzed on a Hewlett–Packard 5890 gas chromatograph coupled to a 5970 mass selective detector equipped with a DB–1 capillary column (30m x 0.2 mm I.D., 0.33–μm film thickness). GC conditions included: helium flow 0.47 ml/min, on column injection.

The column temperature was initially set at 70° C for 5 min increased by 10° C/min until 280° C. Parameters in electron impact (EI) mode were: ion source temperature, 180–250° C; emission current, 300 μA; electron energy, 70 eV.

RESULTS

The separation by GC–MSD of the BSTFA derivatives of alcohols obtained by $NaBH_4$ reduction of the aldehydes other than malondialdehyde, is presented in figure 1. The identification of the nine of these separated products were achieved from the electron impact mass spectra of these derivatives and also from the retention time of the corresponding standards.

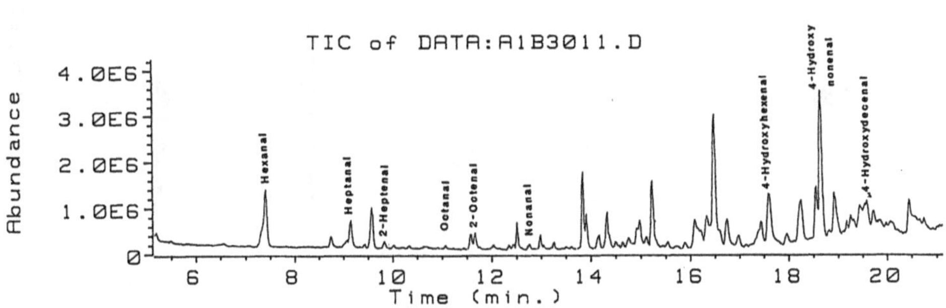

Figure 1. The gas chromatographic analysis of the BSTFA–derivatives otained from the $NaBH_4$ reduction of the aldehydes produced from the peroxidation of arachidonic acid.

It can be noted that 4–hydroxynonenal is the major aldehydic end–product identified from these peroxidations. Experiments are presently in progress to characterize the remaining unidentified derivatives shown in this figure.

The levels of the identified products obtained from the incubation of arachidonic acid alone or with 1 mg/ml UICC crocidolite fibres at 37° C for 6, 21, and 72 hours are shown in figure 2. It can be seen that with the increasing time of incubation, there is a concomitant increase in the aldehydic end–products detected at the end of these time intervals. The presence of crocidolite fibres however, changed the number of the products detected especially after 6 hours of incubation compared to the control mixtures. These fibres also increased the amount of the products formed after 21 and 72 hours of incubation.

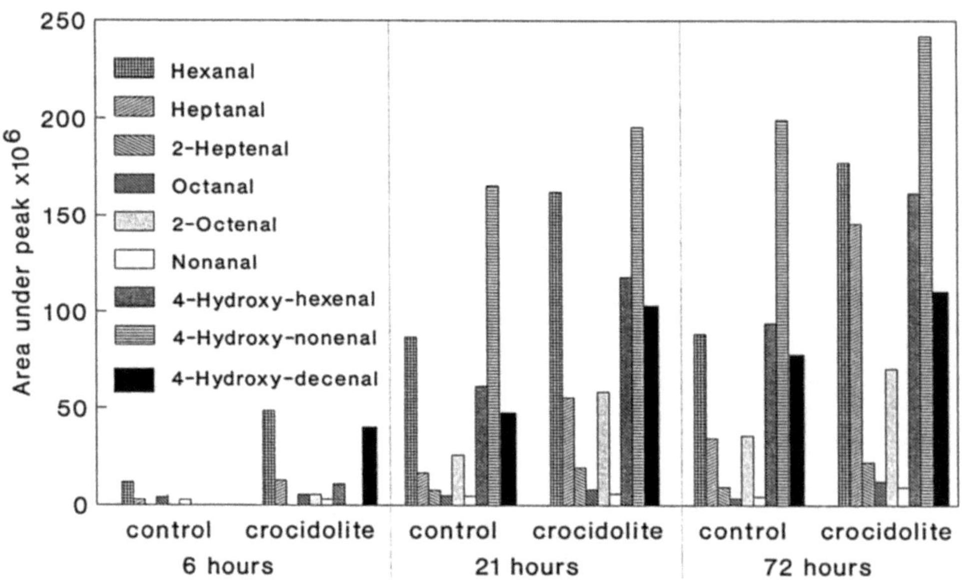

Figure 2 Time curve of the nine identified BSTFA–derivatives of alcohols obtained from the reduction of their corresponding aldehydes produced from the peroxidation of arachidonic acid in the absence (control) and presence of 1 mg/ml UICC crocidolite fibres.

Other types of asbestos fibres did also have this ability. The presence of 1 mg/ml chrysotile or 1 mg/ml amosite fibres increased the level of the aldehydes produced from the peroxidation of arachidonic acid after incubation for 21 hours. This increase was the highest with UICC amosite and the lowest with UICC chrysotile (Figure 3). It can be seen that in these peroxidations, 4–hydroxynonenal was always the highest of the aldehydic products formed followed by hexanal and 4–hydroxydecenal.

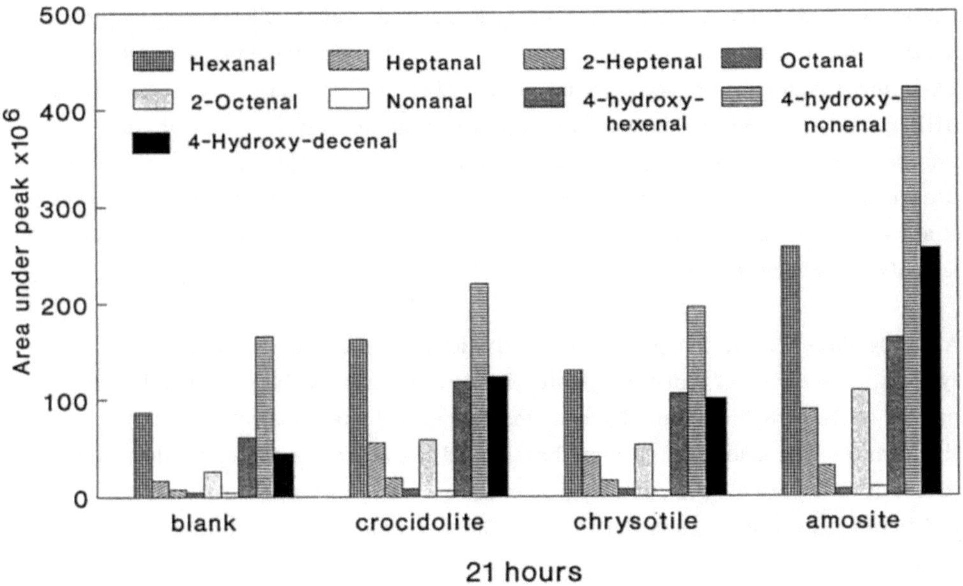

Figure 3. The nine identified BSTFA–derivatives of alcohols obtained from the reduction of their corresponding aldehydes produced from the peroxidation of arachidonic acid in the absence (control) and the presence of 1 mg/ml UICC crocidolite, UICC chrysotile and UICC amosite.

DISCUSSION

The importance of not only the quantification but also the identification of different classes of lipid peroxidation products becomes clearer with the fact that some of these products when tested are shown to have physiological and pathological effects (Horton and Fairhurst, 1987; Benedetti and Comporti, 1987; Esterbauer et al., 1991).

The primary lipid peroxidation products, lipid hydroperoxy acids, for example, are shown to affect the prostaglandin cascade (Hemler et al., 1979) and vascular endothelium (Sasaguri et al., 1984) whilst epoxy–fatty acids are shown to influence the secretion of some hormones (Snyder et al., 1983). The aldehydic secondary lipid peroxidation products are also shown to have different physiological effects at low concentrations and pathological effects at high concentrations. 4–hydroxy–2,3–transnonenal (HNE), for example, at concentrations close to the level in normal cells is found to modulate the expression of *c–myc* and γ–globulin genes in K562 erythroleukemic cells and thus suggest a role in the induction of cell differentiation (Fazio et al., 1992). This aldehydic product is also shown to inactivate β–adrenoreceptors in rat lung membrane preparations by alkylating its sulfhydryl moieties (Leurs et al., 1986). At less than 1 μM concentrations, HNE and other 4–hydroxyalkenals are shown to stimulate *in vitro* the membrane associated phospholipase C– activity in rat liver (Rossi et al., 1988; Rossi et al., 1989; Rossi et al., 1990) and therefore act as a chemoattractant towards rat neutrophils (Rossi et al., 1991) through its ability to activate phospholipase C. They are also found to activate phospholipase D (Natarajan et al., 1993).

Asbestos fibres are not all equal in their activities in *in vitro* and *in vivo* biological systems. The ability of different asbestos fibres in the present study to modulate the pattern of the peroxidation products of arachidonic acid quantitatively is thus of significance. The amount of each product formed from peroxidation might determine the extent of their physiological/pathological effects. This in turn might have an effect in determining the differences in the ability of fibres to produce different pathological effects *in vitro* and *in vivo* systems.

REFERENCES

Benedetti A, Fulceri R, Ferrali M, Ciccoli L, Esterbauer H, Comporti M (1982) Detection of carbonyl functions in phospholipids of liver microsomes in CCl_4– and $BrCCl_3$–poisoned rats. Biochim Biophys Acta 712: 628–638

Benedetti A, Comporti M, Fulceri R, Esterbauer H (1984) Cytotoxic aldehydes originating from the peroxidation of liver microsomal lipids. Identification of 4,5–dihydroxydecenal. Biochim Biophys Acta 792: 172–181

Benedetti A, Comporti M (1987) Formation, reactions and toxicity of aldehydes produced in the course of lipid peroxidation in cellular membranes. Bioelectrochem Bioenerg 18: 187–202

Bucknall T, Edwards HE, Kemsley KG, Moore JS, Phillips GO (1978) The formation of malonaldehyde in irradiated carbohydrates. Carbohydr Res 62: 49–59

Chvapil M, Stankova L, Malshet V (1976) Lipid peroxidation as one of the mechanisms of silica fibrogenicity? I. Study with erythrocytes. Environ Res 11: 78–88

Curzio M, Poli G, Esterbauer H, Biasi F, DiMauro C, Dianzani MU (1986) Detection of carbonyl products of lipid peroxidation in rat pleural exudate. IRCS Med Sci 14: 984–985

Des Rosiers C, Rivest M–J, Boily M–J, Jetté M, Carrobé–Cohen A, Kumar A (1993) Gas chromatographic–Mass spectrometric assay of tissue malondialdehyde, 4–hydroxynonenal, and other aldehydes after their reduction to stable alcohols. Anal Biochem 208: 161–170

Esterbauer H, Cheesman KH, Dianzani MU, Poli G, Slater TF (1982) Separation and characterization of the aldehydic products of lipid peroxidation stimulated by ADP–Fe^{2+} in rat liver microsomes. Biochem J 208: 129–140

Esterbauer H, Lang J, Zadravec S, Slater TF (1984) Detection of malonaldehyde by high–performance liquid chromatography. Methods in Ezymology 105: 319–328

Esterbauer H, Zollner H (1989) Methods for determination of aldehydic lipid peroxidation products. Free Rad Biol Med 7: 197–203

Esterbauer H, Schaur RJ, Zollner H (1991) Chemistry and biochemistry of 4–hydroxynonenal, malonaldehyde and related aldehydes. Free Rad Biol Med 11: 81–128

Fazio VM, Barrera G, Martinotti S, Farace MG, Giglioni B, Frati L, Manzari V, Dianzani MU (1992) 4–hydroxynonenal, a product of cellular lipid peroxidation, which modulates *c–myc* and globin gene expression in K562 erythroleukemic cells. Cancer Res 52: 4866–4871

Fontecave M, Mansuy D, Jaouen M, Pezerat H (1987) The stimulatory effects of asbestos on NADPH–dependent lipid peroxidation in rat liver microsomes. Biochem J 241: 561–565

Gardner HW (1989) Oxygen radical chemistry of polyunsaturated fatty acids. Free Rad Biol Med 7: 65–86

Goodglick LA, Pietras LA, Kane AB (1989) Evaluation of the causal relationship between crocidolite asbestos–induced lipid peroxidation and toxicity to macrophages. Am Rev Respir Dis 139: 1265–1273

Gulumian M, Sardianos F, Kilroe–Smith TA, Ockerse G (1983) Lipid peroxidation in microsomes induced by crocidolite fibres. Chem–Biol Interact 44:111–118

Gupta GSD, Kaw JL (1982) Formation of lipid peroxides in the subcellular fractions of silicotic lungs in rats. Eur J Respir Dis 63: 183–187

Gutteridge JMC (1979) Identification of malondialdehyde as the TBA–reactant formed by bleomycin–iron–radical damage to DNA. FEBS Lett 105: 278–282

Hemler ME, Cook HW, Lands WEM (1979) Prostaglandin synthesis can be triggered by lipid peroxides. Archs Biochem Biophys 193: 340–345

Horton AA, Fairhurst S (1987) Lipid peroxidation and mechanisms of toxicity. CRC Crit Rev Toxicol 18: 27–79

Iguchi H, Kojo S, Ikeda M (1993) Lipid peroxidation and disintegration of the cell membrane structure in cultures of rat lung fibroblasts treated with asbestos. J Appl Toxicol 13: 269–275

Janero DR (1990) Malondialdehyde and thiobarbituric acid–reactivity as diagnostic indices of lipid peroxidation and peroxidative tissue injury. Free Rad Biol Med 9: 515–540

Lang J, Celetto C, Esterbauer H (1985) Quantitative determination of the lipid peroxidation product 4–hydroxynonenal by high–performance liquid chromatography. Anal Biochem 150: 369–378

Leurs R, Rademaker B, Kramer K, Timmerman H, Bast A (1986) The effects of

4–hydroxy–2,3–*trans*–nonenal on β–adrenoreceptors of rat lung membranes. Chem–Biol Interact 59: 211–218

Natarajan V, Scribner WM, Taher MM (1993) 4–hydroxynonenal, a metabolite of lipid peroxidation, activates phospholipase D in vascular endothelial cells. Free Rad Biol Med 15: 365–375.

Norsten–Höög C, Cronholm T (1990) Analysis of aldehydic lipid peroxidation products in rat liver and hepatocytes by gas chromatography and mass spectrometry of the oxime–tert–butyldimethylsilyl derivatives. Anal Chem 189: 131–137

Rossi MA, Garramone A, Dianzani MU (1988) Stimulation of phospholipase C activity by 4–hydroxynonenal; influence of GTP and calcium concentration. Int J Tiss React 10: 321–325

Rossi MA, Garramone A, Dianzani MU (1989) Effect of 4–hydroxy–2,3–*trans*–nonenal, a lipid peroxidation product, on hepatic phospholipase C. Med Sci Res 17: 257–258

Rossi MA, Fidale F, Garramone A, Esterbauer H, Dianzani MU (1990) Effect of 4–hydroxyalkenals on hepatic phosphatidylinositol–4,5–bisphosphate–phospholipase C. Biochem Pharmacol 39: 1715–1719

Rossi MA, Curzio M, Di Mauro C, Fidale F, Garramone A, Esterbauer H, Torrielli M, Dianzani MU (1991) Experimental studies on the mechanism of action of 4–hydroxy–2,3–*trans*–nonenal, a lipid peroxidation product displaying chemotactic activity toward rat neutrophils. Cell Biochem Funct 9: 163–170

Sasaguri Y, Nakashima T, Morimatsu M, Yagi K (1984) Injury to cultured endothelial cells from human umbilical vein by linoleic acid hydroperoxide. J Appl Biochem 6: 144–150

Smith CV, Anderson RE (1987) Methods for determination of lipid peroxidation in biological samples. Free Rad Biol Med 3: 341–344

Snyder GD, Capdevila J, Chacos N, Manna S, Falck JR (1983) Action of luteinising hormone–releasing hormone: involvement of novel arachidonic acid metabolites. Proc Natn Acad Sci USA 80: 3504–3507

Van Kuijk FJGM, Thomas DW, Stephens RJ, Dratz EA (1986) Occurence of 4–hydroxyalkenals in rat tissues determined as pentafluorobenzyl oxime derivatives by gas chromatography–mass spectroscopy. Biochem Biophys Res Commun 139: 144–149

Weitzman SA, Weitberg AB (1985) Asbestos–catalysed lipid peroxidation and its inhibition by desferroxamine. Biochem J 225: 259–262

Yoshino K, Matsuura T, Sano M, Saito S–I, Tomita I (1986) Fluorometric liquid chromatographic determination of aliphatic aldehydes arising from lipid peroxides. Chem Pharm Bull 34: 1694–1700

Zsoldos T, Tigyi A, Montskó T, Puppi A (1983) Lipid peroxidation in the membrane damaging effect of silica–containing dust on rat lungs. Exp Path 23: 73–77

Effect of Chelators on the Surface Properties of Asbestos

Laura Mollo, Elisabetta Merlo, Elio Giamello, Marco Volante, Vera Bolis and Bice Fubini
Dipartimento di Chimica Inorganica, Chimica Fisica e Chimica dei Materiali.
Università di Torino
Via Pietro Giuria 7
10125 Torino
Italy

Introduction

No detectable amounts of iron can be found in simple aqueous suspension of asbestos. In presence of iron chelators, as discovered by Aust and associates (Lund and Aust, 1990; Aust and Lund, 1991; Lund and Aust, 1992; Chao and Aust, 1993 and references therein) considerable amounts of iron are progressively removed from the solid and brought into the solution. It was clearly proved that this mobilized iron was active in DNA damage and lipid peroxidation. The solid may, however, play some role too in asbestos toxicity, acting both as a continuous source of iron ions and as a possible site for deposition of endogenous iron, which under some circumstances may become redox active and consequently toxic. We have therefore set up a thorough investigation on the kind of reactions governing iron release at the solid-liquid interface and on the modifications occurring consequently on the solid. In the present paper we report some results obtained with the three most commonly found asbestos - crocidolite, amosite and chrysotile- and with three different chelators, desferrioxamine B, ferrozine and ascorbate. Desferrioxamine B, a chelator originated from siderophors, because of its strong affinity for Fe(III), is usually used to block any reaction originated from free iron. Ferrozine, at the opposite, because of its nitrogen donors groups, is a typical analytic reagent for Fe(II). Ascorbate is at the same time a chelator and an endogenous reducing agent, thus its reactivity with the inhaled particle may be relevant to fiber toxicity in vivo.

Experimental

Materials. Asbestos: UICC crocidolite, amosite and Canadian chrysotile.

NATO ASI Series, Vol. H 85
Cellular and Molecular Effects
of Mineral and Synthetic Dusts and Fibres
Edited by J. M. G. Davis and M.-C. Jaurand

Chelators: Deferoxamine mesylate, 3-(2-pyridyl)-5,6-diphenyl-1,2,4-triazine-p,p'-disulfonic acid (Ferrozine) and sodium salts of L-ascorbic acid were obtained from Sigma Chemical Co. Spin trap: DMPO (5-5'-dimethyl-1-pyrroline-N-oxide) was obtained by Sigma Chemical Co. *Methods.* Incubation of suspension of the solids in aqueous solutions of chelators: 50 mg of fibers were kept for three days in 5 ml of solution of chelator kept in the dark and continuously stirred. The rate of extraction of Fe at the end of the experiment was very slow by comparison to the first ten hours, in the experimental conditions adopted. Samples were then washed several times and dried up in order to be examined or submitted to a subsequent incubation in chelators solutions. The amounts of extracted iron was determined by measuring the absorbance of the supernatant using an Uvikon 930 dual beam spectrophotometer.
Free radical release: the release of free radicals in simple buffered aqueous solution was performed by means of DMPO as spin trapping agent, by using the formate ion (HCO_2^-) as a target molecule. Hydrogen abstraction following interaction of the formate with the solid surface results in $CO_2^{\bullet -}$ formation and is monitored by the EPR spectrum of the DMPO-$CO_2^{\bullet}$ adduct, following the procedure adopted by Pèzerat and associates (Zalma et al. 1989). The well known EPR spectrum of the adduct was recorded after 30' and 60' and the amount of radicals produced was estimated from the intensity of this spectrum.
EPR spectra of the solid: the spectra originated by Fe(III) in different surroundings, were recorded at room temperature and at 77 K, following what described in previous papers (Astolfi et al., 1991, Fubini et al., 1991). Spectra are shown with magnetic fields increasing from left to right: a corresponding scale for the g values is also given, together with the amplification used in the recording.
All EPR spectra were recorded on a Varian E 109.
High resolution electron microscopy images were obtained with a transmission electron microscope (TEM) Jeol JEM 2000 EX.

Results and Discussion

The amounts of iron removed after the first incubation in desferrioxamine B and ascorbate of the three asbestos are reported in figure 1 as histograms for comparison. The two amphiboles, crocidolite and amosite (where iron is a major component in mineral composition), release, as expected, more iron than chrysotile, (where iron is present only as a substitute for Mg(II)). The amount of iron released does not parallel, however, the actual composition of the solid, as pointed out in another paper by our group in this book (Fubini), where a wider variety of iron containing solids are considered. Mobilization is in fact governed more by the availability of poorly coordinated Fe ions at the interface than by the actual concentration of them in the solid phase.

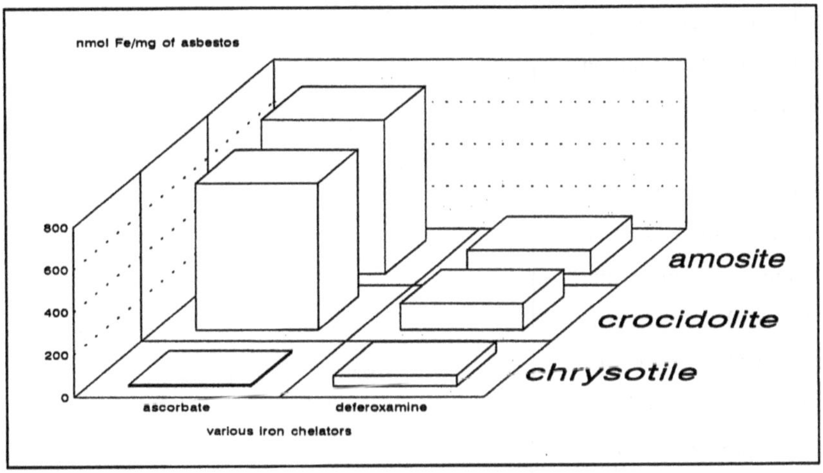

Figure 1 Removal of iron following incubation in an aqueous solution of two chelators for three days.

The amount released obviously also depends upon the chelator: strange enough, with the two amphiboles, ascorbate removes much more iron than desferrioxamine B, in spite of its stability constant being lower (Martell, 1982) than the desferrioxamine B one (pK = 31). This indicates that the mechanism operating at the interface is much more complex than a simple displacement of equilibria of dissolution and complexation. What might happen in a complex biological medium within the cells is thus very difficult to predict. Figure 2 reports the EPR spectra of amosite before and after incubation in the ascorbate solution: profound modification are visible: the overall intensity of the spectrum sligthly decreases and the component falling around g = 4, typical of relatively isolated Fe (III), mostly disappears. This suggests that by acting both as a reductant and as a chelator ascorbate may reduce and extract more easily the ions from the lattice.

The broad spectrum visible after ascorbate chelation may be ascribed to Fe(II)-Fe(III) magnetically interacting, the Fe(II) arising from reduced surface Fe (III). Moreover the weak complex (Martell A.E., 1982) formed is readily dissolved into the solution whereas with the exadentate desferrioxamine B often the chelator is adsorbed irreversibly at some surface site (likely a protruding Fe(III) ion), blocking consequently further dissolution (Weitzman and Graceffa, 1988). The mechanism whereby chrysotile exhibits a trend opposite to amphiboles, (more iron is released with desferrioxamine B than with ascorbate) is not straightforward and requires further investigations. It may be related to the structure of the brucitic layer into which Fe substitutes for Mg. Being easily leached, iron becomes accessible virtually in the same way to both chelators; in such a case the higher stability of desferrioxamine B by respect to the ascorbate complex could account for the amount of iron released. Incubation in chelators deeply affects the capability of releasing free radicals in solution. Figure 3 illustrates the case of chrysotile. A similar behaviour was found in all the three cases examined. As previously reported (Fubini et al., 1993) both desferrioxamine B and ferrozine inactivate the surface and inhibit free radical release. This was taken as evidence for the requirement of both oxidation states of iron at the active site (Fubini et al., 1993) as desferrioxamine B is supposed to extract mainly Fe (III)

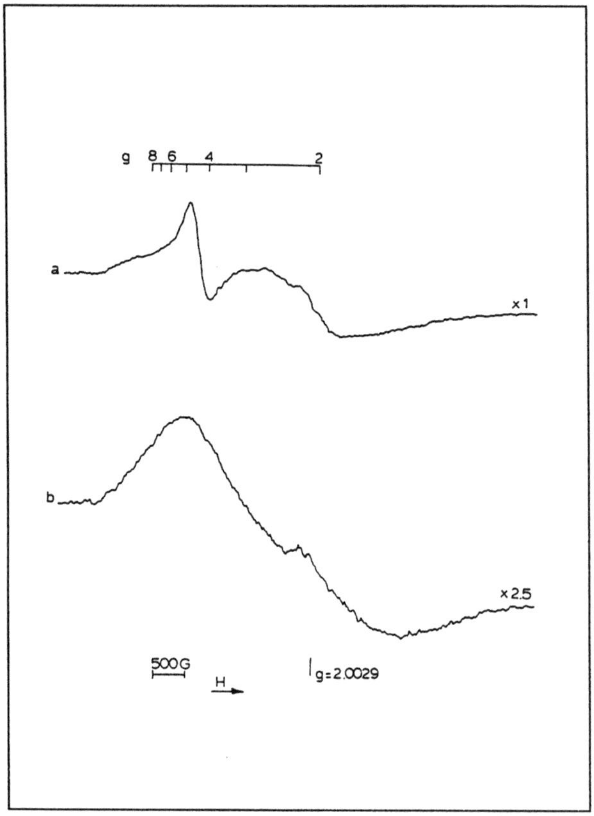

Figure 2. EPR spectra of amosite before (a) and after (b) incubation in ascorbate, recorded at 77 K.

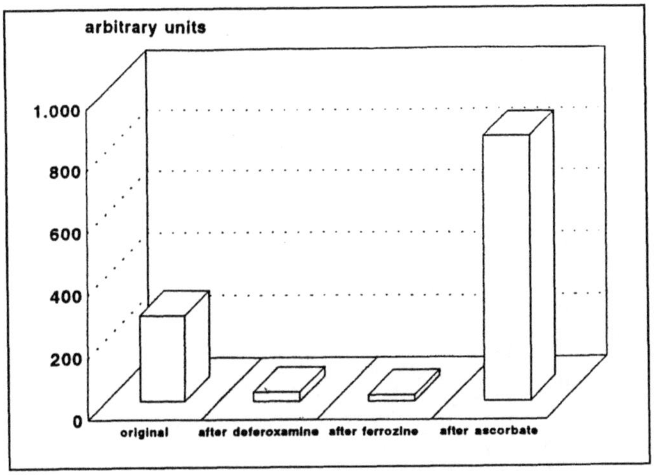

Figure 3. Free radical release of chrysotile before (a) and after (b,c,d) incubation in three chelators.

and ferrozine Fe(II) (Lund and Aust, 1991). The inhibition caused by adsorption of the molecule at the active site (Weitzman and Graceffa, 1984; Weitzman et al. 1988) may also explain the loss of Fenton reactivity (production of OH· from hydrogen peroxide) in desferrioxamine B treated asbestos. Ascorbate induces the opposite effect, as the fibers are activated by it. In such a case the few iron ions left at the surface are reduced so that the number of mixed valencies couples or aggregates, which we have supposed to be the active center for H abstraction, is increased by comparison with the starting material, in spite of the fact that a part of the ions present have been released into the solution.

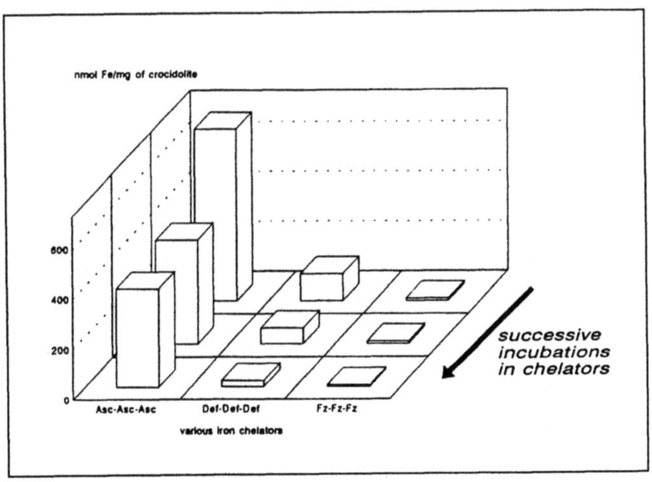

Figure 4. Amounts of iron released upon three successive incubations in the chelating solutions of the same crocidolite fibers.

The possibility, however, that bulk iron might migrate to the surface in order to replace the ions released into the solution has not to be discarded. In order to evaluate the extent of this process we have submitted subsequently, each time after thorough washing and drying, the fibers to the chelating solutions. The results obtained with crocidolite are reported in figure 4. With all the three chelating agents the amounts released decreased following successive incubations, but a substantial part of iron was still present at the surface. This suggests a consistent ion migration within the solid following its extraction from the surface and subsurface layers.

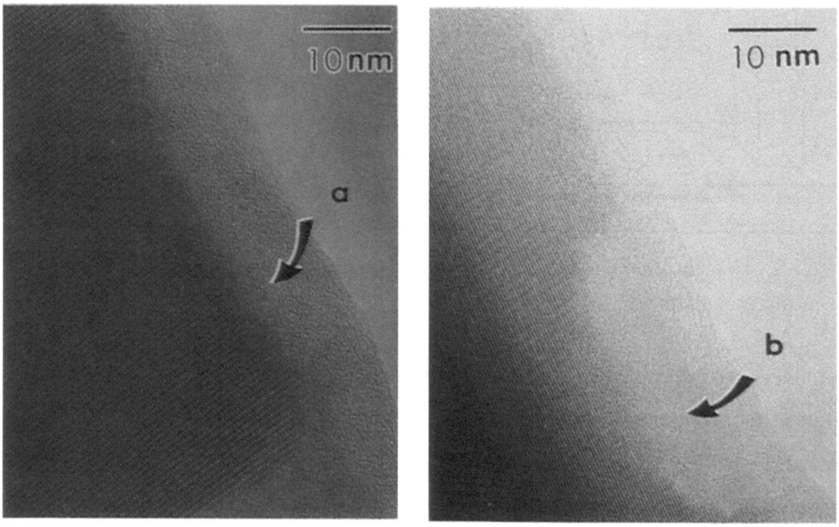

Figure 5. HRTEM image of a crocidolite fiber (a) and of one incubated in desferrioxamine B (b).

In order to detect whether chelators affected the micromorphology of the fibers, both crocidolite and amosite were studied by transmission electron microscopy. Fig. 5 and 6 report for crocidolite and amosite respectively the image of the original fibers and the same following desferrioxamine B treatment. The original crocidolite fibers are surrounded by a regular amorphous layer (~ 12 nm), but are made up of well ordered crystal planes, each plane being evidenced in the high resolution image. In the fiber of the sample examined after incubation the crystal structure is eroded with onset of random irregularities and amorphous patches in the outmost layers. Similar features can be observed for amosite. This process is not merely confined to the surface but takes place via progressive ion migration and seems to affect the micromorphology of the particle. Iron containing fibers may act therefore as a reservoir for active iron, depending upon the kind of endogenous chelators with which they become in contact in vivo. Moreover the progressive depletion of the iron sites at the surface

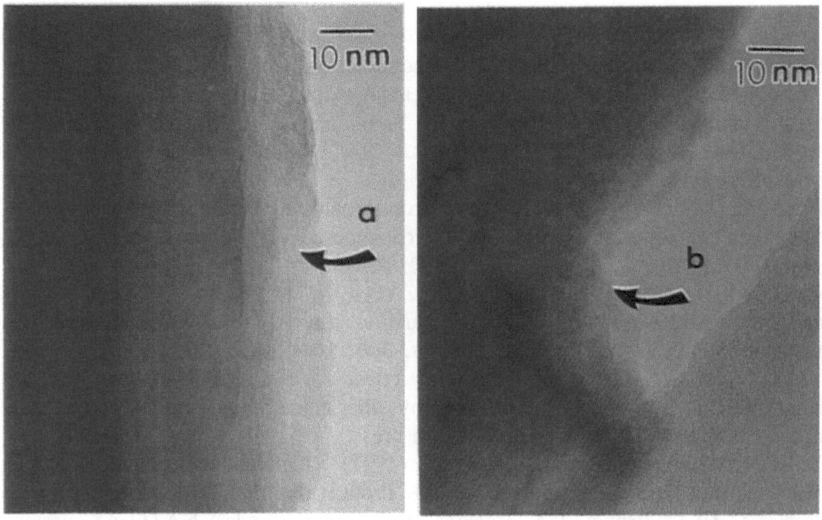

Figure 6. HRTEM image of an amosite fiber (a) and of one incubated in desferrioxamine B (b).

may also provide sites capable of trapping endogenous iron, as recently reported by Aust and coworkers (Lund et al., in press). Under such circumstances a continuous iron cycling might take place at the site of internalization of a fiber, following the mechanisms outlined in figure 7.

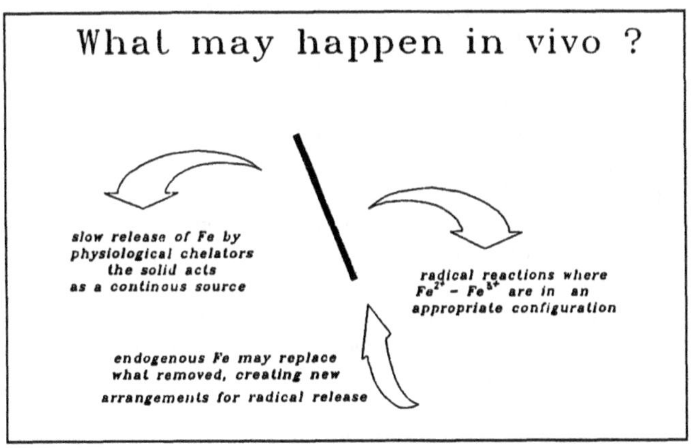

Figure 7.

Acknowledgments

Financial support from MURST ST (60%) is here acknowledged.

References

Astolfi A., Belluso E., Ferraris G., Fubini B., Giamello E., Volante M. (1991) Asbestiform mineral associated with chrysotile from Western Alps (Piedmont-Italy): chemical characteristics and possible related toxicity in *"Mechanisms in Fibre Carcinogenesis"* (R.C. Brown, J.A. Hoskins, N.F. Johnson, eds.), NATO ASI seriesies, Plenum press, New York, p.269-283.

Aust A. and Lund L.G.,(1991). Iron mobilization from crocidolite results in enhanced iron-catalyzed oxygen consumption and hydroxyl radical generation in the presence of cysteine in *"Mechanisms in Fibre Carcinogenesis"* (R.C. Brown, J.A. Hoskins, N.F. Johnson, eds.), NATO ASI seriesies, Plenum press, New York, p.269-283.

Chao C.C. and Aust A., (1993). Photochemical Reduction of ferric Iron By Chelators results in DNA Strand Breaks. *Arch.Biochem. Biophys.* 300: 544-550.

Chao C.C. and Aust A. Effect of long-term removal of iron from asbestos by desferrioxamine B on subsequent mobilization by other chelators and induction of DNA single-strand breaks. *Arch.Biochem.Biophys.* in press.

Fubini B., Bolis V., Giamello E. and Volante M. (1991) Chemical functionalities at the broken fibre surface relatable to free radicals production in *"Mechanisms in Fibre Carcinogenesis"* (R.C. Brown, J.A. Hoskins, N.F. Johnson, eds.) NATO ASI series, Plenum press, New York, p.415-432.

Fubini B., Volante M., Giamello E., Bolis V.(1993). The role of surface chemistry in the pathogenicity of mineral dusts and fibres: some open questions"; in *"Proceedings of the Eighth International Conference on Occupational Lung Diseases"*, J.Hurich, M.Leasage, and A.David eds., ILO, Geneva, Switzerland, Vol. III, p.1180-1185.

Lund L. and Aust A., (1990) Iron mobilization from asbestos by chelators and ascorbic acid. *Arch.Biochem.Biophys.* 278: 60-64.

Lund L. and Aust A., (1991). Iron-catalyzed reactions may be responsible for the biochemical and biological effects of asbestos. *BioFactors*, 3: 83-89.

Lund L. and Aust A., (1991). Mobilization of iron from crocidolite asbestos by certain chelators results in enhanced crocidolite-dependent oxygen consumption. *Arch.Biochem.Biophys.* 287: 91-96.

Lund L. and Aust A., (1992). Iron mobilization from crocidolite asbestos greatly enhances crocidolite-dependent formation of DNA single-strand breaks in ϕX174 RFI DNA. *Carcinogenesis* 13: 637-642.

Lund L.G., Williams M.G., Dodson R.F., Aust A.E. Iron associated with asbestos bodies is responsible for the formation of single-strand breaks in ϕX174 RFI DNA.British J.Ind.Med.,in press.

Martell A.E.,(1982) Chelates of Ascorbic Acid, Formation and Catalytic Properties in *"Advances in Chemistry Series, No.200 Ascorbic Acid: Chemistry Metabolism and Uses"* P.A.Seib and B.M.Tolbert eds. American Chemical Society.

Weitzman S.A., Graceffa P, (1984) Asbestos catalyse hydroxyl and superoxide radical generation from hydrogen peroxide. *Arch. Biochem. Biophys.* 228: 267-274.

Weitzman S.A., Chester J.F., Graceffa P. (1988) Binding of dexferoxamine to asbestos fibres in vitro and in vivo. *Carcinogenesis* 9:1643-1645.

Zalma R., Bonneau L., Jaurand M.C., Guignard J. and Pezerat H. (1987a). Production of hydroxyl radicals by iron solid compounds. *Toxicol. and Environ. Chem.* 13: 171-188; (1987b). Formation of oxy-radicals by oxygen reduction arising from the surface activity of asbestos. *Can. J. Chem.* 652, 338-2341.

NATO ASI Series H

Vol. 1: **Biology and Molecular Biology of Plant-Pathogen Interactions.** Edited by J.A. Bailey. 415 pages. 1986.

Vol. 2: **Glial-Neuronal Communication in Development and Regeneration.** Edited by H.H. Althaus and W. Seifert. 865 pages. 1987.

Vol. 3: **Nicotinic Acetylcholine Receptor: Structure and Function.** Edited by A. Maelicke. 489 pages. 1986.

Vol. 4: **Recognition in Microbe-Plant Symbiotic and Pathogenic Interactions.** Edited by B. Lugtenberg. 449 pages. 1986.

Vol. 5: **Mesenchymal-Epithelial Interactions in Neural Development.** Edited by J. R. Wolff, J. Sievers, and M. Berry. 428 pages. 1987.

Vol. 6: **Molecular Mechanisms of Desensitization to Signal Molecules.** Edited by T M. Konijn, P J. M. Van Haastert, H. Van der Starre, H. Van der Wel, and M.D. Houslay. 336 pages. 1987.

Vol. 7: **Gangliosides and Modulation of Neuronal Functions.** Edited by H. Rahmann. 647 pages. 1987.

Vol. 8: **Molecular and Cellular Aspects of Erythropoietin and Erythropoiesis.** Edited by I.N. Rich. 460 pages. 1987.

Vol. 9: **Modification of Cell to Cell Signals During Normal and Pathological Aging.** Edited by S. Govoni and F. Battaini. 297 pages. 1987.

Vol. 10: **Plant Hormone Receptors.** Edited by D. Klämbt. 319 pages. 1987.

Vol. 11: **Host-Parasite Cellular and Molecular Interactions in Protozoal Infections.** Edited by K.-P. Chang and D. Snary. 425 pages. 1987.

Vol. 12: **The Cell Surface in Signal Transduction.** Edited by E. Wagner, H. Greppin, and B. Millet. 243 pages. 1987.

Vol. 13: **Toxicology of Pesticides: Experimental, Clinical and Regulatory Perspectives.** Edited by L.G. Costa, C.L. Galli, and S.D. Murphy. 320 pages. 1987.

Vol. 14: **Genetics of Translation. New Approaches.** Edited by M.F. Tuite, M. Picard, and M. Bolotin-Fukuhara. 524 pages. 1988.

Vol. 15: **Photosensitisation. Molecular, Cellular and Medical Aspects.** Edited by G. Moreno, R. H. Pottier, and T. G. Truscott. 521 pages. 1988.

Vol. 16: **Membrane Biogenesis.** Edited by J.A.F Op den Kamp. 477 pages. 1988.

Vol. 17: **Cell to Cell Signals in Plant, Animal and Microbial Symbiosis.** Edited by S. Scannerini, D. Smith, P. Bonfante-Fasolo, and V. Gianinazzi-Pearson. 414 pages. 1988.

Vol. 18: **Plant Cell Biotechnology.** Edited by M.S.S. Pais, F. Mavituna, and J. M. Novais. 500 pages. 1988.

Vol. 19: **Modulation of Synaptic Transmission and Plasticity in Nervous Systems.** Edited by G. Hertting and H.-C. Spatz. 457 pages. 1988.

Vol. 20: **Amino Acid Availability and Brain Function in Health and Disease.** Edited by G. Huether. 487 pages. 1988.

NATO ASI Series H

Vol. 21: **Cellular and Molecular Basis of Synaptic Transmission.** Edited by H. Zimmermann. 547 pages. 1988.

Vol. 22: **Neural Development and Regeneration. Cellular and Molecular Aspects.** Edited by A. Gorio, J. R. Perez-Polo, J. de Vellis, and B. Haber. 711 pages. 1988.

Vol. 23: **The Semiotics of Cellular Communication in the Immune System.** Edited by E.E. Sercarz, F. Celada, N.A. Mitchison, and T. Tada. 326 pages. 1988.

Vol. 24: **Bacteria, Complement and the Phagocytic Cell.** Edited by F. C. Cabello und C. Pruzzo. 372 pages. 1988.

Vol. 25: **Nicotinic Acetylcholine Receptors in the Nervous System.** Edited by F. Clementi, C. Gotti, and E. Sher. 424 pages. 1988.

Vol. 26: **Cell to Cell Signals in Mammalian Development.** Edited by S.W. de Laat, J.G. Bluemink, and C.L. Mummery. 322 pages. 1989.

Vol. 27: **Phytotoxins and Plant Pathogenesis.** Edited by A. Graniti, R. D. Durbin, and A. Ballio. 508 pages. 1989.

Vol. 28: **Vascular Wilt Diseases of Plants. Basic Studies and Control.** Edited by E. C. Tjamos and C. H. Beckman. 590 pages. 1989.

Vol. 29: **Receptors, Membrane Transport and Signal Transduction.** Edited by A. E. Evangelopoulos, J. P. Changeux, L. Packer, T. G. Sotiroudis, and K.W.A. Wirtz. 387 pages. 1989.

Vol. 30: **Effects of Mineral Dusts on Cells.** Edited by B.T. Mossman and R.O. Begin. 470 pages. 1989.

Vol. 31: **Neurobiology of the Inner Retina.** Edited by R. Weiler and N.N. Osborne. 529 pages. 1989.

Vol. 32: **Molecular Biology of Neuroreceptors and Ion Channels.** Edited by A. Maelicke. 675 pages. 1989.

Vol. 33: **Regulatory Mechanisms of Neuron to Vessel Communication in Brain.** Edited by F. Battaini, S. Govoni, M.S. Magnoni, and M. Trabucchi. 416 pages. 1989.

Vol. 34: **Vectors asTools for the Study of Normal and Abnormal Growth and Differentiation.** Edited by H. Lother, R. Dernick, and W. Ostertag. 477 pages. 1989.

Vol. 35: **Cell Separation in Plants: Physiology, Biochemistry and Molecular Biology.** Edited by D. J. Osborne and M. B. Jackson. 449 pages. 1989.

Vol. 36: **Signal Molecules in Plants and Plant-Microbe Interactions.** Edited by B.J.J. Lugtenberg. 425 pages. 1989.

Vol. 37: **Tin-Based Antitumour Drugs.** Edited by M. Gielen. 226 pages. 1990.

Vol. 38: **The Molecular Biology of Autoimmune Disease.** Edited by A.G. Demaine, J-P. Banga, and A.M. McGregor. 404 pages. 1990.

NATO ASI Series H

Vol. 39: **Chemosensory Information Processing.**
Edited by D. Schild. 403 pages. 1990.

Vol. 40: **Dynamics and Biogenesis of Membranes.**
Edited by J. A. F. Op den Kamp. 367 pages. 1990.

Vol. 41: **Recognition and Response in Plant-Virus Interactions.**
Edited by R. S. S. Fraser. 467 pages. 1990.

Vol. 42: **Biomechanics of Active Movement and Deformation of Cells.**
Edited by N. Akkas. 524 pages. 1990.

Vol. 43: **Cellular and Molecular Biology of Myelination.**
Edited by G. Jeserich, H. H. Althaus, and T. V. Waehneldt. 565 pages. 1990.

Vol. 44: **Activation and Desensitization of Transducing Pathways.**
Edited by T. M. Konijn, M. D. Houslay, and P. J. M. Van Haastert. 336 pages. 1990.

Vol. 45: **Mechanism of Fertilization: Plants to Humans.**
Edited by B. Dale. 710 pages. 1990.

Vol .46: **Parallels in Cell to Cell Junctions in Plants and Animals.**
Edited by A. W Robards, W. J . Lucas, J . D. Pitts, H . J . Jongsma, and D. C. Spray. 296 pages. 1990.

Vol. 47: **Signal Perception and Transduction in Higher Plants.**
Edited by R. Ranjeva and A. M. Boudet. 357 pages. 1990.

Vol. 48: **Calcium Transport and Intracellular Calcium Homeostasis.**
Edited by D. Pansu and F. Bronner. 456 pages. 1990.

Vol. 49: **Post-Transcriptional Control of Gene Expression.**
Edited by J. E. G. McCarthy and M. F. Tuite. 671 pages. 1990.

Vol. 50: **Phytochrome Properties and Biological Action.**
Edited by B. Thomas and C. B. Johnson. 337 pages. 1991.

Vol. 51: **Cell to Cell Signals in Plants and Animals.**
Edited by V. Neuhoff and J. Friend. 404 pages. 1991.

Vol. 52: **Biological Signal Transduction.**
Edited by E. M . Ross and K . W. A. Wirtz. 560 pages. 1991.

Vol. 53: **Fungal Cell Wall and Immune Response.**
Edited by J. P. Latge and D. Boucias. 472 pages. 1991.

Vol. 54: **The Early Effects of Radiation on DNA.**
Edited by E. M. Fielden and P. O'Neill. 448 pages. 1991.

Vol. 55: **The Translational Apparatus of Photosynthetic Organelles.**
Edited by R. Mache, E. Stutz, and A. R. Subramanian. 260 pages. 1991.

Vol. 56: **Cellular Regulation by Protein Phosphorylation.**
Edited by L. M. G. Heilmeyer, Jr. 520 pages. 1991.

NATO ASI Series H

Vol. 57: **Molecular Techniques in Taxonomy.**
Edited by G . M . Hewitt, A. W. B. Johnston, and J. P. W. Young . 420 pages. 1991.

Vol. 58: **Neurocytochemical Methods.**
Edited by A. Calas and D. Eugene. 352 pages. 1991.

Vol. 59: **Molecular Evolution of the Major Histocompatibility Complex.**
Edited by J. Klein and D. Klein. 522 pages. 1991.

Vol. 60: **Intracellular Regulation of Ion Channels.**
Edited by M. Morad and Z. Agus. 261 pages. 1992.

Vol. 61: **Prader-Willi Syndrome and Other Chromosome 15q Deletion Disorders.**
Edited by S. B. Cassidy. 277 pages. 1992.

Vol. 62: **Endocytosis. From Cell Biology to Health, Disease and Therapie.**
Edited by P. J. Courtoy. 547 pages. 1992.

Vol. 63: **Dynamics of Membrane Assembly.**
Edited by J. A. F. Op den Kamp. 402 pages. 1992.

Vol. 64: **Mechanics of Swelling. From Clays to Living Cells and Tissues.**
Edited by T. K. Karalis. 802 pages. 1992.

Vol. 65: **Bacteriocins, Microcins and Lantibiotics.**
Edited by R. James, C. Lazdunski, and F. Pattus. 530 pages. 1992.

Vol. 66: **Theoretical and Experimental Insights into Immunology.**
Edited by A. S. Perelson and G. Weisbuch. 497 pages. 1992.

Vol. 67: **Flow Cytometry. New Developments.**
Edited by A. Jacquemin-Sablon. 1993.

Vol. 68: **Biomarkers. Research and Application in the Assessment of Environmental Health.** Edited by D. B. Peakall and L. R. Shugart. 138 pages. 1993.

Vol. 69: **Molecular Biology and its Application to Medical Mycology.**
Edited by B. Maresca, G. S. Kobayashi, and H. Yamaguchi. 271 pages. 1993.

Vol. 70: **Phospholipids and Signal Transmission.**
Edited by R. Massarelli, L. A. Horrocks, J. N. Kanfer, and K. Löffelholz. 448 pages. 1993.

Vol. 71: **Protein Synthesis and Targeting in Yeast.**
Edited by A. J. P. Brown, M. F. Tuite, and J. E. G. McCarthy. 425 pages. 1993.

Vol. 72: **Chromosome Segregation and Aneuploidy.**
Edited by B. K. Vig. 425 pages. 1993.

Vol. 73: **Human Apolipoprotein Mutants III. In Diagnosis and Treatment.**
Edited by C. R. Sirtori, G. Franceschini, B. H. Brewer Jr. 302 pages. 1993.

NATO ASI Series H

Vol. 74: **Molecular Mechanisms of Membrane Traffic.**
Edited by D. J. Morré, K. E. Howell, and J. J. M. Bergeron. 429 pages. 1993.

Vol. 75: **Cancer Therapy. Differentiation, Immunomodulation and Angiogenesis.**
Edited by N. D'Alessandro, E. Mihich, L. Rausa, H. Tapiero, and T. R. Tritton. 299 pages. 1993.

Vol. 76: **Tyrosine Phosphorylation/Dephosphorylation and Downstream Signalling.**
Edited by L. M. G. Heilmeyer Jr. 388 pages. 1993.

Vol. 77: **Ataxia-Telangiectasia.** Edited by R. A. Gatti, R. B. Painter. 306 pages. 1993.

Vol. 78: **Toxoplasmosis.** Edited by J. E. Smith. 272 pages. 1993.

Vol. 79: **Cellular Mechanisms of Sensory Processing. The Somatosensory System.**
Edited by L. Urban. 514 pages. 1994.

Vol. 80: **Autoimmunity: Experimental Aspects.**
Edited by M. Zouali. 318 pages. 1994.

Vol. 81: **Plant Molecular Biology. Molecular Genetic Analysis of Plant Development and Metabolism.**
Edited by G. Coruzzi, P. Puigdomènech. 579 pages. 1994.

Vol. 82: **Biological Membranes: Structure, Biogenesis and Dynamics.**
Edited by Jos A. F. Op den Kamp. 367 pages. 1994.

Vol. 83: **Molecular Biology of Mitochondrial Transport Systems.**
Edited by M. Forte, M. Colombini. 420 pages. 1994.

Vol. 84: **Biomechanics of Active Movement and Division of Cells.**
Edited by N. Akkaş. 587 pages. 1994.

Vol. 85: **Cellular and Molecular Effects of Mineral and Synthetic Dusts and Fibres.**
Edited by J. M. G. Davis, M.-C. Jaurand. 448 pages. 1994.

Vol. 86: **Biochemical and Cellular Mechanisms of Stress Tolerance in Plants.**
Edited by J. H. Cherry. 616 pages. 1994.

Springer-Verlag and the Environment

We at Springer-Verlag firmly believe that an international science publisher has a special obligation to the environment, and our corporate policies consistently reflect this conviction.

We also expect our business partners – paper mills, printers, packaging manufacturers, etc. – to commit themselves to using environmentally friendly materials and production processes.

The paper in this book is made from low- or no-chlorine pulp and is acid free, in conformance with international standards for paper permanency.

MIX
Papier aus verantwortungsvollen Quellen
Paper from responsible sources
FSC® C105338

If you have any concerns about our products,
you can contact us on
ProductSafety@springernature.com

In case Publisher is established outside the EU,
the EU authorized representative is:
Springer Nature Customer Service Center GmbH
Europaplatz 3, 69115 Heidelberg, Germany

Printed by Libri Plureos GmbH
in Hamburg, Germany